T0321789

Blockchain, IoT, and AI Technologies for Supply Chain Management

Supply chain management, often known as SCM, refers to the extensive variety of operations that are required to plan, monitor, and coordinate the movement of a product from its raw materials to its finished state in the most time- and money-efficient manner possible. How the supply chain is managed has an impact not only on the quality of the product and the service but also on the distribution, costs, and overall customer experience. Supply chain management is a massive undertaking that needs firms to reevaluate the method in which they operate their supply chains.

Blockchain, IoT, and AI Technologies for Supply Chain Management discusses the problems and difficulties that the facilitators of the supply chain confront, in addition to the possible solutions to such problems and difficulties. This book will be the only one of its kind to address the impact of COVID-19 on supply chain systems involving different stakeholders such as producers, dealers, and manufacturers, and will provide a foundation for future research opportunities that will allow for the unrestricted expansion and prosperity of business. It will serve as a foundation for academics, scientists, and educationists interested in the use of modern technologies in the field of supply chain management, such as the Internet of Things (IoT), Artificial Intelligence (AI), and Blockchain. In addition to those engaged in research, undergraduate and postgraduate students in higher education can also use this publication as a reference book. This book also presents a multifaceted perspective for the general public, including topics such as computer science, the food business, hotel management, fashion, medical, inventory management, and agricultural spheres.

Innovations in Intelligent Internet of Everything (IoE)
Series Editor: Fadi Al-Turjman

Computational Intelligence in Healthcare
Applications, Challenges, and Management
Meenu Gupta, Shakeel Ahmed, Rakesh Kumar, Chadi Altrjman

Blockchain, IoT, and AI Technologies for Supply Chain Management
Priyanka Chawla, Adarsh Kumar, Anand Nayyar, and Mohd Naved

For more information about the series, please visit: https://www.routledge.com/Innovations-in-Intelligent-Internet-of-Everything-IoE/book-series/IOE

Blockchain, IoT, and AI Technologies for Supply Chain Management

Edited by
Priyanka Chawla
Adarsh Kumar
Anand Nayyar
Mohd Naved

CRC Press
Taylor & Francis Group
Boca Raton London New York

CRC Press is an imprint of the
Taylor & Francis Group, an **informa** business

First Edition published 2023
by CRC Press
6000 Broken Sound Parkway NW, Suite 300, Boca Raton, FL 33487-2742

and by CRC Press
4 Park Square, Milton Park, Abingdon, Oxon, OX14 4RN

CRC Press is an imprint of Taylor & Francis Group, LLC

Library of Congress Cataloging-in-Publication Data
Names: Chawla, Priyanka, editor. | Kumar, Adarsh, editor. | Nayyar, Anand, editor.
Title: Blockchain, IoT and AI technologies for supply chain management / edited by Priyanka Chawla, Adarsh Kumar, Anand Nayyar and Mohd Naved.
Description: 1 Edition. | Boca Raton, FL : CRC Press, 2023. |
Series: Innovations in Intelligent Internet of Everything IoE | Includes bibliographical references and index.
Identifiers: LCCN 2022045906 (print) | LCCN 2022045907 (ebook) | ISBN 9781032206400 (hardback) | ISBN 9781032206417 (paperback) | ISBN 9781003264521 (ebook)
Subjects: LCSH: Business logistics. | Blockchains (Databases) | Internet of things. | Artificial intelligence.
Classification: LCC HD38.5 .B5356 2023 (print) | LCC HD38.5 (ebook) | DDC 658.8--dc23/eng/20220922
LC record available at https://lccn.loc.gov/2022045906
LC ebook record available at https://lccn.loc.gov/2022045907

ISBN: 978-1-032-20640-0 (hbk)
ISBN: 978-1-032-20641-7 (pbk)
ISBN: 978-1-003-26452-1 (ebk)

DOI: 10.1201/9781003264521

Typeset in Times
by MPS Limited, Dehradun

Contents

Preface

Supply chain management (SCM) is the wide range of activities needed to schedule, track, and conduct the flow of a product from materials to production to delivery in the most cost-effective way. While the supply chains of yesterday were centered on the availability, movement, and expense of physical objects, the supply chains of today are focused on the management of data, resources, and goods integrated into solutions. The management of the supply chain influences the quality of the product and service, distribution, prices and customer experience. In the present situation, SCM is a huge challenge and requires businesses to revalidate the way their supply chain is run. Technologies such as IoT, blockchain, augmented reality, robotics, and cloud computing have emerged which can change the traditional working approach. Indeed, the blockchain in the IoT industry has grown from USD$30 million to $113 million since 2018 and is expected to grow to over $3 billion annually with a compound annual growth rate (CAGR) of nearly 93 percent by 2024. By 2025, the economic impact of industrial IoT applications may be somewhere between USD$4 trillion and $11 trillion on global GDP. This economic impact of IoT is primarily due to the visibility and remote control capabilities it offers to all industries, resulting in increased operating efficiency and protection. The pandemic has exposed weaknesses and fragility through most, if not all, sectors and industries in the global supply chains. A blockchain-based supply chain management system is built on a decentralized distributed ledger that provides an irrefutable record of all data relating to shipment status, track status, storage conditions, and more. To boost supply and demand, AI makes it easier to access and evaluate a virtually unlimited source of data at once. AI can also be used to further minimize human and labor costs, thus reducing the time taken in early to later stages.

This book provides a holistic view of the supply chain concept from different dimensions. The book covers the issues and challenges that are faced by the facilitators of the supply chain, along with their potential solutions. The book has thrown a light on the impact of COVID-19 in the progress of already-existing supply chain projects as well as the newly proposed ones. Since this type of book will be the only of its kind to address the impact of COVID-19 on supply chain systems involving different stakeholders such as producers, dealers, and manufacturers, it provides a foundation for future research opportunities for unlimited growth and prosperity of business. This book provides a foundation for researchers, scientists, and educationists towards the application of advanced technologies such as IoT, AI, and blockchain in the area of supply chain management to track assets accurately and for upgrading supply chain business operations. Since the delivery chains are both dynamic and distributed these days, involving a great number of parties, this book provides a multidimensional view for the masses ranging from computer science, food industry, hotel management, apparel, medical, inventory management, and agriculture domain. It can be used as a reference book for higher-education UG and PG students apart from the research scholars. This book is comprised of 12 Chapters including comprehensive coverage to IoT, AI, and Blockchain in supply chain management.

Chapter 1, titled "**Fundamentals of IoT, Artificial Intelligence, and Blockchain Approaches for Application in Supply Chain Management**" elaborates that supply chains have a significant impact not only on society but also on the performance of businesses and the environment. Supply chains have a greater potential to develop and support a sustainable future if contemporary technology is used appropriately and used in the appropriate ways. Businesses have the ability to cut their emissions of greenhouse gases by combining technology such as the Internet of Things, artificial intelligence, and blockchain. Chapter 2, titled "**Blockchain, IoT, AI Technologies for Supply Chain Management: Bibliometric Analysis,**" explains that the world is always changing, and businesses are doing their best to keep up. Technologies such as the Internet of Things are used by artificial intelligence, particularly for the development of more complex supply chain management systems. There has been a rise in the number of research projects conducted in the areas of blockchain, IoT, AI, and SCM, respectively. Chapter 3, titled **"Complete Scenario for Supply Chain Management Using IoT and 5G,"** explains that 4G and 5G technologies are able to offer smart services such as smart homes, smart energy, and smart health by analyzing data collected from the Internet of Things. By using accessible interactive dashboards, inefficiencies and errors in the supply chain may be corrected in a time frame that is very close to being considered real time. Chapter 4, titled **"Impact of Artificial Intelligence on Agriculture Value Chain Performance: Agritech Perspective,"** elaborates that agriculture is one of the necessary activities that play a significant part in maintaining all of the activities that humans engage in. The most significant issues that the sector is now facing are an expanding population and increased competition for available resources. These issues may be overcome with the use of technological advances, such as information and communication technology, artificial intelligence, machine learning, and blockchain technology. Chapter 5, titled **"Applications of Artificial Intelligence of Things in Green Supply Chain Management: Challenges and Future Directions,"** explains that Internet of Things and artificial intelligence are two of the most exciting new technologies that will emerge in the next several decades. The purpose of this chapter is to investigate the role that artificial intelligence and the Internet of Things, often known as hybrid technology and described as artificial intelligence of things (AIoT), play in green supply chain management (GSCM). This chapter lays out a plan for the deployment of an IoT-based green supply chain for businesses, with a particular focus on the healthcare industry. Chapter 6, titled **"Effects on Supply Chain Management due to COVID-19,"** elaborates on the global pandemic scenario with COVID-19 that has had an impact on the management of the supply chain in every country across the globe. There is a great need for innovation in the supply chain management of many different enterprises. In the past, many businesses and manufacturers were unsuccessful because they did not have sufficient knowledge at the appropriate moment. To have a better understanding of the current state of the market, a survey was carried out. It has been discovered that some businesses are making a loss as a result of the pandemic; yet, after implementing certain advancements, these businesses are now making incredible profits. Chapter 7, titled **"Artificial Intelligence from Vaccine Development to Pharmaceutical Supply Chain Management in Post COVID-19 Period,"** discusses that on March 11, 2020, the World Health Organization classified the SARS-CoV-2 coronavirus outbreak as a pandemic across the world. The real-time decision-making process was carried out with the assistance of AI algorithms to meet the obstacles associated with vaccine production, storage, logistics, and safety concerns. Machine learning (ML) and artificial intelligence (AI) based methodologies will be presented as applications for vaccine development. During the process of supplying pharmaceuticals, we will discuss the use of AI to increase end-to-end visibility, demand forecasting, and maintaining the integrity of the supply chain. Chapter 8, titled **"Blockchain for SCM: A Prospective Study Based on A Panel of Literature Reviews,"** explains that the industrial and service sectors are two areas that might be significantly altered by blockchain technology. The management of supply chains is one of a number of different domains that may be influenced by developing technology. The newly popular ideas of "smart manufacturing," "cyber physical systems," and the "Internet of Things" would make it possible for intelligent communication to take place between machines that are linked to one another. The integration of cutting-edge technology into the management of supply chains will result in an increase in the openness and accessibility of the information that is shared. Chapter 9, titled **"A Blockchain-Based Framework for Circular Plastic Waste Supply Chain Management in India: A Case Study of Kolkata, India,"** discusses the use of blockchain technology in the management of supply chains that has the potential for exponential expansion. It may be worthwhile to put into action in an industry in which supply chain networks are still experiencing difficulties. The industry of managing plastic garbage is not held in particularly high regard across the board. The SCN of this sector may be improved by turning it into a closed loop, which will increase cashflow. The development of a waste management system that is both effective and efficient is another prerequisite for sustainable cities.

Chapter 10, titled "**Supply Chain Management-Based Transportation System Using IoT and Blockchain Technology,**" discusses that the modern corporate environment is characterized by a particularly high degree of rivalry due to the more demanding and dynamic nature of the market. This is because supply networks need to be able to keep up with the ever-evolving external environment. The incorporation of blockchain technology into the management of supply chains may result in increased visibility and traceability, in addition to higher levels of both operational efficiency and security. Chapter 11, titled "**Perspective Analysis of Three Types of Services on a Queueing-Inventory System with a Sharing Buffer for a Two Class of Customers,**" explains the (s,Q) ordering strategy for the replenishment products stored in the inventory system that only uses a single server. In accordance with the probability p, the HNC(HNC) takes precedence over the LNC(LNC). The hypothesis of the Markov chain is used to examine the steady state of the model that is being investigated. The current quantity of goods in the inventory, as well as the number of low- and high-priority items in the queue, are all included in the joint probability distribution. Chapter 12, titled "**Ensuring Provenance and Traceability in a Pharmaceutical Supply Chain Using Blockchain and Internet of Things,**" discusses the intergration of blockchain and IoT into the pharmaceutical supply chain for better record management and also discusses ecosystem, visbility technologies, and temperature sensors in a cold supply chain.

This book acts as a bridging information resource between basic concepts and advanced-level contents from technical experts to blockchain, AI, and supply chain practitioners. This book facilitates the research group to study and publish novel work towards the advancement of emerging technologies in applications of supply chain management with AI and blockchain integrated into it. The content is aimed at students at the graduate and post-graduate levels from different engineering disciplines.

Priyanka Chawla
Adarsh Kumar
Anand Nayyar
Mohd Naved

About the Editors

Dr. Priyanka Chawla is working as an Associate Professor in the Department of Computer Science and Engineering, NIT Warangal, India. She earned her PhD degree from Thapar University, Patiala, Punjab, India. Dr. Chawla has had a rich experience of Industry and Academia of around 20 years. Her research interests lie in the area of Internet of Things, Blockchain, Artificial Intelligence, Big data, Data Science and Sustainable development. She has several publications in reputed national and international journals and conferences. She is on the editorial board of many reputed journals. She is also the reviewer for many reputed journals and the member of the organizing committee for several conferences and workshops. She has successfully completed various certifications in the field of education. She is a member of reputed professional bodies like ISTE, ISC etc.

Dr. Adarsh Kumar is an associate professor in the School of Computer Science with University of Petroleum & Energy Studies, Dehradun, India. He received his master's degree (M. Tech) in software engineering from Thapar University, Patiala, Punjab, India, and received his PhD degree from Jaypee Institute of Information Technology University, Noida, India, followed by a post-doc from Software Research Institute, Athlone Institute of Technology, Ireland. From 2005 to 2016, he has been associated with the Department of Computer Science Engineering & Information Technology, Jaypee Institute of Information Technology, Noida, Uttar-Pardesh, India, where he has worked as an assistant professor. His main research interests are cybersecurity, cryptography, network security, and ad-hoc networks. He has many research papers in reputed journals, conferences, and workshops. He participated in one European Union H2020 sponsored research project and he is currently executing two research projects sponsored from UPES SEED division and one sponsored from Lancaster University.

Dr. Anand Nayyar received his PhD (Computer Science) from Desh Bhagat University in 2017 in the area of Wireless Sensor Networks, Swarm Intelligence and Network Simulation. He is currently working in School of Computer Science-Duy Tan University, Da Nang, Vietnam as **Professor, Scientist, Vice-Chairman (Research) and Director- IoT and Intelligent Systems Lab**. A Certified Professional with **125+ Professional certificates** from CISCO, Microsoft, Amazon, EC-Council, Oracle, Google, Beingcert, EXIN, GAQM, Cyberoam and many more. Published more than **150+ Research Papers** in various High-Quality ISI-SCI/SCIE/SSCI Impact Factor Journals cum Scopus/ ESCI indexed Journals, 70+ Papers in International Conferences indexed with Springer, IEEE and ACM Digital Library, 40+ Book Chapters in various SCOPUS, WEB OF SCIENCE Indexed Books with Springer, CRC Press, Wiley, IET, Elsevier with Citations: **8000+, H-Index: 46 and I-Index: 165**. Member of more than 60+ Associations as Senior and Life Member including IEEE, ACM. He has authored/co-authored cum Edited **40+ Books of Computer Science**. Associated with more than 500+ International Conferences as Programme Committee/Chair/Advisory Board/Review Board member. He has 18 Australian Patents, 4 German Patents, 2 Japanese Patents, 11 Indian Design cum Utility Patents, 1 USA Patent, 3 Indian Copyrights and 2 Canadian Copyrights to his credit in the area of Wireless Communications, Artificial Intelligence, Cloud Computing, IoT and Image Processing. Awarded **38 Awards for Teaching and Research**—Young Scientist, Best Scientist, Best Senior Scientist, **Asia Top 50 Academicians and Researchers**, Young Researcher Award, Outstanding Researcher Award, Excellence in Teaching, Best Senior Scientist Award, **DTU Best Professor and Researcher Award- 2019, 2020-2021, 2022** and many more. **He is listed in Top 2% Scientists as per Stanford University (2020, 2021, 2022)**. He is acting as Associate Editor for *Wireless Networks* (Springer), Computer Communications (Elsevier), *International Journal of Sensor Networks (IJSNET)* (Inderscience), *Frontiers in Computer Science, PeerJ Computer Science, Human*

Centric Computing and Information Sciences (HCIS), IET-Quantum Communications, IET Wireless Sensor Systems, IET Networks, IJDST, IJISP, IJCINI, and *IJGC.* He is acting as Editor-in-Chief of IGI-Global, USA Journal titled ***International Journal of Smart Vehicles and Smart Transportation (IJSVST).*** He has reviewed more than 2500+ Articles for diverse Web of Science and Scopus Indexed Journals. He is currently researching in the area of Wireless Sensor Networks, Internet of Things, Swarm Intelligence, Cloud Computing, Artificial Intelligence, Drones, Blockchain, Cyber Security, Network Simulation, Big Data and Wireless Communications.

Dr. Mohd Naved is a passionate researcher and educator with 16 years of experience and proven track record of quality research publications and leading teams for the research and overall management of the educational institution. He is a senior member of IEEE and is associated with multiple leading research organizations. He is a machine learning consultant and researcher, currently teaching in Amity University (Noida) for various degree programs in analytics and machine learning. He is actively engaged in academic research on various topics in management as well as on 21st-century technologies. He has published 60+ research articles in reputed journals (SCI/Scopus Indexed/peer reviewed). He has 16 patents in AI/ML and is actively engaged in commercialization of innovative products.

Contributors

Hamed Alqahtani
Department of Information Systems
King Khalid University
Abha, Saudi Arabia

Tomina Anoop
M.Sc Scholar, Tesside University
Middlesbrough, United Kingdom

Mahnoor Bano
Department of Computer Science
University of Agriculture Faisalabad
Faisalabad, Pakistan

Nurbahar Bora
Atatürk University Social Sciences Institute
Erzurum, Turkey

Ahmed Mateen Buttar
Department of Computer Science
University of Agriculture Faisalabad
Faisalabad, Pakistan

Abhishek Dadhich
School of Allied Health Sciences and
    Management
New Delhi, India

Priyanka Dadhich
Department of Computer Science and
    Engineering
Delhi Technical Campus
Guru Gobind Singh Indraprastha University
New Delhi, India

Abhijit Das
Department of Information Technology
Institute of Leadership
Entrepreneurship and Development (iLead)
Kolkata, India
and
Department of Data Science and Cyber
    Security
Institute of Leadership
Entrepreneurship and Development (iLead)
Kolkata, India

Adrija Das
A.K. Choudhury School of Information
    Technology
University of Calcutta
Kolkata, India

Ankita Das
Consortium of Researchers for Sustainable
    Development (C.R.S.D.)
Agra, India
and
Department of Data Science and Cyber
    Security
Institute of Leadership
Entrepreneurship and Development (iLead)
Kolkata, India

Santanu Das
Seshadripuram First Grade College
Yelahanka New Town
Bangalore, India

Arokiaraj David
Department of Management Studies
St. Francis Institute of Management and
    Research
Mumbai, India

Biswajit Debnath
Department of Chemical Engineering
Jadavpur University
Kolkata, India
and
Consortium of Researchers for Sustainable
    Development (C.R.S.D.)
Agra, India

Sadaf Fatima
Department of Business Administration
Aligarh Muslim University
Aligarh, India

C. Ganeshkumar
Indian Institute of Plantation Management
Bengaluru, India

Ankur Gupta
Department of Computer Science and
   Engineering
Vaish College of Engineering
Rohtak, Haryana, India

Shipra Gupta
School of Management
Graphic Era Hill University
Dehradun, Uttarakhand, India

T. Harikrishnan
Department of Mathematics
Guru Nanak College
Chennai, India

Abdellah Houssaini
Faculty of Economics and Management
University Ibn Tofail
Kenitra, Morocco

K. Jeganathan
Ramanujan Institute for Advanced Study
   in Mathematics
University of Madras
Chennai, India

P.P. Joby
Computer Science and Engineering
St. Joseph's College of Engineering and
   Technology
Palai, Kerala, India

Amna Khalid
Department of Computer Science
University of Agriculture Faisalabad
Faisalabad, Pakistan

Mohammad Faiz Khan
Security Forces Hospital
Ministry of Interiors
Abha, Saudi Arabia

Vijay Kumar
Physics Department
Graphic Era Hill University
Dehradun, Uttarakhand, India

Kukati Aruna Kumari
Prasad V. Potluri Siddhartha Institute of
   Technology
Vijaywada, Andhra Pradesh, India

Anna N. Kurian
Computer Science and Engineering
St. Joseph's College of Engineering and
   Technology
Palai, Kerala, India

K. Prasanna Lakshmi
Department of Mathematics with Computer
   Applications
Ethiraj College for Women
Chennai, India

Allen Mathew
Financial Analyst, X L Dynamics
Kochi, Kerala, India

D. Nagarajan
Department of Mathematics
Rajalakshmi Institute of Technology
Chennai, India

Arshi Naim
Department of Information Systems
King Khalid University
Abha, Saudi Arabia

Arpit Namdev
University Institute of Technology RGPV
Bhopal, Madhya Pradesh, India

Praful Nandankar
Government College of Engineering
Nagpur, India

Esra Ozmen
Ankara Hacı Bayram Veli University
Ankara, Turkey

V. Padmavathi
Department of Information Technology
A. V. C. College of Engineering
Mayiladuthurai, Tamilnadu, India

Jay Kumar Pandey
Shri Ramswaroop Memorial University
Barabanki, Uttar Pradesh, India

Vandana B. Patil
Dr. D.Y. Patil Institute of Engineering
Management, and Research
Pimpri-Chinchwad, Maharashtra, India

Mostafa Qandoussi
Faculty of Economics and Management
Laboratory of Economics and Management of
   Organizations
University Ibn Tofail
Kenitra, Morocco

R. Saminathan
Department of Computer Science and
   Engineering
Annamalai University
Chidambaram, Tamilnadu, India

Jeganthan Gomathi Sankar
BSSS Institute of Advanced Studies
Bhopal, India

S. Selvakumar
Ramanujan Institute for Advanced Study in
   Mathematics
University of Madras
Chennai, India

1 Fundamentals of IoT, Artificial Intelligence, and Blockchain Approaches for Applications in Supply Chain Management

V. Padmavathi
Department of Information Technology, A. V. C. College of Engineering, Mayiladuthurai, Tamilnadu, India

R. Saminathan
Department of Computer Science and Engineering, Annamalai University, Chidambaram, Tamilnadu, India

CONTENTS

DOI: 10.1201/9781003264521-1

1.1 INTRODUCTION

The Internet of Things (IoT) [1] is a network of physical objects or people that are equipped with software, electronics, networks, and sensors to collect and exchange data. The Internet of Things aims to extend Internet connectivity beyond traditional devices like computers, smartphones, and tablets to more mundane products like toasters. As a result of the Internet of Things, virtually everything becomes "smart," utilizing the power of data collection, AI algorithms, and networks to improve aspects of our life. Things in the Internet of Things include people with diabetes monitor implants, animals with tracking devices, and so on.

IoT technology enables supply chain members to maximize their competitive advantage by attaining visual management and intelligent management throughout the supply chain, enhancing supply chain transparency, and facilitating information sharing.

Every day, trucks, ships, and humans must deliver, track, and account for millions of commodities. The Internet of Things connects these products, assets, and people across the supply chain, enabling efficiencies and optimizing operations that save companies time and money every year. IoT [2] devices have the potential to have a substantial impact on all aspects of the supply chain, including warehouse management, transportation and logistics, and last-mile delivery to the end consumer, as shown in Figure 1.1. Manufacturers may increase operational efficiency, reduce theft and counterfeiting, and deliver outstanding customer service by responding to data from IoT devices in their supply chain; allows a manufacturer to see their whole supply chain in real time without having to rely on a third party to scan packaging at any point along the way; writes data from IoT sensors straight to their decentralized blockchain environment, ensuring that the data cannot be tampered with or hacked and is completely correct; and allows the end-user to scan the product to check that they're getting the actual thing and that it's in the best possible condition as specified by the manufacturer.

Artificial intelligence is already incredibly powerful, and it's only getting better. From social media to self-driving cars, the speed with which robots can be educated to replicate or even outperform humans is becoming increasingly important. Although still in the early phases of development, enterprise applications based on advanced technologies such as machine learning and AI [3] are beginning to generate distinctive and novel business strategies. In the logistics and supply chain industries, these technologies have proven to be game changers. According to Gartner, the value of machine automation in supply chain operations is expected to triple in the next five years. By 2022, it is expected that annual Industrial IoT (IIOT) investment by increasing businesses will reach $600 billion.

AI-enhanced technologies are being utilized across supply chains to boost efficiency; reduce the effects of a global labor shortage; and discover better, safer ways to move goods from one point to another. AI applications can be found across the supply chain [4], from the production floor to the front door. AI is being employed in the supply chain and logistics business because organizations have realized that AI has the power to solve the difficulties of running a global logistics network. When used effectively, AI may assist businesses in becoming smarter, making more agile decisions, and anticipating challenges.

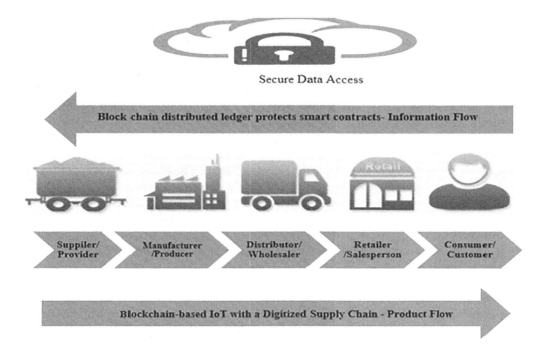

FIGURE 1.1   Blockchain-Based IoT with Supply Chain Process.

Consumer expectations for on-time and intact delivery are being exceeded by AI-enabled proactive solutions, which are also improving service quality. They're also sharpening their talents through automating compliance. As a result, there are fewer problems and lower costs across the logistics network. The most intriguing feature of AI is its seemingly infinite potential. When algorithms are combined with technologies like the Internet of Things (IoT) [5–9], machine learning (ML), and predictive analytics, they become even more powerful. Because of increased access to data, companies now have a better grasp of their worldwide logistics networks. Transparency is important because it improves people's perceptions of supply chain management and logistics.

Figure 1.2 depicts architecture of IoT systems. It is a three-layered structure and consists of an application layer, network layer, and perception layer. All layers in a supply chain can access the same data via blockchain, potentially reducing communication and data transmission issues. More time can be spent on delivering goods and services, either enhancing quality or cutting costs, or both. Blockchain technology can be used by businesses to track all types of transactions more securely and transparently. The impact on supply chain operations might be massive. Companies

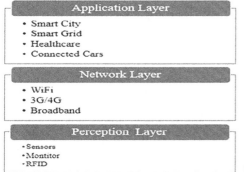

FIGURE 1.2   The IoT System's Three-Layer Architecture.

can utilize blockchain to track a product's history from conception to present-day location. The transaction is securely documented every time a product changes hands, creating a permanent record of the transaction from beginning to end. With the use of this advanced technology, parties collaborating on a single shared platform might substantially reduce the time delays, additional costs, and human error that are typically associated with transactions. By minimizing the number of middle people in the supply chain, fraud risks are lowered. Finally, detailed records assist businesses in identifying the source of fraud when it occurs.

The following are the objectives of this chapter:

- To study the background and review the extensive literature matching the concept of IoT fundamentals, blockchain, AI highlights, and supply chain management.
- To propose a novel privacy protection technique based on blockchain technology. The McEliece Cryptography system is used as base for secure data flow between suppliers and customers.
- In addition, the security and privacy implications of blockchain-based IoT applications are examined.

Organization of Chapter: The chapter is structured as follows: Section 1.2 addresses related studies as well as a survey summary. Section 1.3 looks at data security and privacy in the context of blockchain technology and the integration of blockchain, the Internet of Things, and artificial intelligence in supply chain management. Section 1.4 describes the architectural design of IoT, AI, and blockchain-based SCM with security mechanisms. Section 1.5 discusses the various case study applications. Finally, Section 1.6 concludes the chapter with a future scope.

1.2 FUNDAMENTS OF IOT, ARTIFICIAL INTELLIGENCE, AND BLOCKCHAIN APPROACHES

The digital transformation is being accelerated by three key technologies: Blockchain, IoT, and AI. We believe these technologies will converge, enabling for the emergence of new business models: In the future, autonomous agents (sensors, cars, machines, trucks, cameras, and other IoT devices) will be able to (1) create a digital twin with IoT, (2) send and receive money using blockchain technology on their own, and (3) make autonomous decisions as independent economic agents using AI and data analytics. This convergence, we believe, will also stimulate the development of self-contained business models and the digital transformation of industrial firms.

1.2.1 ESSENTIALS OF IoT

The Internet of Things is a network of networked devices that can collect and share data about their own operations as well as their surroundings. Any device that has a two-way data link, such as connected sensors, thermostats, cars, biometric devices, and luminaires, could be part of the IoT.

The three basic levels of an IoT system are the device layer, network layer, and platform layer. The device layer contains the things that engage in the Internet of Things. The network layer contains everything you need to connect devices to each other and to the platform layer [10–14]. An IoT system has four basic components, shown in Figure 1.3.

1. **Sensors/Devices:** Sensors and devices are necessary for collecting real-time data from the environment. All of this data could be challenging in some way. It could be something as simple as a temperature sensor or something as complex as a video stream.

   A gadget could have a range of sensors that do more than just sense. A smartphone, for example, is a device having multiple sensors, including GPS and a camera, yet it is unable to detect these features.

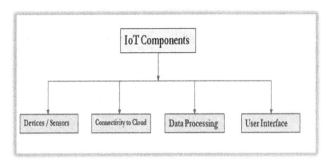

FIGURE 1.3 Four Basic Components of IoT.

2. **Data Transmission:** All data is transmitted to a cloud infrastructure. A number of communication methods should be used to connect the sensors to the cloud. Examples of communication methods include mobile or satellite networks, Bluetooth, Wi-Fi, WAN, and others.

3. **Data Processing:** The software processes the information after it has been captured and transmitted to the cloud. This strategy may consist of merely checking the temperature or readings from devices such as air conditioners or heaters. However, some tasks, such as recognizing objects using computer vision on video, can be rather tough.

4. **User Interface:** The information must be accessible to the end-user in some way, which can be done by sending them an email or text message, or by setting up alerts on their phones. An interface that actively monitors the user's IoT equipment may be required, on occasion. For example, the user has a camera installed in their home. They want to be able to access video recordings and all streams using a web server.

1.2.2 Highlights of Artificial Intelligence

1.2.2.1 Data Access in Real Time

To improve on traditional corporate systems that use outmoded batch planning methodologies, new AI systems must address the stale data problem. Most supply chains today attempt to carry out plans based on data that is several days old, but this results in poor decision making, which either under-optimizes the supply chain or needs human user involvement. Without real-time data, AI technology merely makes wrong decisions faster.

1.2.2.2 Information from the Public (Multi-Party)

Any AI, deep learning, or machine learning algorithm must be able to access data from outside the organization or, more importantly, be granted access to data relevant to your trading community. Unless the AI tool can see forward-most demand and downstream supply, as well as any relevant limits and capacity in the supply chain, the results will be no better than a traditional planning system. Unfortunately, this lack of awareness and access to real-time community data is the standard in nearly all supply chains. This, of course, must change if an AI tool is to be successful.

1.2.2.3 Participating in a Game

You can acquire a machine that can play master level chess for a few hundred dollars. They have some AI, but they largely compete with people via brute force calculations, which entail scanning hundreds of thousands of locations. You must be able to scan 200 million spots per second to overcome a world champion utilizing brute force and well-established techniques [15].

1.2.2.4 Speech Recognition

By the 1990s, computer-based speech recognition had advanced enough to be used for limited purposes. As a result, United Airlines has replaced the flight information keyboard tree with a system that uses voice recognition to recognize flight numbers and city names. It's very comfortable. On the other hand, certain computers can be programmed by voice, but most users prefer the keyboard and mouse for more convenience.

1.2.3 FUNDAMENTAL OF BLOCKCHAIN APPROACHES

Blockchain is a system of interconnected blocks that contains transaction history and other user data. It is based on the concept of a decentralised distributed digital ledger. This technology enables the network's user nodes to conduct cryptographically secure and anonymous money transactions, with the transactions being authenticated and approved by all users in a public manner. It's a cutting-edge technology that's becoming increasingly popular as a result of the use of digital currency. Although blockchain has a promising future in online transactions, it is prone to a variety of security and vulnerability flaws.

A peer-to-peer architecture underpins blockchain technology. Because it's decentralized and made up of a series of blocks, it's called blockchain. Since Satoshi Nakamoto's original concept was derived and implemented in Bitcoin, blockchain has been a hot topic among scholars. Its attributes have also increased the scope of its applicability [16–20]. Because it saves all of the nodes' calculations in each of them, it's also known as distributed ledger technology. The network's reliability is not an issue because the ledger is shared. Hash code, which is a one-of-a-kind and unchangeable value generated by a sophisticated mathematical hash function, is also included in the blocks. Transactions are not done in the traditional way, especially with real user IDs and addresses, so there are numerous options for anonymizing both senders and recipients. The system is somewhat autonomous because there is no central authority. As a result of these factors, the blockchain concept has evolved into a new technology that can be used in a variety of industries.

1.3 THE INTERSECTION OF BLOCKCHAIN, INTERNET OF THINGS (IOT), AND ARTIFICIAL INTELLIGENCE IN SUPPLY CHAIN MANAGEMENT

1.3.1 IMPROVING DATA STANDARDIZATION, PRIVACY, SECURITY, AND SCALABILITY

Blockchain enables decentralized aggregation of large amounts of data generated by IoT devices, allowing benefits to be shared more evenly among participants in supply chain exchanges [21]. As shown in Figure 1.4, the confluence of IoT, AI, and blockchain technology approaches in supply chain management includes data standardization, privacy, security, scalability, traceability, and quality.

1.3.1.1 Standardization of Data

Smart objects generate a large amount of data. This data must be managed, processed, transported, and stored correctly. Standardization is necessary for true device and application interoperability [22]. The right standards, issued at the right time in a technology's development, can help ensure interoperability, generate trust in the system, and make it easier to use. Standardization's ultimate goal is to reduce all features to a single scale without distorting the variances in value ranges. Data normalization is the process of rescaling the attributes to have a mean of 0 and a variation of 1.

1.3.1.2 Privacy

Special considerations are needed to protect personal information from disclosure of things in the Internet of Things environment [23–25]. You can give almost every physical or logical entity or object a unique identifier and the ability to autonomously navigate communications over the Internet or similar networks.

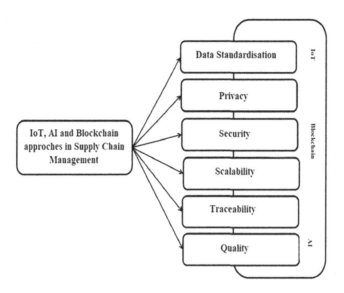

FIGURE 1.4 The Confluence of IoT, AI, and Blockchain Technology Approaches in Supply Chain Management.

1.3.1.3 Security

IoT, AI, and Blockchain are three new technologies that can boost productivity and help supply chain exchange partners provide the integrity they need. The combination of Blockchain, IoT, and AI technology is thought to have a huge revolutionary influence across a variety of industries, as it shows a possible avenue for controlling IoT devices. The use of blockchains and IoT allows users to create a more resilient, responsive, and distributed peer-to-peer system with the ability to engage with supply chain exchange partners in a "trustless," safe, and real-time manner. More importantly, blockchain-based solutions [26] have the potential to change the IoT's potential benefits by bridging the device-data interoperability gap while maintaining security, privacy, and dependability.

1.3.1.4 Scalability

The blockchain can be managed using either a centralized database (such as MySQL) or distributed hash table (DHT) technology. As a result, (off-chain) transactions complete faster and are more scalable. Off-chain solutions also have the potential to interoperate with enterprise infrastructures that store non-essential data. In addition, supply chain blockchain-enabled IoT applications adapt to IoT network features such as heterogeneity, dynamic topology, complexity, scalability, throughput, and memory size [27]. These solutions seek to improve scalability by modifying the basic parts of blockchain transactions by using new or especially lightweight network protocols for IoT devices, editable blockchain enhancement techniques, and directed acyclic graphs (DAGs) to increase block size. Increasing the block size of a public blockchain improves scalability and provides additional storage and processing capabilities, but it can also slow down the block propagation speed of your network.

1.3.2 USING A BLOCKCHAIN-BASED IDENTITY FOR AUTHENTICATION

When the blockchain is implemented, users will be able to create their own identities and publicly verifiable information for them. Users can save public data on the blockchain, removing the requirement for third parties to handle identity data [15]. A local blockchain is used to register ordinary nodes for authentication. Smart contracts are utilized on cluster head nodes to validate

FIGURE 1.5   Blockchain-Based Identity for Authentication.

registration and authentication requests made by normal nodes. The public chain stores information about the registered nodes.

According to the W3C, "Verifiable Credentials reflect assertions made by an issuer in a tamper-evident and privacy-respecting manner." Verifiable credentials, in essence, allow for the digital watermarking of claims data using a combination of public key cryptography and privacy-preserving methods to avoid correlation [28,29]. As a result, physical credentials can be safely transformed to digital, and holders of such credentials can choose publish particular information from their credentials without disclosing the actual data, and other parties can instantaneously validate this data without having to contact the issuer.

Distributed identifiers are globally unique and persistent identifiers. The ID owner has full control over them. Distributed IDentifier (DID) does not depend on a centralized registry, authority, or identity provider. When an organization provides verifiable credentials, it also provides a public DID. The blockchain, which is an immutable record of data, stores the same public DID. Distributed identifiers are globally unique and persistent identifiers. The ID owner has full control over them. DID does not depend on a centralized registry, authority, or identity provider. When an organization provides verifiable credentials, it also provides a public DID. The blockchain, which is an immutable record of data, stores the same public DID. If you want to verify the authenticity/validity of your credentials, you can search they DID on the blockchain to see who issued it without contacting the publisher. The blockchain acts as a distributed registry of verifiable data. A "phone book" that anyone can use to find out which organization a particular public DID belongs to.

The owner, publisher, and verifier are three participants in blockchain-based identity management. When using blockchain technology for identity management, it is important to remember that there are three different parties involved: identity owner, issuer, and verifier [30–34]. The ID holder's personal credentials can be issued by the ID issuer. The identity issuer may be a trusted third party B. Local government (user). When issuing a card, the card issuer confirms the legality of the personal data (name, date of birth, etc.) on the card. Figure 1.5 shows the identity based on blockchain for authentication.

1.3.3   FIVE STAGES OF SUPPLY CHAIN MANAGEMENT

The control of the flow of products and services from the point of origin to the point of consumption by the end-user is referred to as supply chain management. The regulation of the flow of goods and services from the point of manufacture to the point of consumption is referred to as supply chain management. Transportation and storage of raw materials utilized in work-in-progress, inventory, and fully furnished objects are also included.

FIGURE 1.6   The Stages of Supply Chain Management.

The following actors are briefly introduced in Figure 1.6, which depicts a simplified version of such a process:

    i. **Suppliers/Provider:** Raw materials like as seeds and nutrients, as well as pesticides, chemicals, and other substances.
   ii. **Manufacturer/Producer:** Typically, the farmer is responsible for everything from seeding to harvesting. Raw resources are transformed into finished goods.
  iii. **Distributor/Wholesaler:** The distributor is responsible for conveying the producer's output (e.g., the product) from the processor to retailers.
  iv. **Retailer/Dealer:** A retailer is responsible for selling items, whether they are sold in small neighbourhood stores or enormous supermarkets.
   v. **Consumer/Customer:** The consumer is the final link in the chain.

The graphic above depicts the flow of goods, services, and information from the producer to the consumer [35]. The passage of a product from the producer to the manufacturer, who then transfers it to the distributor for distribution, is depicted in the diagram. The distributor then delivers it to a wholesaler or retailer, who distributes the items to a variety of locations where customers can easily purchase them.

Supply chain management is basically a combination of supply and demand management. Examine the entire chain using a variety of tactics and methods and work efficiently at each stage. Each entity involved in the process must strive to reduce costs and help organizations improve long-term performance while creating value for stakeholders and consumers [36]. This method helps reduce charges by eliminating unnecessary costs, transfers, and processing.

The following are the main advantages of supply chain management:

- Improves client service and relationships.
- Develops more efficient distribution systems for in-demand items and services.
- Enhances efficiency and business processes.
- Lowers the cost of storage and transportation.
- Reduces both direct and indirect expenditures.
- Assists in the timely delivery of the correct items to the correct location.
- Supports the successful execution of just-in-time stock models by improving inventory management.
- Assists businesses in adjusting to globalisation, economic turmoil, rising customer demands, and other changes.
- Assists businesses in reducing waste, lowering costs, and increasing efficiency throughout the supply chain.

1.3.4 Areas of Expertise in Supply Chain Management

- The five aspects that make up supply chain management are supply planning, production planning, inventory planning, capacity planning, and sales planning.
- Supply planning identifies the most efficient way to cover the demand generated by demand planning. The goal is to create a supply-demand balance that meets the company's financial and service goals.
- Production planning deals with the company's production and manufacturing modules. It takes into account resource allocation for employees, materials, and manufacturing capacity.
- The production/supply plan includes the following steps:
  - Collaboration and Supplier Management
  - Production Planning
- Inventory planning analyzes the optimal quantity and timing of inventory to meet sales and production requirements.
- Capacity planning estimates the number of production workers and equipment needed to meet product demand.
- The movement of items from the supplier or manufacturer to the POS is monitored by sales and network plans. Packaging, inventory, warehousing, supply chain, and logistics are all procedures under the control of sales.

A supply chain is required that is connected from beginning to end, across the organization and beyond, to flourish in a rapidly developing global market [37–39]. These five steps should be followed to accomplish connected supply chain planning.

1.3.4.1 Make the Transition to Real-Time Supply Chain Planning

Companies that plan using ERP systems and spreadsheets rely almost entirely on historical data, leaving little room for change in the case of demand or supply disruptions. For example, a company can anticipate the number of products it will sell in the current quarter using data from the previous year.

1.3.4.2 Supply Chain and Enterprise Planning Should Be Combined

The second phase is to connect formerly compartmentalised supply chain planning to sales and operations planning, as well as finance planning. By synchronizing their short-term operational planning with their bigger business planning operations, companies can make real-time changes to inventory estimations and supplies.

1.3.4.3 Get Ready for the Demands of the End Consumers

Consumer packaged goods manufacturers face a perpetual challenge in anticipating what consumers want and when they want it. For example, detects client demand signals by providing end-to-end intelligence across the supply chain, as well as beyond an established network of wholesalers and retailers.

1.3.4.4 Utilize Real-Time Data at Every Stage of the Supply Chain

Supply chain planning often includes different suppliers, channels, customers, and pricing schemes, so leverage real-time data at all stages of the supply chain, especially if spreadsheets are the primary planning tool. The model is huge and can become unmanageable in some cases.

1.3.5 Applications in Supply Chain Management Using Blockchain Approaches

Blockchain allows all participants in a supply chain to have access to the same data, potentially minimizing communication and data transfer problems. Less time can be spent confirming data and more time can be spent delivering goods and services – improving quality, lowering costs, or doing both.

Blockchain allows all participants in the supply chain to access the same data, which may minimize communication and data transfer issues. You can spend less time reviewing your data and more time delivering goods and services to improve quality, reduce costs, or both. Blockchain has the potential to improve supply chain transparency while reducing costs and risks. Below are some of the key benefits that blockchain supply chain innovation can offer.

Key Potential Benefits
- Improved traceability of the material supply chain to ensure compliance with business requirements.
- Reduce losses from gray market/counterfeit transactions.
- Reduce paperwork and management costs by improving visibility and compliance of outsourced contract manufacturing.

Possible Secondary Benefits
- Enhance your company's reputation by maintaining transparency about the materials used in your products.
- Increase the reliability of the data provided and the trust of the public.
- Reduce the risk of PR disasters due to supply chain failures.
- Participate in discussions with stakeholders.

The company digitizes physical assets; provides decentralized, invariant records of all transactions; and allows them to trace assets across the supply chain, from manufacturing to distribution to end user. End-to-end tracking is visible and accurate. This greater supply chain transparency benefits both firms and customers.

1.4 IOT, ARTIFICIAL INTELLIGENCE, AND BLOCKCHAIN-BASED SUPPLY CHAIN MANAGEMENT: ARCHITECTURAL DESCRIPTION

1.4.1 PROPOSED SUPPLY CHAIN STRUCTURE AND DATA PRESERVATION APPROACH

Figure 1.7 depicts the whole architecture of the proposed privacy protection technique based on blockchain technology. The block diagram for the proposed work includes three processes: supply chain, AI, and IoT [40]. In the supply chain process, several segments are linked to one another in various areas of the chain, such as service, inventory, and cost, showing the whole product structure and data will be transmitted via blockchain technology.

A supplier of raw materials such as seeds, fertilizers, pesticides, chemicals, and other substances. A farmer who usually oversees the entire process from seed to harvest. Raw materials are converted to finished products. The distributor is responsible for transferring the producer's products (such as products) from the processor to the retailer. Retailers are responsible for selling their products, whether they are sold in small stores in the neighborhood or in large supermarkets.

The consumer is the last link in the chain. Blockchain helps prevent fraud on high-value products, such as diamonds and medicines, by increasing the transparency of the supply chain [41–43]. By mitigating or eliminating the effects of counterfeit goods, blockchain helps companies understand how raw materials and finished goods pass through each subcontractor, reducing lost profits from counterfeit and gray market transactions and helps building confidence in end-market consumers.

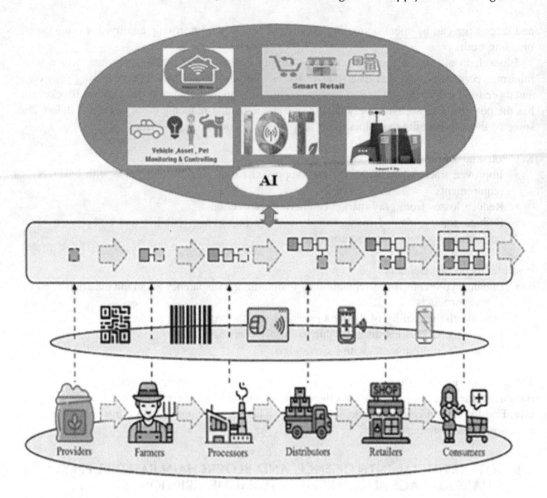

FIGURE 1.7 Proposed Architecture.

1.4.2 McEliece Cryptosystem

McEliece's cryptosystem, which dates from the 1970s and is based on the syndrome decoding issue, is an example of a code-based cryptosystem. McEliece's technique has a fast encryption and decryption time, which is beneficial for executing quick blockchain transactions. McEliece's cryptosystem, on the other hand, necessitates the storage and execution of enormous matrices that serve as public and private keys. Such matrices often take up between 100 kilobytes and several megabytes, which can be a limitation when dealing with resource-constrained systems.

For encryption and decryption, the McEliece algorithm is used in Figure 1.8.

     a. Consider plain text a, which is embedded with the weight vector e.
     b. The Cipher text b is calculated using the formula $b = a * G + e$
     c. The equation that generates the encrypted text is as follows: $y_1 = b * P^{-1}$

The encrypted text is given by y_1.
The decryption procedure is as follows

     a. By extracting the first four components of y_1, which is represented by X_0
     b. The decryption procedure is carried out. $x = S^{-1} * y^1$
     c. Finally, x outputs plain text.

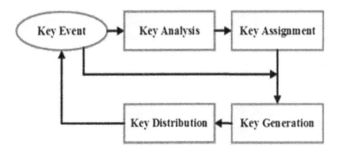

FIGURE 1.8 Key Management Process.

As a result, utilizing appropriate keys, the McEliece algorithm is used to encrypt and decode the data.

Figure 1.9 shows a comparison of execution times for the asymmetric techniques. According to the graph, the recommended McEliece technique completes the encryption procedure in 1.2 seconds. The RSA technique, on the other hand, completes the encryption operation in 2.8 seconds [44]. As a result, the proposed McEliece algorithm requires less time to encrypt data than the existing RSA technique. The encryption time of the proposed McEliece is compared to symmetric and asymmetric encryption algorithms. Overall, the results show that the McEliece algorithm encrypts in a very short amount of time.

The suggested algorithm has the following advantages:

- Both the RSA and McEliece algorithms use a key length of 1024-bits.
- Despite having the same number of rounds for the encryption process, the suggested McEliece algorithm takes less time to encrypt than the existing RSA technique.
- This is because the number of encryption rounds required is smaller than the number of rounds required by the RSA method.

A comparison of the execution times consumed by the asymmetric approach during the decryption process is shown in Figure 1.10. According to statistics, the decryption process for the proposed McEliece algorithm takes 1.3 seconds. On the other hand, with the RSA method, it takes 3.6 seconds to decrypt the message [45]. As a result, the proposed McEliece algorithm takes less time to decode the data than the RSA algorithm. Compared to existing algorithms like RSA, the proposed McEliece method takes less time to decrypt [46]. This is because the McEliece method requires less blocks to decrypt than other algorithms.

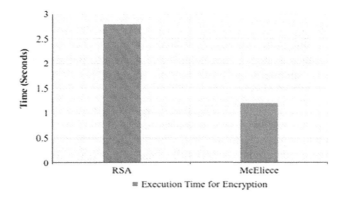

FIGURE 1.9 Asymmetric Methods Execution Time (Encryption) Comparison.

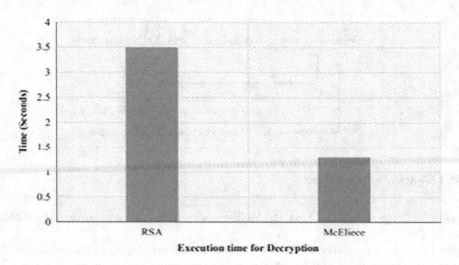

FIGURE 1.10   Asymmetric Methods Execution Time (Decryption) Comparison.

TABLE 1.1
Algorithms in Summary

Algorithms	Key Length (Bits)	Execution Time (Seconds)	
		Encryption	Decryption
RSA	1024	2.8	3.6
McEliece	1024	1.2	1.3

The execution times for the symmetric and asymmetric techniques are compared in Table 1.1. The table includes a list of algorithms, as well as their corresponding key lengths (in bits) and encryption and decryption process times. Both symmetric and asymmetric encryption approaches are compared to the proposed McEliece decryption time. The McEliece symmetric algorithm is compared to the RSA asymmetric approach and the results demonstrate that the McEliece method requires relatively little time to decrypt.

1.4.3   OPTIMIZED SECURE DATA TRANSMISSION

In supply chain management, blockchain technology is a viable alternative for secure information exchange. However, maintaining security at all layers of the blockchain [47] is so important that the McEliece cryptography system is becoming more and more popular. A secure flow of data between manufacturers, suppliers and customers is critical to being able to respond to market fluctuations.

Secure data flow between manufacturers, suppliers, and customers is essential. The value of information or assets that a supplier owns, owns, accesses, or processes under a contract cannot be quantified. This is a major obstacle to the entire supply chain. The main advantage of this procedure is that only identified people can access the shared data using the same key issued by the sender. This makes the exchange of information very secure and allows stakeholders to function in any corporate environment.

1.4.4 Security Issues in Supply Chain Management for Information Exchange

Reducing supply chain links is responsible for reducing the risks associated with recipients and shipping. Supply chain transportation and other weather, border security, economic collapse, and natural disaster capabilities make it easy to connect supply chain technology with as much data as possible with automated supplier analysis and improves supply chain reliability and efficiency, and thanks to sellers, delivery partners, and more, the business is less exposed to unforeseen disasters. Supply chain management is used in a variety of professions such as health care and logistics, and information in data formats is highly sensitive as moved from intermediate-to-intermediate levels [48]. Researchers use a variety of optimization tools and technologies in logistics to address many areas where security is a major concern. Network optimization tools model the entire supply chain [49], including forecast data, manufacturing, product and asset data, inventory and distribution rates, features, and costs. Site-level supply chain modeling with other planning tools does not fall into the category of network optimization techniques.

1.5 CASE STUDIES

1.5.1 Case Study on Supply Chain Traceability

Every stage of the procedure might be considerably speeding up with blockchain technology. From raw materials to completed items, every transaction signaling a movement of commodities would be documented. Documentation would be generated, updated, read, or confirmed by parties on the blockchain, giving visibility of the whole supply chain. A complete audit trail would be generated, which could be utilized to protect consumers from counterfeit goods while also increasing company trust in the authenticity and quality of items, influencing sourcing decisions. Payments might be made effortlessly between stakeholders throughout the process in finance, based on agreements. As an addition, linked sensors and smart devices might measure the state of containers and other information can be stored to inform final.

1.5.2 Case Study on Blockchain-Based Agri-Food Supply Chain Management in China

The primary goal of agri-food supply chain management is to limit opportunism brought on by knowledge asymmetry. To control the unpredictability of the agri-food quasi-organization, a traditional Chinese agri-food supply chain management uses a contract system and a trust mechanism. In the situation of asymmetric information, however, improving transaction efficiency and maintaining agri-food supply chain stability is nearly difficult. Today, Blockchain, IoT and Big Data are transforming the agri-food supply chain into a massive smart network that will break down information barriers.

1.5.3 Case Study on Blockchain-Based Supply Chain for Transaction Process Automation

The transaction problem in supply chain management is extremely important to stakeholders. The exchange of transactions is critical for the supply chain's efficient transportation and logistical activities. The present supply chain management system has various flaws in terms of security and confidence in the transaction process. As a result, information is transferred in a semi-digital and paper-based manner. In this article, we examine a supply chain trust issue and offer a new method based on blockchain technology for addressing the issue and automating the entire payment process using a smart contract. Validation using a case study is carried out.

1.6   CONCLUSION AND FUTURE SCOPE

The blockchain technology will ensure data security at every level of the system. An attacker can't update or give bogus data without breaking the chain. The blockchain is a new technology that is increasingly being used in security systems and other fields. Furthermore, this chapter looked at the many security issues, obstacles, flaws, and attacks that are preventing more people from using blockchain technology from a variety of angles. To make the suggested architecture more dependable, safe, and scalable, blockchain components are added. Finally, the security and privacy implications of blockchain IoT applications are examined, as well as methods for improving the scalability and throughput of such systems. Blockchain, the Internet of Things, and Artificial Intelligence are cutting-edge technologies that will usher in a new era of digital transformation and disrupt a variety of industries. The convergence of these three technologies will result in the birth of new business models: Autonomous agents will operate as profit centres on their own. 1) that utilise IoT in supply chain to build a digital twin, 2) use blockchain technology in supply chain to send and receive money autonomously, and 3) and use AI in supply chain to make autonomous decisions as independent economic agents. The influence of supply chains on society, the environment, and financial success is immense, as is the establishment of autonomous business models and corporate digital revolutions. With the right application of modern technologies, supply chains have more potential to construct and support a sustainable future. By merging technologies such as IoT, AI, and Blockchain, businesses can reduce greenhouse gas emissions, optimize routes, remove waste, ensure efficient transactions with suppliers, improve worker safety, and limit hazards. A supply chain is made up of many firms from one end to the other. These contracts are vulnerable to forgery. The traceability of changing data is one of the major issues in supply chain systems. When compared to traditional database management systems, the current cost of blockchain solutions is simply too high, impeding commercial adoption. Current systems rely on the faith of persons involved in the supply chain's operation; this reliance on trust must be eliminated. Better blockchain technology, a mechanism to prevent any node from joining the supply chain network without permission, and the McEliece cryptosystem for a more added security layer to IoT communication systems without the use of any sophisticated or complicated systems are all included in the system.

The semantic-based technique of access control was taken to the next level of security in the future, with the history of the user's request and the development of fine-grained policies. With the proposed system, which will be implemented with heterogeneous devices in a real social network, other elements such as trust will be investigated.

REFERENCES

[1] Biswas, K., & Muthukkumarasamy, V. (2016). Securing smart cities using blockchain technology. In *High Performance Computing and Communications, IEEE 14th International Conference on Smart City*, pp. 1392–1393.

[2] Fernandez-Carames, T. M., & Fraga-Lamas, P. (2018). A review on the use of ~blockchain for the internet of things. *IEEE Access*, 6(6), 32979–33001.

[3] Song, J. C., Demir, M. A., Prevost, J. J., & Rad, P. (2018). Blockchain design for trusted decentralized IoT networks. In *2018 13th Annual Conference on System of Systems Engineering (SoSE)*. IEEE, pp. 169–174.

[4] Singh, S., Sanwar Hosen, A. S. M., & Yoon, B. (2021). Blockchain security attacks, challenges and solutions for the future distributed IoT network. *IEEE Access*, 9, 13938–13959, DOI: 10.1109/ACCESS.2021.3051602

[5] Agrawal, R., Verma, P., Sonanis, R., Goel, U., & De, A. (2018). Continuous security in IoT using blockchain. IEEE International Conference on Acoustics, Speech and Signal Processing (ICASSP), Samsung Research Institute Bangalore, IEEE. 6423–6427.

[6] Roy, S., Ashaduzzaman, Md, Hassan, M., & Chowdhury, A. (2018). Blockchain for IoT security and management: Current prospects, challenges and future directions. 5th International Conference on Networking, Systems and Security (NSysS), IEEE, DOI: 978-1-7281-1325-8/18/$31.00

[7] Krishnan, N., & Roopesh Jenu, K. (2018). Blockchain based security framework for IoT implementations. In *International CET Conference on Control, Communication and Computing*, July 5.

[8] Saxena, S., Bhushan, B., & Abdul Ahad, M. (2021). Blockchain based solutions to secure IoT: Background, integration trends and a way forward. *Journal of Network and Computer Applications*, 181, 103050, ISSN, 1084–8045, DOI: 10.1016/j.jnca.2021.103050

[9] Nartey, C. (2021). On blockchain and IoT integration platforms: Current implementation challenges and future perspectives. *Hindawi Wireless Communications and Mobile Computing*, Article ID 6672482, DOI: 10.1155/2021/6672482

[10] Abu-elkheir, M., Hayajneh, M., & Ali, N. A. (2013). Data management for the Internet of Things: Design primitives and solution. Sensors, 13(11), pp. 15582–15612, DOI: 10.3390/s131115582

[11] Kosta, B. P., & Naidu, P. S. (2021). Design and implementation of a strong and secure lightweight cryptographic hash algorithm using elliptic curve concept: SSLHA-160. *(IJACSA) International Journal of Advanced Computer Science and Applications*, 12(2), 15582–15612.

[12] Krishnan, K. N., Jenu, R., Joseph, T., & Silpa, M. L. (2010). Blockchain based security framework for IoT implementations. In *International CET Conference on Control, Communication, and Computing (IC4)*. IEEE, Thiruvananthapuram, pp. 425–429. DOI: 10.1109/CETIC4.2018.8531042.

[13] Roy, S., Ashaduzzaman, M. D., Hassan, M., & Chowdhury, A. R. (2018). Blockchain for IoT security and management: Current prospects, challenges and future directions. In *5th International Conference on Networking, Systems and Security (NSysS).*, IEEE, Dhaka, Bangladesh, pp. 1–9, DOI: 10.1109/NSysS.2018.8631365.

[14] Saxena, S., Bhushan, B., & Ahad, M. A. (2021). Blockchain based solutions to secure IoT: Background, integration trends and a way forward. *Journal of Network and Computer Applications*, 181, 103050. DOI: 10.1016/j.jnca.2021.103050

[15] Singh, S., Sanwar Hosen, A. S. M., & Yoon, B. (2021). Blockchain security attacks, challenges, and solutions for the future distributed IoT network. *IEEE Access*, 9, 13938 –13959, DOI: 10.1109/ACCESS.2021.3051602

[16] Bermeo-Almeida, O., Cardenas-Rodriguez, M., Samaniego-Cobo, T., Ferruzola-Gomez, E., & Cabezas-Cabezas, R. (2018). Blockchain in agriculture: A systematic literature review. In *Proceedings of the 4th International Conference*, vol. 883, pp. 44–56.

[17] Liang, X., Shetty, S., Tosh, D., Kamhoua, C., Kwia, K., & Njilla, L. (2017). ProvChain: A blockchain-based data provenance architecture in cloud environment with enhanced privacy and availability. In *Proceedings of the CCGRID*, pp. 468–477.

[18] Kim, T. H., Solanki, V. S., Baraiya, H. J., Mitra, A., & Shah, H. (2020). A smart, sensible agriculture system using the exponential moving average model. *Symmetry*, 12(3), 457.

[19] Anupama, H. S., DurgaBhavani, A., & Fayaz, A. B. A. Z. (2020). Smart farming: IoT based water managing system. *International Journal of Innovative Technology and Exploring Engineering*, 9(4), 2383–2385.

[20] Caro, M. P., Ali, M. S., Vecchio, M., & Giaffreda, R. (2018). Blockchain-based traceability in Agri-Food supply chain management: A practical implementation. In *IoT Vertical and Topical Summit on Agriculture*, pp. 1–4.

[21] Tian, F. An agri-food supply chain traceability system for China based on RFID & blockchain technology. In *Proceedings of the 13th International Conference on Service Systems and Service Management, ICSSSM, China.* pp. 1–6.

[22] Tian, F. A Supply chain traceability system for food safety based on HACCP, blockchain & internet of things. In *Proceedingsof the* 14th *International Conference on Services Systems and Services Management, ICSSSM*, pp. 1–6.

[23] Fu, Y., & Zhu, J. (2019). Big production enterprise supply chain endogenous risk management based on blockchain. *IEEE Access*, 7(8626088), 15310–15319.

[24] Fernández-Caramés, T. M., & Fraga-Lamas, P. A. (2018). Review on the use of blockchain for the Internet of Things. *IEEE Access*, 6, 32979–33001.

[25] Liang, G., Weller, S. R., Luo, F., Zhao, J., & Dong, Z. Y. Distributed blockchain-based data protection framework for modern power systems against cyber-attacks. *IEEE Transactions on Smart Grid*, 10(3), 3162–3173. DOI: 10.1109/TSG.2018.2819663

[26] Perboli, G., Musso, S., & Rosano, M. (2018). Blockchain in logistics and supply chain: A lean approach for designing real-world use cases. *IEEE Access*, 6, 62018–62028.

[27] Khan, P. S., & Byun, Y. C. (2020). IoT-Blockchain enabled optimized provenance system for food industry 4.0 using advanced deep learning. *Sensors*, 20(10), 2990. DOI: 10.3390/s20102990

[28] Pham, T. D., & Yan, H. (2017). Tensor Decomposition of Gait Dynamics in Parkinson's Disease. *IEEE Transactions on Biomedical Engineering*, 65(8), 1820–1827, DOI: 10.1109/TBME.2017.2779884

[29] Lindholm, B., Nilsson, M. H., Hansson, O., & Hagell, P. (2016). External validation of a 3-step falls prediction model in mild Parkinson's disease. *Journal of Neurology*, 263, 2462–2469, DOI: 10.1007/s00415-016-8287-9

[30] Delval, A. (2014). Why we should study gait initiation in Parkinson's disease. *Neurophysiologie Clinique/Clinical Neurophysiology*, 44, 69–76, DOI: 10.1016/j.neucli.2013.10.127

[31] Liang, J., Qin, Z., Xiao, S., Zhang, J., Yin, H., & Li, K. (2020). Privacy-preserving range query over multi-source electronic health records in public clouds. *Journal of Parallel and Distributed Computing*, 135, 127–139, DOI: 10.1016/j.jpdc.2019.08.011

[32] Liu, Y., Zhang, Y., Ling, J., & Liu, Z., (2018). Secure and fine-grained access control on e-healthcare records in mobile cloud computing. *Future Generation Computer Systems*, 78(3), 1020–1026, DOI: 10.1155/2017/6426495

[33] Akbarnejad, A., & Baghshah, M. S. (2017). A probabilistic multi-label classifier with missing and noisy labels handling capability. *Pattern Recognition Letters*, 89, 18–24, DOI: 10.1016/j.patrec.2017.01.02.

[34] Pei-Hao, C., Rong-Long, W., De-Jyun, L., & Jin-Siang, S. (2013). Gait Disorders in Parkinson's disease: Assessment and Management. *International Journal of Gerontology*, 7(4), 189–193, DOI: 10.1016/j.ijge.2013.03.005

[35] Hathaliya, J. J., & Tanwar, S. (2020). An exhaustive survey on security and privacy issues in Healthcare 4.0. *Computer Communication*, 153, 311–335. DOI: 10.1016/j.comcom.2020.02.018

[36] Omar, A. A., Bhuiyan, M. Z. A., Basu, A., Kiyomoto, S., & Rahman, M. S. (2019). Privacy-friendly platform for healthcare data in cloud based on blockchain environment. *Future Generation Computer Systems*, 95, 511–521. DOI: 10.1016/j.future.2018.12.044

[37] Prashanth, R., Dutta, R. S., Mandal, P. K., & Ghosh, S. (2016). High accuracy detection of early Parkinson's disease through multimodal features and machine learning. *International Journal of Medical Informatics*, 90, 13–21. DOI: 10.1016/j.ijmedinf.2016.03.001

[38] Hui-Ling, C., Chang-Cheng, H., Xin-Gang, Y., & Su-Jing, W. (2013). An efficient diagnosis system for detection of Parkinson's disease using fuzzy k-nearest neighbor approach. *Expert Systems with Applications*, 40(1), 263–271. DOI: 10.1016/j.eswa.2012.07.014

[39] AI, M. K. A., Musaed, A., Kashfia, S., & Saiful, I. M. (2017). Cloud based framework for Parkinson's disease diagnosis and monitoring system for remote healthcare applications. *Future Generation Computer Systems*. 66, 36–47. DOI: 10.1016/j.future.2015.11.010

[40] Hausdorff, J. M. (2009). Gait dynamics in Parkinson's disease: Common and distinct behavior among stride length, gait variability, and fractal-like scaling. *Chaos*, 19, 026113.

[41] Zeng, W., & Wang, C. (2016). Parkinson's disease classification using gait analysis via deterministic learning. *Neuroscience Letters*, 633, 268–278.

[42] Cong, F. (2015). Tensor decomposition of EEG signals: A brief review. *Journal of Neuroscience Methods*, 248, 59–69.

[43] Lin, K., Pankaj, S., & Wang, D. (2018). Task offloading and resource allocation for edge-of things computing on smart healthcare systems. *Computers & Electrical Engineering*, 95(72), 511–521. DOI: 10.1016/j.compeleceng.2018.10.003

[44] Bro, R., & Kiers, H. A. L. (2003). A new efficient method for determining the number of components in PARAFAC models. *Journal of Chemometrics*, 17, 274–286.

[45] Despotovic, V., Skovranek, T., & Schommer, C. (2020). Speech based estimation of Parkinson's disease using Gaussian processes and automatic relevance determination. *Neuro Computing*, 401, 173–181.

[46] Goyal, J., Khandnor, P., & Aseri, T. C. (2020). Classification, prediction, and monitoring of Parkinson's disease using computer assisted technologies: A comparative analysis. *Engineering Applications of Artificial Intelligence*, 96, 103955.

[47] Armañanzas, R., Bielza, C., Chaudhuri, R. K., Martinez-Martin, P., & Larrañaga, P. (2013). Unveiling relevant non-motor Parkinson's disease severity symptoms using a machine learning approach. *Artificial Intelligence in Medicine*, 58(3), 195–202, ISSN 0933-3657, DOI: 10.1016/j.artmed.2013.04.002

[48] Sgantzos, K., & Grigg, I. (2019). Artificial intelligence implementations on the blockchain. Use cases and future applications. *Future Internet*, 11, 170. DOI: 10.3390/fi11080170.

[49] Lin, K., Pankaj, S., & Wang, D. (2016). Task offloading and resource allocation for edge-of things computing on smart healthcare systems. *Computers & Electrical Engineering*, 72, 348–360. DOI: 10.1016/j.compeleceng.2018.10.003

2 Blockchain, IoT, and Artificial Intelligence Technologies for Supply Chain Management: Bibliometric Analysis

Esra Ozmen
Ankara Hacı Bayram Veli University, Ankara, Turkey

Nurbahar Bora
Atatürk University Social Sciences Institute, Erzurum, Turkey

CONTENTS

2.1 INTRODUCTION

Business firms follow different strategies to have competitive advantage in the market. With the development of technology, various systems and applications are utilized by enterprises for having competitive advantage, survival, decision planning, management, and so on. Moreover, with the emergence of Industry 4.0, it is seen that various concepts and systems have also emerged in the use of technology by enterprises; for example, improved supply chain management systems providing material, information, and money flow; artificial intelligence imitating human intelligence; the Internet of Things, which enable physical objects to be connected with each other and with larger systems are new concepts that have entered our life with the development of technology. In this chapter, a bibliometric analysis study has been conducted with 1558 metadatasets.

The data were collected in January 2022, via Scopus database. All research studies published between 1995 and 2021 and including Blockchain, IoT, AI, or SCM keywords are listed. With the bibliometric analysis it is possible to produce scientific maps that reveals relationships in inter-disciplinary fields. This original study also aimed to contribute related literature, which requires holistic bibliometric analysis covering these four areas. For data visualization, analysis was per-formed using the VOSviewer tool. It is plausible to say that this study will be useful in terms of providing an overview of the blockchain, IoT, AI, and SCM literature and may guide researchers who want to work in these fields. In addition, this research presents the analysis of Blockchain, IoT, AI, and SCM research from a holistic perspective using bibliometric visualization techniques. It contributes to researchers, institutions, and scientists who want to work in these fields, to reveal and evaluate the effect of developing technological processes. This study is limited to the Scopus database. Those who want to work in this field can examine the databases and find WoS, IEEE Xplore, Google Scholar, etc.

In this chapter, it is aimed to contribute to the literature by subjecting the studies in the fields of Blockchain, IoT, AI, and SCM to a holistic bibliometric analysis. The execution of the study was carried out in four main steps. Firstly, it was aimed to examine the current status of the studies in the field of blockchain, IoT, AI, or SCM. Then, the databases with the studies on these fields were searched. The data collection process was determined by deciding on the publication range of the studies in the Scopus database. The bibliometric analysis step of the collected data was started by using the VOSviewer tool. Details such as country, source, and keywords were discussed in the analysis phase. The results were presented in light of the findings and various suggestions were included.

Within the scope of this chapter, the following objectives are expected to be achieved:

- To provide a systematic perspective on past, present, and future developments in Blockchain, IoT, AI, and SCM technologies.
- To fill the gap in the literature by providing bibliometric analysis of interdisciplinary fields.
- To guide future research on the course of technology and to contribute to the academic literature.

Organization of Chapter: The chapter is organized as: Section 2.2 elaborates the background of the study. Section 2.3 highlights the related works. Section 2.4 elaborates on the methodology. Section 2.5 outlines the findings. Section 2.6 stresses the discussion. Section 2.7 concludes the chapter with future scope.

2.2 BACKGROUND OF THE STUDY

2.2.1 BLOCKCHAIN

There are many definitions in the literature about blockchain. For example, Beck [1] (2018) defined it as a database that allows secure and consistent transactions by many nodes in the network. Zheng et al. [2] (2017) defined blockchain as a data ledger. While the transactions approved in this data ledger are stored in block lists, the data ledger grows as each block is added. In another definition of blockchain, it is expressed as a part of a distributed software system to ensure data integrity [3]. According to Swan [4] (2015), blockchain is essentially a distributed database system that records transaction data and information, secured by cryptography and managed by a consensus mecha-nism. Looking at the definitions in the literature, blockchain can be technically expressed as a combination of a distributed database and cryptographic algorithms. Blockchain technology has evolved over time: Blockchain 1.0 in money transfer and payment systems in 2009, Blockchain 2.0, known as smart contracts in 2013, and Blockchain 3.0 in non-financial applications as of 2020 [5]. With the aforementioned development of blockchain, various classifications have been made. In current classifications, there are public, private, and consortium blockchain categories [6].

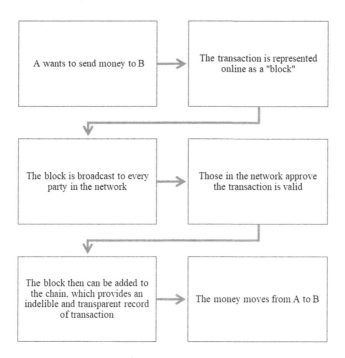

FIGURE 2.1   Blockchain Working Logic.

Blockchain offers various advantages such as preventing data loss and destruction, high level of trust, and transparency. In addition to these advantages, there are also disadvantages such as performance and user privacy [7]. The blockchain working logic conveyed by Crosby et al. [8] (2016) is presented in Figure 2.1.

2.2.2   Internet of Things (IoT)

The introduction of the IoT concept into our lives was proposed in the Massachusetts Institute of Technology (MIT) Auto-ID laboratories in the early 1990s. However, the first IoT application "Trojan Room coffee pot" was developed in 1999 [9]. In the same year, the world's first Internet-controlled device, a remote toaster, was developed [10]. However, the concept we call IoT is located in the middle of the concepts of object, human, and Internet today. The Internet of Things is defined as a network formed by connecting objects in the simplest sense [11]. In other words, it can also mean that objects have communication with each other. The Internet of Things also refers to the integration of the virtual world with objects, which has emerged with the development of technology [12]. RFID, NFC, and wireless technologies have been extremely effective in the development of the Internet of Things [13]. On the other hand, embedded intelligence, connection, and interaction steps in the development of this technology should not be forgotten [14]. An example of a high-level IoT architecture is presented in Figure 2.2 [15].

2.2.3   Artificial Intelligence

Along with the industrial revolution, many developments have occurred in the field of technology. Technology has helped to do countless jobs that humanity has been challenged or unable to do. Artificial intelligence is among the technological innovations that started to do the work done manually by humans. Artificial intelligence is the field of science and technology created with computer programs and intelligent systems to perform various tasks and tasks that require human

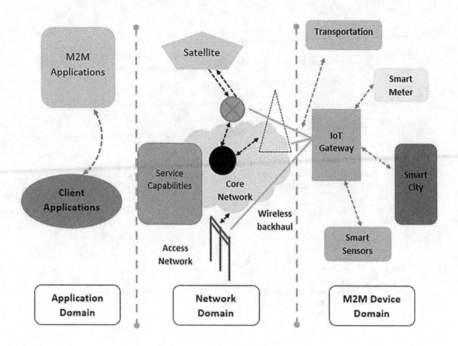

FIGURE 2.2   An Example of a High-Level IoT Architecture.

intelligence [16]. Artificial intelligence, which is used more and more every day, is the most important source of innovation today. Many places such as hotels, health care, restaurants, etc. can be given as examples of fields where artificial intelligence services are employed [17,18].

2.2.4   SUPPLY CHAIN MANAGEMENT (SCM)

Supply Chain is defined as a set of elements including suppliers, logistics services, manufacturers, distributors, and retailers, and there is a flow of materials, products, and information among those components [19]. Supply chain management, on the other hand, is the management of materials and products from the supply of basic raw materials to the final product stage; it is a management philosophy that focuses on how to exploit the process, competitive advantages, technology, and capabilities. Supply chain management has some objectives such as increasing customer satisfaction, shortening the cycle time, reducing costs, and product defects [20]. Supply chain management also includes processes such as customer relationship management, order processing, demand management, purchasing, and product development [21,22]. The interrelationship of blockchain, IoT, AI, and SCM technologies is explained in Figure 2.3 [23].

2.3   RELATED WORKS

There are bibliometric studies in many fields in the literature. Related studies were reached by querying the keywords "blockchain," "IoT," "AI," "SCM," and "bibliometric analysis" in WoS, Scopus, Science Direct, and Google Scholar research databases. In this section, related studies in the literature are summarized.

Firdaus et al. [24] (2019) analyzed blockchain research activities with 1119 publications in the Scopus database between 2013 and 2018 using bibliometric methods. The countries with the highest number of publications in the field of blockchain were the United States, China, and Germany respectively, while Singapore and Switzerland are the countries with the least number of publications. On the other hand, in terms of the number of citations, it is seen that the publications

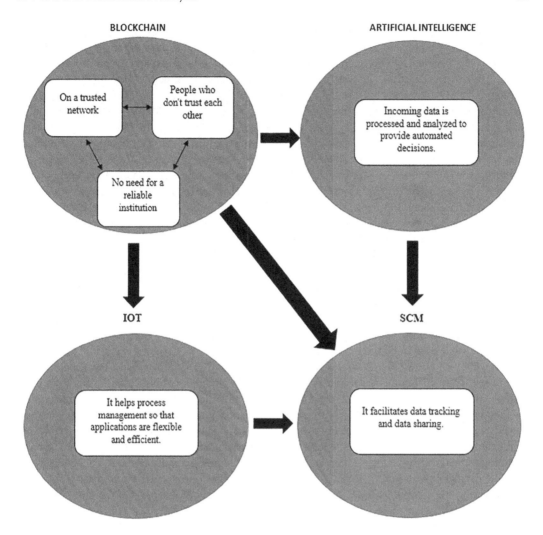

FIGURE 2.3 The Interrelationship of Blockchain, IoT, AI, and SCM Technologies.

originated from Singapore and Switzerland receive higher citations. In terms of keywords used in the publication by authors, it is observed that blockchain, Bitcoin, privacy, smart contract, and cryptocurrency were determined as the most used keywords. "Data privacy," "digital storage," "data security," "big data," and "distributed databases" were determined as the most used categories in solving problems in blockchain. Miau and Yang [25] (2018) analyzed 801 publications and blockchain technologies in the Scopus database related to blockchain technology with bibliometric methods. The countries with the most publications in the blockchain space are the United States, United Kingdom, Germany, Switzerland, and China, respectively. "Bitcoin," "blockchain," and "electronic money" are the most used keywords by the authors who publish these posts. It is stated that the number of literatures on blockchain has increased. Blockchain research in the WoS database was conducted by Alam et al. [26] (2022). In the bibliometric analysis, China and the USA were determined as the countries that contributed the most to blockchain research. It is seen that the blockchain is not only about informatics, but also operates in the fields of telecommunications, commerce, trust, reputation, and law.

Guo et al. [27] (2021) analyzed the data collected from WoS database and CiteSpace with VOSviewer analysis tools. It has been determined that the countries with the most publications in the field of blockchain are China, the USA, and the United Kingdom, respectively. The USA is the most cited country compared to other countries. The most popular journals are *IEEE Access, IEEE Journal of the Internet of Things, Next Generation Computer Systems,* and *Sensor and Applied Science Basel.* In addition, it was concluded that "smart contract," "Bitcoin," "security," "Ethereum," and "cryptography" are the most used keywords, respectively. Luo et al. [28] (2021) performed bibliometric analysis using the blockchain technology SciMat tool with 2993 publications in the WoS database. They concluded that China, USA, South Korea, England, and Australia are the countries that broadcast the most, respectively. On the other hand, it is stated that there are five groups of data sets: "Data Security and Privacy Protection," "New Algorithms," "Features and Applications of Blockchain," "New Technology Research," and "Cloud Technology Research." Leong et al. [29] (2021) performed a bibliometric analysis of IoT technology studies on the WoS database using VOSviewer. The most publications in the field of IoT have been seen as The Technological Forecasting and Social Change. While the most cited journals were *Technological Forecasting* and *Social Change*, the country with the highest number of publications in IoT studies was the USA. It is also found that "Internet of Things," "model," "technology," "adoption," and "user acceptance" are the mostly used keywords by the authors. Katoch [30] (2021) discussed studies in the fields of IoT, SCM, and logistics in the Scopus database. When the journals were analyzed according to the number of publications, *The International Journal of RF Technologies Research* took the first place. The country with the highest number of publications in these fields was the USA. It has been seen that "RFID," "SCM," "IoT," "logistics," and "Industry 4.0" are the most used keywords.

In the study of Raza [31] (2022), research focused on RFID applications in the supply chain were investigated on Scopus database. According to the findings, while China is the most productive country, USA is the most cited country in this field. *The International Journal of Production Economics* became the most widely published journal. When the keywords are examined, it is seen that the most used keywords are "supply chain management," "radio frequency identification," and "supply chains." It is stated that machine learning, IoT, and blockchain technologies benefit from supply chain and RFID technologies. Examining the IoT technology in the WoS database, Wang et al. [32] (2021) reported that the journal with the highest number of publications in the study was IEEE Access; it has been a sensor magazine. On the other hand, he stated that China is the most cited country. Internet of Things, wireless sensor networks, management, and security appear to be the most used keywords. In the bibliometric analysis study conducted by Niu et al. [33] (2016) about artificial intelligence on the WoS database. *Expert Systems with Applications* was the journal with the most publications in the study. When the journals were examined in terms of the number of citations, artificial intelligence took the first place. The country that produces the most in this field is the USA. When the keywords are examined, there are "artificial intelligence (AI)," "artificial neural network (ANN)," "genetic algorithm (GA)," "expert system (IT IS)," and "optimization." Gao and Ding [34] (2022) performed bibliometric analysis about artificial intelligence in WoS, Scopus, and Derwent Innovation Index (DII) databases. They reported that USA, China, England, Germany, and Spain have higher of publications. In the bibliometric analysis, areas of expertise were determined by focusing on three clusters named random forest, deep learning, and machine learning.

Zeba et al. [35] (2021) discussed the applications of artificial intelligence in the manufacturing sector in the WoS database. Bibliometric analysis was performed using VOSviewer and WordStat analysis tools. The most used keywords in 1979–2010 were "artificial intelligence," "production system," "neural networks," "flexible manufacturing system," "expert systems," "petri nets," "control system," and "decision support." In 2011–2019, it is stated as artificial intelligence, neural networks, production system, machine learning, smart manufacturing, cyber-physics, genetic algorithms, and deep learning. The journal that ranked first in terms of citations in 2019 was *International Journal of Production Research.* Riahi et al. [36] (2021) conducted bibliometric analysis using RStudio mapping software in their artificial intelligence-based supply chain research in the Scopus database. *International Journal of Production Research* ranks first in terms of number of journals. The countries with the most publications are the United States and the United Kingdom. Sharma et al. [37] (2022) demonstrated artificial intelligence research in SCM. The keyword supply chain management ranked first. In SCM and AI research, it was concluded that there is a relationship with smart agent and inventory control issues. It was concluded that "supplier selection" in SCM and "genetic algorithms" in artificial intelligence are widely used. Zekhnin et al. [23] (2020) covered SCM searches in the Scopus, Elsevier, Emerald, Taylor & Francis, Springer, IEEE, and Google Scholar databases. It has been seen that the USA is the first country in terms of publishing research in the field of SCM. *Supply Chain Management: An International Journal* is the most popular one having the highest number of studies about the topic. In addition, "SMC," "supply chain," and "blockchain" are the most used keywords. Marty [38] (2022) used BibExcel and Gephi software for supply chain management research in Scopus and WoS databases. It has been seen that the USA ranks first in terms of the number of publications. *International Journal of Production Economics* took the first place in terms of publication quantity. It is seen that the keywords "supply chain management," "integration," "sector," and "logistics" are the most common ones.

Kamran et al. [39] (2020) revealed the concept of BIot by analyzing blockchain and iodine research with bibliometric methods, with 151 publications in the WoS database between 2008 and 2019 (April). In BIot research, the journals with the most publications are listed as *IEEE Access, Sensors, International Journal of Distributed Sensor Networks, Future Generation Computer Systems,* and *The International Journal of eScience.* The countries with the most publications in this field were China and the USA, respectively. When the author keywords are examined, it is seen that the results of "security," "smart contracts," "computing," "privacy," and "smart city" are reached, respectively. In the study conducted by Szum [40] (2021), VOSviewer analysis tool was used with 1019 publications in WoS, Scopus and IEEE Xplore databases between 2012 and 2021 and the author analyzed IoT-based smart city research with bibliometric methods. The countries with the most broadcasts in the field of IoT-based smart city were determined as India, United States, and China, respectively. *IEEE Internet of Things Journal* ranks first among the most widely published journals, with *Sensor* second and *IEEE Access* third. Author keywords were analyzed in five cluster titles. The words specified are "IoT application domains in smart cities," "IoT architecture for smart cities," "energy," "security-privacy," and "data."

Çiğdem [41] (2021) used the R analysis tool in her study with 8036 publications in the WoS database between 1991 and 2021. Author analyzed research on digital transformation of the supply chain using bibliometric methods. China ranked first with the most publications in this field, and USA took the second place. USA ranks first in terms of number of citations, while China ranks second. *International Journal of Production Economics* ranks first in terms of citations and publications, while *International Journal of Production Research* ranks second. When the author keywords are examined, it is seen that the keywords "RFID," "IoT," "blockchain," "sustainability," and "big data" are used intensively, respectively. Zhang et al. [42] (2020) used R and VOSviewer analysis tools with 777 publications in the Scopus database in (2012-2019) and analyzed big data and sustainable supply chain management research with bibliometric methods. The first country with the most publications in these fields was the USA, while the second was China. *Sustainability* ranks first among the most popular magazines, while *Lecture Notes in Computer Science* is second. *Advances in Intelligent System and Computing* ranks third. "Big data," "sustainability," "supply chain management," "big data analysis," and "supply chains" were used for author keywords, respectively.

In summary, it is understood that the number of publications in the field of blockchain is quite high in the USA and China. *IEEE Access* is the most popular journal; blockchain and Bitcoin seem to be the most used keywords. In the field of IoT, it is also seen that the number of publications made in the USA and China is quite high. It can be inferenced that *The Technological Forecasting and Social Change* and *IEEE Access* are among the most popular journals, and IoT and RFID are the most used keywords. *Applied Expert Systems: International Journal of Production Research* has become the most popular journal in the field of artificial intelligence with its numerous publications in the USA. It is seen that "artificial intelligence" is the most commonly used keyword. It is seen that the number of publications in the USA in the field of SCM is high, *Supply Chain Management: An International Journal* is the most popular journal, and "SCM" and "supply chain" are the most used keywords.

2.4 METHODOLOGY

In order to fill the gap in the literature about Blockchain, IoT, AI, and SCM with a holistic bibliometric analysis and provide the interaction between developing technologies from a holistic perspective, a bibliometric analysis was conducted in this study. This chapter focus on the following research questions:

- Which countries, journals, and author keywords are included in Blockchain, IoT, AI, and SCM research?
- Which countries and journals are mostly cited in Blockchain, IoT, AI, and SCM researches?
- How are these areas related to each other over time?

2.4.1 RESEARCH DESIGN

Bibliometric analysis is one of the quantitative research methods that helps to analyze publications. With this analysis, it is possible to clearly display and visualize the interactions and performances of different disciplines [27]. Lists of countries, authors, institutions, journals, keywords, etc. help to visualize the relationships between them using scientific maps [43].

2.4.2 DATA COLLECTION

Within the scope of this research, data collection was carried out on 23 January 2022 via the Scopus database. In this context, research published in English languages between the years 1995 and 2021, having the keywords "Blockchain," "IoT," "AI," and "SCM" in part of the "title," "abstract," and "keywords" were queried. The listed 1558 researches were incorporated in this study. The metadata set was obtained in ".csv" format and the data collection process was completed. Scopus, which has been in existence since 2004, is a database that holds close to 69 million records, serving many researchers in the social, health, life, and physical sciences [43]. The query used in the data collection process is given below. The steps followed in the study are shown in Figure 2.4.

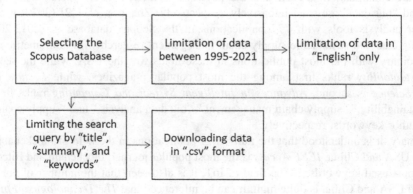

FIGURE 2.4 Data Collection Process.

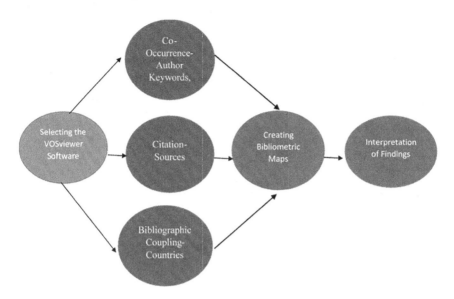

FIGURE 2.5 Data Analysis Process.

Query: TITLE-ABS-KEY (((("Blockchain" OR "IoT" OR "Internet of Things" OR "Artificial Intelligence" OR "AI") AND ("SCM" OR "Supply Chain Management"))))

2.4.3 DATA ANALYSIS

There are many analysis tools with different features for the realization of bibliometric analysis. Within the scope of this research, bibliometric analysis was performed using the open source VOSviewer software developed by Leiden University (Netherlands) for the analysis of the data [44]. With VOSviewer, it is possible to create nodes according to the size of the data, repetition of nodes, and data visualization according to the level of interaction between nodes [45]. The bibliometric analysis of the data was completed with Co-Occurrence-Author Keywords, Citation-Sources, and Bibliographic Coupling-Countries. The data analysis process is presented in Figure 2.5.

2.5 FINDINGS

The findings obtained within the scope of the study were discussed in three categories: the author, the journals cited, and the countries where the study was conducted.

2.5.1 CO-OCCURRENCE/AUTHOR KEYWORDS

When the publications on the use of Blockchain, IoT, and artificial intelligence technologies in supply chain management are examined, "co-occurrence-author keywords" were chosen to determine the most used author keywords in the VOSviewer analysis tool. The minimum number of repetitions for each keyword is set at five. In total, 147 out of 3252 keywords were found to be related to each other. The analysis process was provided by including only one of the same expressions with the same meaning and spelling. The analysis process was completed by visualizing 108 data as seven clusters. Table 2.1 shows the 20 most repeated keywords.

When Table 2.1 is examined, it is seen that "block chain" is the most used keyword with the highest connection strength (853). "Supply chain management" is second place, and "Internet of

TABLE 2.1
Related Keywords Used in Studies

Keywords	Occurrences	Total Link Strength
Blockchain	489	853
Supply Chain Management	414	722
Internet of Things	215	335
Industry 4.0	92	160
Rfid	77	97
Traceability	63	162
Smart Contract	58	144
Sustainability	53	124
Logistics	49	110
Smart Contracts	41	97
Big Data	38	75
Artificial Intelligence	37	60
Cloud Computing	33	63
Security	30	79
Machine Learning	29	61
Distributed Ledger	28	69
Transparency	27	74
Trust	23	64
Ethereum	23	60
Distributed Ledger Technology	18	45

Things" in third place. Although "Industry 4.0" is the fourth most used keyword, it is seen that it ranks fifth in terms of connection strength (160). Figure 2.6 includes the most commonly used keywords related to each other in studies.

The author keywords that emerged as a result of the bibliometric analysis are presented in Figure 2.6. It seems that "blockchain" ranks first as the keyword with the most used formations (489)

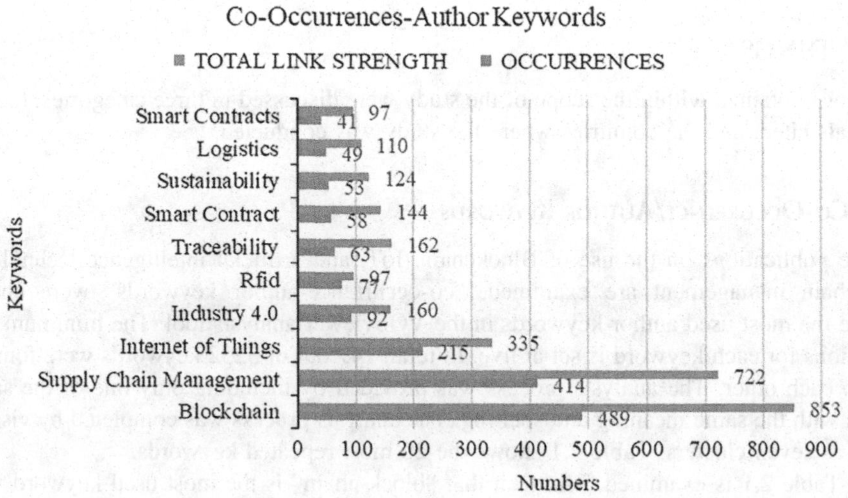

FIGURE 2.6 Related Author Keywords in the Top 10.

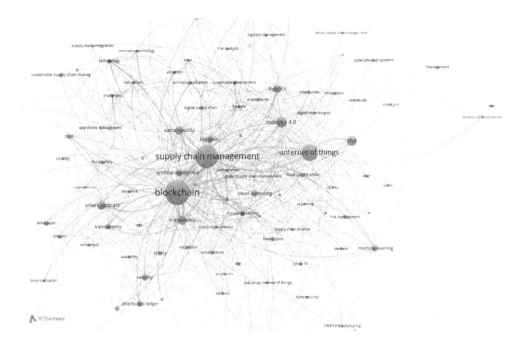

FIGURE 2.7 "Co-Occurrence-Author Keywords" Analysis (Network Visualization) in Supply Chain Management of Blockchain, IoT, and Artificial Intelligence Technologies.

and the highest connection strength (853), while "Supply Chain Management" ranks second as the keyword with occurrences (414) and connection strength (722). On the other hand, "Internet of Things" ranks third as the keyword with occurrences (215) and connection strength (335), while "Industry 4.0" occurrences (92) ranks fifth in terms of connection strength (160), although it is the fourth keyword. "RFID" is the fifth keyword for occurrences (77), while it ranks eleventh in terms of link strength (97).

Figure 2.7 shows "network visualization" links for co-author keywords. It is clear that there are strong ties between the keywords questioned for the four disciplines. Large or small circles indicate strong and weak relationships among others keywords. A total of 108 keywords are divided into seven clusters. It is understood that there is an intense interaction between the clusters. The keywords "supply chain management" (cluster 1), Fig 2.7 "blockchain" (cluster 2), and "Internet of Things" (cluster 1) are represented by the largest circles.

In cluster 1 (28 keywords), "game theory," "Industry 4.0," "IoT," "RFID," "tedarik zinciri yönetimi," etc. keywords are included. Cluster 2 includes (19 keywords), "3D Printing," "AHP," "drones," "TOPSIS," etc. keywords. In Cluster 3, 16 keywords such as "Bitcoin," "blockchain," "cryptocurrency," "Ethereum," etc. are displayed. Cluster 4 has 14 keywords: "cybersecurity," "interoperability," "machine learning," and "sensors". In Cluster 5 (12 keywords), "information technology," "innovation," "technology adoption," "technology management," etc. keywords are included. Cluster 6 caps 10 keywords such as "artificial intelligence," "big data," "cloud computing," and "e-commerce." And in Cluster 7 (9 keywords), "traceability," "transparency," "trust," "visibility," etc. are represented. The clusters and frequently used words obtained as a result of the analysis are listed below.

Total 108 Products (7 Clusters)

1. **CLUSTER 1 (Items 28):** case study, coordination, cyber-physical systems, digital supply chain, digital technologies, digital transformation, EPC, food supply chain, game theory,

green supply chain management, healthcare, industry 4.0, IoT, logistics, logistics management, management, performance, procurement, RFID, risk analysis, simulation, smart cities, smart factory, smart grid, smart logistics, smart supply chain management, supply chain management, wireless sensor networks.

2. **CLUSTER 2 (Items 19):** 3D printing, additive manufacturing, agriculture supply chain, AHP, partnership, COVID-19, decentralization, demasel, digital economy, digitization, drones, information sharing, integration, risk management, supply chain finance, supply chain resilience, supply chain risk management, sustainable supply chain, topsis.

3. **CLUSTER 3 (Items 16):** Bitcoin, blockchain, consensus, cryptocurrency, data integrity, distributed ledger, Ethereum, hyper notebookmaterial, IPFS, privacy, product traceability, flexibility, scalability, security, smart contract.

4. **CLUSTER 4 (Items 14):** cyber security, data analytics, data management, data mining, high-throughput video coding(HEVC), industrial Internet of things, information security, interoperability, machine learning, resource, screen content coding (SCC), sensors, smart manufacturing, standards.

5. **CLUSTER 5 (Items 12):** adoption, barriers, challenges, circular economy, information technology, innovation, sustainable development, technology, technology adoption, technology management, utaut, value chain.

6. **CLUSTER 6 (Items 10):** agriculture, artificial intelligence, big data, cloud computing, e-commerce, ERP, food safety, network analysis, operations management, optimization.

7. **Cluster 7 (Items 9):** design science research, supply chain integration, supply chain performance, sustainable, sustainable supply chain management, traceability, transparency, trust, visibility.

2.5.2 ATTRIBUTION/SOURCES

Taking into account the relevant publications on the use of Blockchain, IoT, and Artificial Intelligence technologies, in supply chain management, the "common citation sources," a selection was made to determine the most cited journals in the VOSviewer analysis tool.

When the cited journals are examined, the number of repetitions of the publications has been determined as at least five. In total, 56 out of 769 journals were found to be interrelated. Only one of the same expressions with the same meaning and spelling was included in the process to do the analysis. The analysis was completed by visualizing 56 journals as 12 clusters. The top 20 most cited journals are presented in Table 2.2.

In Table 2.2, *International Journal of Production Research* ranks first as the most cited journal. *International Journal of Information Management* second, *Sustainability* (Switzerland) third, *IEEE Access* fourth, and *Supply Chain Management* is fifth.

In Figure 2.8, the most cited journals as a result of the bibliometric analysis are shown. *International Journal of Production Research* ranks first with the highest number of citations (2918) and total link strength (241), and fourth with the number of publications (30). It seems that *International Journal of Information Management* is second in citation count (1436), third in total link strength (154), and and 19th in terms of number of publications (12). *Sustainability* (Switzerland) ranks third with the number of citations (974), second with the total connection strength (239), and first with the number of publications (69). *IEEE Access* ranks fourth (951) in terms of number of citations, fifth (127) in terms of total link strength, and third (10) in terms of number of publications. *Supply Chain Management* ranks fifth in citation count (880), fourth in total link strength (138), and 25th according to the number of publications (10).

"Density Visualization" links of cited journals are shown in Figure 2.9. When examined in general, it is understood that there are strong connections between the journals questioned for the four disciplines. More or less density indicates strong and weak relationships between other journals. While it is seen that there is an intense interaction between the four clusters, it is

TABLE 2.2
Most Cited Journal List

Source	Documents	Quotes	Total Connection Strength
International Journal of Production Research	30	2918	241
International Journal of Information Management	12	1436	154
Sustainability (Switzerland)	69	974	239
IEEE Access	31	951	127
Supply chain management	10	880	138
Computer and Industrial Engineering	11th	829	79
Transportation Research Part E: Logistics and Transportation Study	11th	524	77
Technological Forecasting and Social Change	13	498	53
International Journal of Production Economics	16	465	69
Cleaner Production Magazine	9	258	11th
Procedia Manufacturing	7	208	13
Industrial Management and Data Systems	9	187	35
Sensors (Switzerland)	6	187	7
Electronics (Switzerland)	5	185	16
International Journal of Logistics Research and Applications	6	180	26
European Journal of Operations Research	6	162	5
Applied Sciences (Switzerland)	9	151	27
Computer Science Lecture Notes (Including Artificial Intelligence Subseries Lecture Notes and Bioinformatics Lecture Notes)	30	132	2
Journal of Business Logistics	6	132	9
Computers in Industry	5	130	16

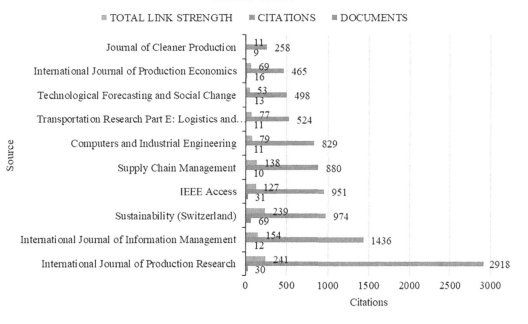

FIGURE 2.8   Top 10 Most Cited Journals.

FIGURE 2.9 Journals that are not related to each other are listed below. Those, Eaı/Springer Innovations in communication and Computing, Proceeding of the International, Intelligence Systems Reference, 2020 13 th International Colloquium, Journal of Physics: Conference, Wit Transactions on Information, Applied Mechanics and Material, Iclem 2014: System Planning.

understood that there is no interaction between the eight clusters. Interactive logs between the four clusters are presented in Figure 2.10.

"Network visualization" links of linked and cited journals are shown in Figure 2.10. It is understood that there are strong connections between the journals questioned for the four disciplines. Large or small circles indicate that other cited journals have strong and weak relationships. Among a total of 56 journals, *Sustainability* (Switzerland) cluster 1, *International Journal of Information*

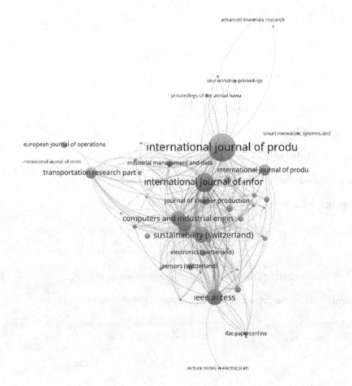

FIGURE 2.10 "Co-Citation-Sources" Analysis (Network Visualization) in Blockchain Supply Chain Management, IoT, and Artificial Intelligence Technologies.

Management" cluster 2, *Computer and Industrial Engineering* cluster 3, and *International Journal of Production Research* (cluster 6) are marked with large circles.

In the cluster 1 (16 magazines), *Electronics (Switzerland), Supply Chain Management, "Sustainability (*Switzerland*), Technological Foresight and Social Change*, etc. are the most cited journals. In the cluster 2 (12 magazines), *International Journal of Information Management, Applied Sciences (*Switzerland*), IEEE International Conference on Industrial Engineering and Engineering Management, Ifac-Papersonline*, etc. are seen as the most cited journals. The most cited journals in the cluster 3 (11 journals) are *Computer and Industrial Engineering, Industrial Management and Data Systems, Journal of Business Logistics, Computer Science Lecture Notes (Including Artificial Intelligence Sub-Series Lecture Notes and Bioinformatics Lecture Notes)*. In the cluster 6, nine magazines, like *International Journal of Manufacturing Research, European Journal of Operations Research, International Journal of Manufacturing Economics, Transportation Research Part E: Logistics and Transportation Review*, etc. are the most cited journals. The clusters and frequently used words obtained as a result of the analysis are listed below.

Total 56 Products (12 Clusters)

1. **CLUSTER (items 16):** communications in computer and information science, computing in industry, electronics (Switzerland), IEEE transactions on engineering, advances in knowledge, international logistics research journal, international supply chain management journal, cleaner production journal, institute lecture notes, operations and supply chain management, ongoing manufacturing, production planning and control, sensors (Switzerland), supply chain management,sustainability (Switzerland), technological forecasting and social change.

2. **CLUSTER (items 12):** ACM international conference minutes, smart systems and developments in computing, appliedsciences (Switzerland), CEUR workshop progression, IEEE access, IEEE international conferences, online impact articles, international knowledge management journal, lop conference series: earth and environmental science, lecture notes on commercial computing, lecture notes on electrical engineering, continuation of annual Hawaii international meeting systems science conferences.

3. **CLUSTER (Items 11):** computers and industrial engineering, industrial management and data systems, international journal of advanced computer sciences and application, journal of business logistic, journal of enterprise information management, lecture notes in computer science (including subseries lecture notes in artificial intelligence and lecture notes in bioinformatics), lecture notes in mechanical engineering, lecture notes in networks and systems, Procedia computer sciences, proceeding of the international conference on industrial engineering and operations management, wireless personal communications.

4. **CLUSTER (Items 9):** advanced materials research, annals of operations research, European journal of operational research, international journal of production economics, international journal of production research, international journal of recent technology, pervasive health pervasive computing technologies for healthcare, smart innovation systems and technology, transportation research part logistics and transportation review.

5. **CLUSTER (Item 1):** 2020 IEEE 13th International colloquium of logistic.

6. **CLUSTER (Item 1):** applied mechanics and materials.

7. **CLUSTER (Item 1):** springer innovation.

8. **CLUSTER (Item 1):** journal of physics, conference.

9. **CLUSTER (Item 1):** proceeding of the international.

10. **CLUSTER (Item 1):** operations related to information.

11. **CLUSTER (Item 1):** ICLEM item 2014; system planning.

12. **CLUSTER (Item 1):** intelligent systems reference library.

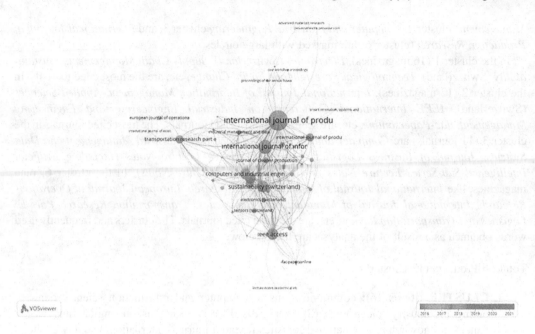

FIGURE 2.11   Scale Showing the Increase in Publications by Years.

In addition, distribution analysis was made according to publication years and is presented in Figure 2.11. It is seen that there has been an increase in the number of studies in these areas since 2017 and this increase continues exponentially in our country from 2019 to 2020.

2.5.3   BIBLIOGRAPHIC LINK/COUNTRY

"Bibliography unification countries" was chosen to determine the most cited journals in the VOSviewer analysis tool for related publications for the use of blockchain, IoT, and artificial intelligence technologies in supply chain management.

When examining the countries working together, the number of repetitions of the publications was determined as a minimum of five. In total, 53 out of 114 countries turned out to be related to each other. The analysis process was provided by including only one of the same expressions with the same meaning and spelling. The analysis process was completed by visualizing 53 countries in three clusters. The first 20 countries ranked according to the number of publications are shown in Table 2.3.

In terms of number of publications, China is first, India is second, United States is third place, England is fourth, and in fifth place is Germany. In Figure 2.12, a visualized version of the publication numbers of countries with bibliometric pairs is mapped.

A visualized version of the citations of countries with bibliometric pairs is shown in Figure 2.13. "United States" ranks first with 6617 citations, "China" ranks second with 3543 citations, "United Kingdom" ranks third with 3442 citations, "India" ranks fourth with 2421 citations, and "Germany" ranks fifth with 1391 citations.

A visualized version of the total link strengths of countries with bibliometric pairs is shown in Figure 2.14. "United States of America" ranks first with 68755 total connection strength, "United Kingdom" ranks second with 66157 total connection strength, "India" ranks third with 65997 total connection strength, and "China" ranks fourth with a total connection strength of 61798, "Australia" ranks fifth with a total connection strength of 33585.

Countries with interaction between the three clusters are presented in Figure 2.15. It is clear from the figure that there are strong links between countries for the four disciplines. Large or small

---

TABLE 2.3

Bibliographic Link: Countries

Country	Documents	Quotes	Total Connection Strength
Chinese	303	3543	61798
India	247	2421	65997
United States of America	178	6617	68755
United Kingdom	129	3442	66157
Germany	78	1391	27123
Australia	70	1030	33585
Canada	58	1106	23922
South Korea	57	877	15241
Italy	56	1128	30012
France	55	1328	30035
Hong Kong	44	1330	15959
United Arab Emirates	39	546	22320
Malaysia	37	431	18512
Taiwan	37	746	15530
Morocco	32	134	6600
İran	29	250	18260
Russian Federation	29	548	5846
Japan	26	432	4851
Brazil	25	636	15570
Greece	25	433	10981

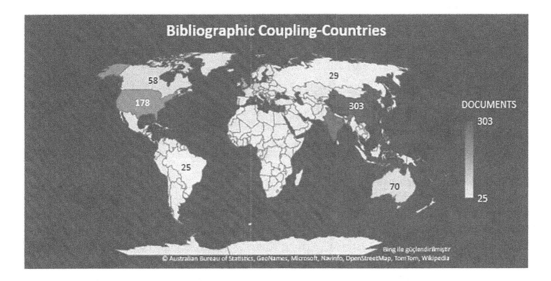

FIGURE 2.12   Number of Publications by Countries with Top 20 Bibliometric Pairs.

circles indicate strong and weak relationships between other journals. Fifty-three countries are divided into three clusters. It is understood that there is an intense interaction between the clusters. The first cluster has the highest density compared to the other clusters. The countries "United States of America", "India", "United Kingdom", and "China" are represented by the largest circles.

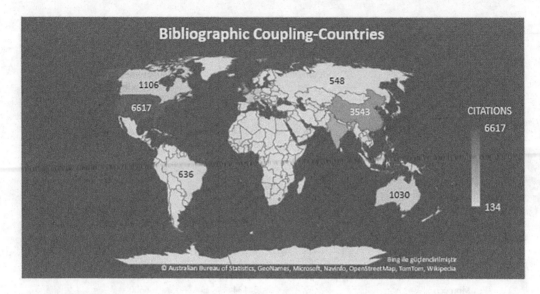

FIGURE 2.13   Number of Citations of Countries with Bibliometric Pairs in the Top 20.

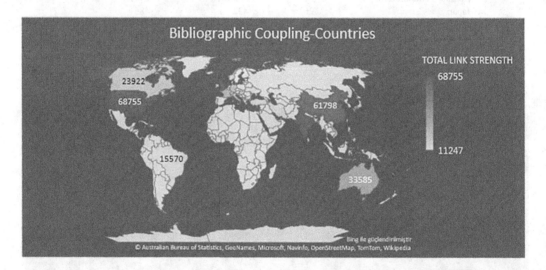

FIGURE 2.14   Total Connectivity of Countries with Bibliometric Pairs in the Top 20.

In the first cluster 35 countries such as "United States", "India", "United Kingdom", "Germany", "Italy", "Japan", "Turkey" etc. countries are grouped. Within the second cluster 9 countries including "China", "Australia", "Hong Kong", "Saudi Arabia" etc. are heaped together. In the third cluster 9 countries such as "Canada", "Malaysia", "Austria", "Hungary", etc. are shown. The clusters and frequently used words obtained as a result of the analysis are listed below.

Total 53 Products (3 Clusters)

1. **CLUSTER (Items 35):** Bangladesh, Belgium, Brazil, Chile, Croatia, Czech Republic, Denmark, Egypt, Finland, France, Germany, Greece, India, Iran, Ireland, Italy, Japan, Luxembourg, Morocco, Netherlands, Norway, Poland, Qatar, Romania, Russian Federation, South Africa, South Korea, Spain, Sweden, Switzerland, Turkey, United Arab Emirates, United Kingdom, United States.

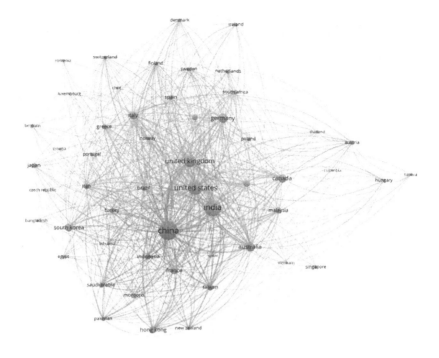

VOSviewer

FIGURE 2.15 "Bibliographic Consolidation-Countries" Analysis in Supply Chain Management of Blockchain, IoT, and Artificial Intelligence Technologies (Network Visualization).

   2. **CLUSTER (Items 9):** Australia, China, Indonesia, Lithuania, New Zealand, Poland, Saudi Arabia, Taiwan.
   3. **CLUSTER (Items 9):** Austria, Canada, Colombia, Hungary, Malaysia, Thailand, Tunisia, Vietnam.

"Density Visualization" of countries with bibliometric pair quality is shown in Figure 2.16. In general, it is seen that there is a density in the middle part. The density in is quite high in the "United Kingdom," "India," and "China." On the other hand, "France," "Canada," "Malaysia," "Australia," "Germany," "Italy," "Turkey," "İran," "Brazil," "SouthKorea," "HongKong," and "Austria" have a medium density compared to other countries.

2.6   DISCUSSION

Based on the research questions, "co-occurrence-author keywords" was selected for analysis by the most frequently used author keyword when the related publications in blockchain, IoT, artificial intelligence, and SCM research were examined. In the bibliometric map, the large nodes represent the most used author keywords, and small nodes represent the least used keywords. The "blockchain" keyword in the cluster is indicated by the largest node. This shows that "blockchain" has a very high impact and is in a leading position compared to the other three areas. The keyword "smart contract" ranks second in the cluster. This expression used by Guo et al. [27] (2021), "smart contract," is the most used author keyword in blockchain bibliometric analysis research. However, in the studies of Miau and Yang [25](2018) the most used keyword is "Bitcoin." According to the results obtained from the findings, the keyword coming after blockchain in second place is "supply chain management." It is represented by a cluster and a large node. It is in the same cluster as "Internet Things." This reveals the conclusion that SCM and IoT studies coexist. Katoch [30] (2021) shows similarities

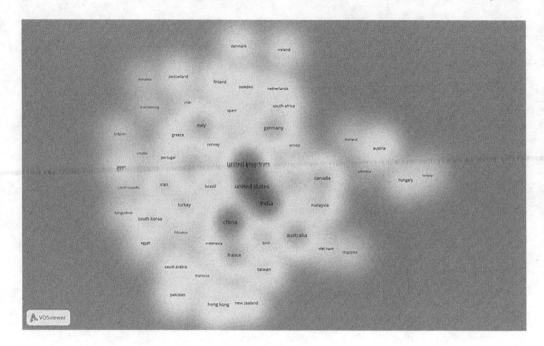

FIGURE 2.16 "Bibliographic Coupling-Countries" Analysis in Blockchain Supply Chain Management (Density Visualization), IoT, and Artificial Intelligence Technologies.

with the results of the most used author as a result of bibliometric analysis in the fields of IoT, SCM, and logistics.-It is also consistent with the result of IoT and blockchain in Raza's [31] (2022) bibliometric analysis study. It is widely used in supply chain and RFID technology. "Artificial intelligence" is included in the cluster and is represented by a small node among the keywords "supply chain management" and "blockchain". As can be seen from here, it is seen that there is a close interaction between the two sides and the keywords "supply chain management", "artificial intelligence", and "blockchain". Sharma et al. [37] (2022) states that SCM is the most important study discipline in artificial intelligence applications and this relationship can be explained. Also, while blockchain technology helps to create a reliable environment in business processes, this can be explained by the fact that artificial intelligence makes processes smart by analyzing business processes [46].

When the related publications about Blockchain, IoT, AI and SCM researches were examined, the "Co-Citation-Source" was chosen for the analysis of the most cited journals. It has been observed that there has been an increase in the number of studies carried out in these areas since 2017 and this increase continues exponentially in 2019-2020. The "International Journal of Production Research" (pink) represents the biggest node for all four disciplines. Other publications are seen to be clustered close to each other. Similar to this result, Zeba et al. [35], as a result of the bibliometric analysis of artificial intelligence applications in the manufacturing sector in 2011-2019, ranked "International Journal of Production Research" first in terms of the number of citations. When the "International Journal of Production Research" is examined in terms of the number of publications, it ranks fourth place within the scope of this study. According to Riahi et al. [36] (2021) "*International Journal of Production Research*" was ranked first in supply chain bibliometric analysis based on artificial intelligence, while Katoch [30] (2021) ranked it second in bibliometric studies for the fields of IoT, SCM and logistics. Unlike this result, Zekhnini et al. [23] (2020) stated that he ranked "*International Journal of Production Research*" second in his bibliometric study for SCM research, while Marty [38] (2021) ranked it third. According to the findings, it is seen that the *IEEE Access* journal ranks third in terms of the number of publications and fourth in terms of the number of citations. Guo et al. [27] (2021) ranked it first in terms of the number of publications and citations in the bibliometric analysis of blockchain technology. Similarly, Wang et al. [32] (2021) when the related publications about Blockchain, IoT, AI and SCM researches were examined, the "Co-Citation-Source" was chosen for the analysis of the most cited journals ranked it first in terms of the number of publications in his bibliometric analysis on IoT. It can be said that these differences are due to the fact that the data are collected in different databases and the publications are limited to years. In addition, the number of citations and publications change from day to day, the analysis of different disciplines, and the production of too many publications can cause this difference.

When relevant publications are made, blockchain, IoT, artificial intelligence, and SCM studies are examined, "bibliographic matching countries" selection was made for the analysis of countries with bibliometric pairs. In the bibliometric map, large nodes represent the countries with the highest total connection strength, and small nodes represent the countries with the lowest total connection strength. It is seen that the United States ranks third in terms of number of publications and first in terms of number of citations and total link strength. China ranked first in the number of publications and second in the number of publications and quotes, and fourth in terms of total connection strength. India ranked second in number of posts, fourth in citations, and third in overall link strength. It seems that the United Kingdom is in the fourth place in terms of the number of publications and the third in terms of the number of publications and quotes, and second in terms of total connection strength. When the literature is examined, it is concluded that the countries that contribute the most to blockchain, IoT, AI, and SCM studies are China and the USA [23,26,29,37]. It is seen that similar results were obtained in this study. In addition, the fact that the United States, United Kingdom, and India are in the same cluster as the countries that dominate the technology may also lead to the conclusion that they are in cooperation. On the other hand, the fact that China, which is a powerful country, is not included in this cluster can be said to be an indication of its technological independence.

2.7 CONCLUSION AND FUTURE SCOPE

In this study, metadatasets available in Blockchain, IoT, Artificial Intelligence, and SCM research are used. Within the scope of the study, the links between the author keywords, the most cited journals and the country relations, which are the bibliometric pair, were tried to be revealed. The aforementioned links were revealed by analyzing bibliometric maps. In addition, the aim is to contribute to the literature by examining these four areas holistically. Considering the trends in technology, it was attempted to present a perspective to those who want to work in these fields.

This study is limited to the Scopus database. Those who want to work in this field can examine the databases and find WoS, IEEE Xplore, Google Academician, etc. They will be able to apply to programs and assist future research for field research.

REFERENCES

[1] Beck, R. (2018). Beyond bitcoin: The rise of blockchain world. *Computer*, 51(2), 54–58.

[2] Zheng, Z., Xie, S., Dai, H., Chen, X., & Wang, H. (2017, June). An overview of blockchain technology: Architecture, consensus, and future trends. In *2017 IEEE International Congress on Big Data (BigData congress)* (pp. 557–564). IEEE.

[3] Tama, B. A., Kweka, B. J., Park, Y., & Rhee, K. H. (2017, August). A critical review of blockchain and its current applications. In *2017 International Conference on Electrical Engineering and Computer Science (ICECOS)* (pp. 109–113). IEEE.

[4] Swan, M. (2015). *Blockchain: Blueprint for a new economy*. Sebastopol, CA: O'Reilly Media.

[5] Cheng, J. C., Lee, N. Y., Chi, C., & Chen, Y. H. (2018). Blockchain and smart contract for digital certificate. In *2018 IEEE International Conference on Applied System Invention (ICASI)* (pp. 1046–1051). IEEE.

[6] Puthal, D., Malik, N., Mohanty, S. P., Kougianos, E., & Das, G. (2018). Everything you wanted to know about the blockchain: Its promise, components, processes, and problems. *IEEE Consumer Electronics Magazine*, 7(4), 6–14.

[7] Gatteschi, V., Lamberti, F., Demartini, C., Pranteda, C., & Santamaria, V. (2018). To blockchain or not to blockchain: That is the question. *IT Professional*, 20(2), 62–74.

[8] Crosby, M., Pattanayak, P., Verma, S., & Kalyanaraman, V. (2016). Blockchain technology: Beyond bitcoin. *Applied Innovation*, 2(6–10), 71.

[9] Jia, X., Feng, Q., Fan, T., & Lei, Q. (2012, April). RFID technology and its applications in Internet of Things (Iot). In *2012 2nd International Conference On Consumer Electronics, Communications and Networks (CECNet)* (pp. 1282–1285). IEEE.

[10] Welbourne, E., Battle, L., Cole, G., Gould, K., Rector, K., Raymer, S., ... & Borriello, G. (2009). Building the internet of things using RFID: The RFID ecosystem experience. *IEEE Internet Computing*, 13(3), 48–55.

[11] Roman, R., Najera, P., Lopez, J. (2011). Securing the internet of things. *Computer*, 44 (9), 51–58.

[12] Uckelmann, D., Harrison, M., & Michahelles, F. (2011). An architectural approach towards the future internet of things. In *Architecting the Internet of Things* (pp. 1–24). Berlin, Heidelberg: Springer.

[13] Aazam, M., Khan, I., Alsaffar, A. A., & Huh, E. N. (2014). Cloud of things: Integrating internet of things and cloud computing and the issues involved. In *Proceedings of 2014 11th International Bhurban Conference on Applied Sciences & Technology (IBCAST) Islamabad, Pakistan*, 14th–18th January, 2014 (pp. 414–419). IEEE.

[14] Tan, L., & Wang, N. (2010). Future internet: The internet of things. In *3rd International Conference on Advanced Computer Theory and Engineering (ICACTE)* (Vol. 5, pp. 376–380).

[15] Hakiri, A., Berthou, P., Gokhale, A., & Abdellatif, S. (2015). Publish/subscribe-enabled software defined networking for efficient and scalable Iot communications. *IEEE Communications Magazine*, 53(9), 48–54.

[16] PK, F. A. (1984). What is Artificial Intelligence?. *"Success is no accident. It is hard work, perseverance, learning, studying, sacrifice and most of all, love of what you are doing or learning to do"*, 65.

[17] Rust, T., & Huang M. (2014). the service revolution and the transformation of marketing science. *Marketing Science*, 33(2), 206–221.

[18] Fluss, D. (2017). The AI revolution in customer service. *Customer Relationship Management*, January. 21(1), 38.

[19] Kopczak, L. R. (1997). Logistics partnership and supply chain restructuring. survey results from the US computer industry. *Production and Operations Management*, 6(3), 226–247.

[20] Kehoe, D., & Boughton, N. (2001). Internet based supply chain management: A classification of approaches to manufacturing planning and control. *International Journal of Operations & Production Management*.

[21] Ozdemir, A. I. (2004). Development, processes and benefits of supply chain *management. Journal of Erciyes University Faculty of Economics and Administrative Sciences*, (23).

[22] Croxton, K. L., Garcia-Dastugue, S. J., Lambert, D. M., & Rogers, D. S. (2001). The supply chain management processes. *The International Journal of Logistics Management*, 12(2), 13–36.

[23] Zekhnini, K., Cherrafi, A., Bouhaddou, I., Benghabrit, Y., & Garza-Reyes, J. A. (2021). Supply chain management 4.0: A literature review and research framework. *Benchmarking*, 28(2), 465–501. 10.1108/BIJ-04-2020-0156

[24] Firdaus, A., Razak, M. F. A., Feizollah, A., Hashem, I. A. T., Hazim, M., & Anuar, N. B. (2019). The rise of "blockchain": Bibliometric analysis of blockchain study. *Scientometrics*, 120(3), 1289–1331. 10.1007/s11192-019-03170-4

[25] Miau, S., & Yang, J. M. (2018). Bibliometrics-based evaluation of the blockchain research trend: 2008–March 2017. *Technology Analysis and Strategic Management*, 30(9), 1029–1045. 10.1080/09537325.2018.1434138

[26] Alam, S., Zardari, S., & Shamsi, J. (2022). Comprehensive three-phase bibliometric assessment on the blockchain (2012–2020). Library Hi Tech. 10.1108/LHT-07-2021-0244

[27] Guo, Y. M., Huang, Z. L., Guo, J., Guo, X. R., Li, H., Liu, M. Y., Ezzeddine, S., & Nkeli, M. J. (2021). A bibliometric analysis and visualization of blockchain. *Future Generation Computer Systems*, 116, 316–332. 10.1016/j.future.2020.10.023

[28] Luo, J., Hu, Y., & Bai, Y. (2021). Bibliometric analysis of the blockchain scientific evolution: 2014-2020. *IEEE Access*, 9, 120227–120246. Marty, J. (2021). Consumer/user/customer integration in Supply Chain Management: a review and bibliometric analysis. *Supply Chain Forum: An International Journal*. 10.1080/16258312.2021.1984168

[29] Leong, Y. R., Tajudeen, F. P., & Yeong, W. C. (2021). Bibliometric and content analysis of the internet of things research: A social science perspective. In *Online Information Review* (Vol. 45, Issue 6, pp. 1148–1166). Emerald Group Holdings Ltd. 10.1108/OIR-08-2020-0358

[30] Katoch, R. (2021). IoT research in supply chain management and logistics: A bibliometric analysis using vosviewer software. *Materials Today: Proceedings*. 10.1016/j.matpr.2021.08.272

[31] Raza, S. A. (2022). A systematic literature review of RFID in supply chain management. *Journal of Enterprise Information Management*, 35(2), 617–649. 10.1108/JEIM-08-2020-0322

[32] Wang, J., Lim, M. K., Wang, C., & Tseng, M. L. (2021). The evolution of the Internet of Things (IoT) over the past 20 years. *Computers and Industrial Engineering*, 155. 10.1016/j.cie.2021.107174

[33] Niu, J., Tang, W., Xu, F., Zhou, X., & Song, Y. (2016). Global research on artificial intelligence from 1990-2014: Spatially-explicit bibliometric analysis. *ISPRS International Journal of Geo-Information*, 5(5). MDPI AG. 10.3390/ijgi5050066

[34] Gao, H., & Ding, X. (2022). The research landscape on the artificial intelligence: A bibliometric analysis of recent 20 years. *Multimedia Tools and Applications*. 10.1007/s11042-022-12208-4

[35] Zeba, G., Dabić, M., Čičak, M., Daim, T., & Yalcin, H. (2021). Technology mining: Artificial intelligence in manufacturing. *Technological Forecasting and Social Change*, 171. 10.1016/j.techfore.2021.120971

[36] Riahi, Y., Saikouk, T., Gunasekaran, A., & Badraoui, I. (2021). Artificial intelligence applications in supply chain: A descriptive bibliometric analysis and future research directions. In *Expert Systems with Applications* (Vol. 173). Elsevier Ltd. 10.1016/j.eswa.2021.114702

[37] Sharma, R., Shishodia, A., Gunasekaran, A., Min, H., & Munim, Z. H. (2022). The role of artificial intelligence in supply chain management: Mapping the territory. *International Journal of Production Research*. 10.1080/00207543.2022.2029611

[38] Marty, J. (2021, October). Consumer/user/customer integration in Supply Chain Management: A review and bibliometric analysis. *Supply Chain Forum: An International Journal*, 1–16. Taylor & Francis.

[39] Kamran, M., Khan, H. U., Nisar, W., Farooq, M., & Rehman, S. U. (2020). Blockchain and internet of things: A bibliometric study. *Computers and Electrical Engineering*, 81. 10.1016/j.compeleceng.2019.106525

[40] Szum, K. (2021). IoT-based smart cities: A bibliometric analysis and literature review. *Engineering Management in Production and Services*, 13(2), 115–136. 10.2478/emj-2021-0017

[41] Çiğdem, Ş. (2021). A bibliometric analysis of digitalization in supply chains. *Gaziantep University Journal of Social Sciences*, 20(2), 657–677.

[42] Zhang, X., Yu, Y., & Zhang, N. (2020). Sustainable supply chain management under big data: A bibliometric analysis. *Journal of Enterprise Information Management*, 34(1), 427–445. 10.1108/JEIM-12-2019-0381

[43] Moral-Muñoz, J. A., Herrera-Viedma, E., Santisteban-Espejo, A., & Cobo, M. J. (2020). Software tools for conducting bibliometric analysis in science: An up-to-date review. *El Profesional de la Informa-ción*, 29(1), 1699–2407. e290103. 10.3145/epi.2020.ene.03

[44] Van-Erck, N., & Waltman, L. (2010). Software survey: VOSviewer, a computer program for bibliometric mapping. *Scientometric*, 84, 523–538.

[45] Donthu, N., Kumar, S., Mukherjee, D., Pandey, N. veLim, W. M. (2021). How to conduct a bibliometric analysis: An overview and guidelines. *Journal of Business Research*, 133, 285–296.

[46] Singh, P., & Singh N. (2020). Blockchain with IoT and AI: A review of agriculture and healthcare. *International Journal of Applied Evolutionary Computation*, 11(4), 13–27.

3 Complete Scenario for Supply Chain Management Using IoT and 5G

Jay Kumar Pandey
Shri Ramswaroop Memorial University, Barabanki, Uttar Pradesh, India

Vandana B. Patil
Dr. D.Y. Patil Institute of Engineering, Management, and Research, Pimpri-Chinchwad, Maharashtra, India

Kukati Aruna Kumari
Prasad V. Potluri Siddhartha Institute of Technology, Vijaywada, Andhra Pradesh, India

Santanu Das
Seshadripuram First Grade College, Yelahanka New Town, Bangalore, India

Arpit Namdev
University Institute of Technology RGPV, Bhopal, Madhya Pradesh, India

Praful Nandankar
Government College of Engineering, Nagpur, India

Ankur Gupta
Department of Computer Science and Engineering, Vaish College of Engineering, Rohtak, Haryana, India

CONTENTS

DOI: 10.1201/9781003264521-3

3.1   INTRODUCTION

IoT refers to Internet-enabled access and control of everyday equipment and devices. Everything you need to know about IoT is covered in this chapter, from the basics to the more advanced topics like biometrics and security cameras to the devices themselves [1,2].

3.1.1   INTERNET OF THINGS (IoT)

Let's take a closer look at our mobile device, which includes features like GPS tracking, a mobile gyroscope, adaptive brightness, voice recognition, and face recognition, among others. What if all of these components communicate with each other to create a better overall environment? Consider how the GPS location and direction are used to adjust the phone's brightness. Things embedded with sensors, electronics, and software that can communicate within the IoT, plans that converse with each other without human interaction [3]. In IoT, "Things" refers to everything that can be accessed or connected via the Internet (Figure 3.1).

An Internet of Things system is an analytics system and advanced automation that incorporates artificial intelligence, sensors, networking, and cloud messaging to provide systems for services or products. Internet of Things creates a system with greater visibility, control, and efficiency. For every IoT echo system, the workings of IoT are unique (architecture). However, the fundamental working principles are the same. IoT devices, like home electronics, smartphones, and digital watches, all connect to the IoT platform via secure wireless networks [4]. To transfer the most valuable data, a broad variety of devices and platforms are used by platforms to gather data for analysis.

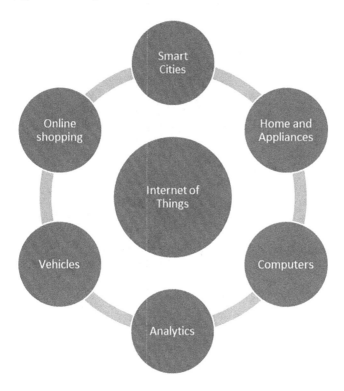

FIGURE 3.1 IoT.

3.1.2 FEATURES OF IoT

Among the many IoT features, it relies on are connectivity, analysis, integration, active partici-pation, and a host of other elements (Figure 3.2). Here are a few examples:

A. **Connectivity:** The term "connectivity" refers to the ability to connect all of the IoT devices to an IoT platform, whether that platform is a server or a cloud. For reliable, secure, and bidirectional communication between IoT devices and the cloud, high-speed messaging is required after the devices are connected [5].

B. **Analyzing:** After connecting all the required items, the following stage is to do real-time analysis of the acquired information and apply the results to the development of useful business information. As long as we have a clear comprehension of the data received from all of these sources, we call our system intelligent.

C. **Integrating:** Additionally, IoT integrates multiple models to enhance the overall user experience [6].

D. **Artificial Intelligence:** Using data, IoT makes things smarter and enhances people's lives. The coffee machine, for example, can order your preferred brand of coffee beans from the retailer if the machine's beans are running low. Sensing: IoT sensor devices monitor their surroundings and report on any changes they notice. The Internet of Things converts inert networks into active ones. There will be no Internet of Things if there were no sensors [7].

E. **Active Engagement:** IoT has made it easier for technology, goods, and services to communicate with one another.

F. **Management of Endpoints:** It is critical that all IoT systems have endpoint manage-ment; otherwise, the entire system will fail. A coffee machine, for example, can order the beans it needs from a retailer on its own, but what happens when we're not home for a few days and the machine orders beans from a retailer? This results in the IoT system failing. As a result, endpoint management is required [8].

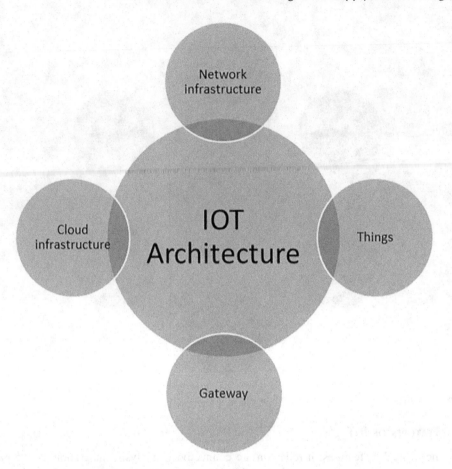

FIGURE 3.2   IoT Architecture.

3.1.3 Advantages of IoT

IoT facilitates several advantages in the business sector daily. The following are a few of its advantages: [9]

    A. The efficient use of resources: The more we understand how different devices work and how they function, the better we'll be able to use resources and keep track of them.

    B. Reduce human effort: As IoT devices interact and communicate with one another and execute a range of functions on our behalf, we may lessen our reliance on human labor.

    C. Make most of your time: It saves time because it reduces the amount of human effort required. By using an IoT platform, you can save a significant amount of time.

    D. Boost safety measures: All of these things can be made more secure and efficient if they are connected in a system that is interconnected.

3.1.4 Disadvantages of IoT

There are many advantages to the Internet of Things, but there are also several drawbacks. The following are a few of the IoT's difficulties:

    A. Security: Despite security precautions, the system gives minimal control and may be exploited to launch a variety of network assaults since IoT is a network of linked devices.

B. **Privacy:** IoT system provides a wealth of personal information even if the user does not actively participate in the process.

C. **Complexity:** The design, development, maintenance, and enabling of a large IoT system is extremely difficult.

3.1.5 APPLICATIONS OF IoT

A game-changing invention has allowed us to stay in touch with friends and family around the world via computers and smartphones. IoT is the next generation of networks that connects things with the capability of sensing, controlling, and communicating with one another [10].

A. **Home and office automation:** The use of smart sensors in smart home applications is becoming increasingly popular. A simple mobile application can be used to set up and connect virtually any smart device to the Internet [11].

B. **Embedded electronics:** Around a decade ago, the first wearable smart devices appeared as smartwatches, and since then, many more features have been added. Messages can now be read on our wearable and smartwatches, notifications from other apps can be shown, we can track our whereabouts, we can keep track of our workouts, and we can constantly monitor our health. Wearable technology can now be used for much more than just these basic functions thanks to the Internet of Things. Some of the biggest names in smart clothing are working on custom operating systems and applications just for their products. It has been reported that many smartwatches saved people's lives in emergencies. Among other IoT devices, smart wearable has become popular because of their potential for life-saving applications. Caregivers will be alerted if a patient's vital signs drop or their blood sugar levels fluctuate, and parents can keep track of where their children are at all times. Wearable technology enables doctors and medical professionals to constantly monitor the health of their patients in real-time. It's expected that future smart wearable, such as watches and fitness bands, will be better able to connect to other IoT devices in the home and other environments. These smart wearables will be able to perform more tasks and receive notifications more quickly if they are paired with smartphone apps [12].

C. **Healthcare:** The healthcare industry is taking advantage of the Internet of Things' potential to save lives. Patient care can be improved with IoT implementation from bedside devices to real-time diagnosis to accessing medical records and patient information from across multiple departments. The IoT will create life easier for medical professionals, recover data accuracy (thereby reducing data errors), boost productivity, and shorten procedure times. Doctors can keep an eye on their patients' health from afar and make treatment recommendations when it's warranted. With IoT devices, there will be less data loss and human error. The vast majority of modern medical devices are network-capable, allowing for secure data access (in the future, all devices will have the capability to connect to the network). Smart Internet of Things (IoT) devices allow for round-the-clock patient monitoring. Medical personnel will be alerted immediately if a patient's vital signs change. After assessing the patient remotely using smart IoT devices, doctors can prescribe medication. Hospitalization may not be necessary in many cases [13].

D. **Driverless cars:** AI and IoT smart sensor technology have accelerated the development of self-driving cars. Early autonomous vehicles (partial automation) will help drivers avoid collisions and warn them about the road and their vehicle's condition [14]. This includes things like cruise control assistance (parking and line-changing assistance), efficient fuel/energy management, and more. AI can predict certain scenarios on the road based on the huge amount of data we collect from thousands of vehicles (using millions of sensors and cameras units). This data can then be used to improve the safety and efficiency of future vehicles. AI and IoT are making self-driving cars and connected car concepts safer to use on public roads in the future (AI). As part of IoT in automobiles, smart sensors constantly collect information about the vehicle, the road, other vehicles, and the road conditions. To help the vehicle decide in the event of sudden road changes, the system includes camera units, proximity sensors, RADARs, and RF antenna arrays. Sharing information between vehicles and smart objects is possible via radio frequency (RF) technology [15,16].

E. **Smart farming and agriculture:** Increasing the human population necessitates an increase in the number of food crops and vegetables that can produce. Using the Internet of Things [17], agriculturalists and researchers can discover more efficient and cost-effective methods of increasing output. Conventional farming and agriculture are not appealing to the younger generation in developed countries. The absence of support staff could hurt productivity, so the authorities must come up with new solutions to this problem. Agriculture and farming can be more efficient with fewer workers thanks to the Internet of Things [13], which is one of the most promising solutions in this area. With the help of smart sensor technology and automation, agriculture can be improved at each stage and the amount of manual labor reduced [18,19].

F. **Analytics of big data:** Big data analytics relies heavily on data, which many businesses see as their most valuable asset for advancing their business strategies. Data can come from a variety of places, including machines, the environment; plants, animals, and even people. The Internet of Things uses a broad range of sensors to collect data from several sources. Machine learning and artificial intelligence will help big data analytics enhance its decision-making algorithm by evaluating data from millions of smart sensors. There is a lot of data that is needed for autonomous driving technology, for example. To improve the self-driving algorithm's ability to deal with any situation that may arise while driving, these sophisticated sensors collect data on engine behaviors, field data, maps, and camera feeds [15,20].

3.1.6 INTRODUCTION OF 5G TECHNOLOGIES

The 5th generation of mobile technology is referred to as "5G" technology. Mobile telecommunication standards 5G represent the next major phase beyond 4G standards. Product engineering, documentation, electronic transactions, and more are all supported by 5G technologies [19]. As customers become more knowledgeable about mobile phone technology, they'll be on the lookout for a comprehensive package that includes all of the most useful extras. As a result, the leading cell phone companies are always on the lookout for new technology to out-innovate their rivals. In an ideal world, a 5G telecommunications network would be able to address the issues that will arise once the 4G model is widely adopted. For millimeter waves (10 to 1 mm), OFDM provides wide-area coverage, high throughput for millimeter waves, and a 20 Mbps data rate across distances of up to 2 km. The recent explosion in wireless Internet usage can best be addressed by using the millimeter-wave band. Wireless WWW (World Wide Web) services may be supported by these standards. WWW allows for a dynamic, ad-hoc wireless network to be built with a channel bandwidth of between 5 and 20 MHz, preferably up to 40 MHz (DAWN). Intelligent antennas (such as switched beam and adaptive array antennae) and the flexible modulation technique are used to provide bidirectional high bandwidth, allowing the transmission of massive amounts of broadcasting data in gigabytes, 60,000 connections, and 25 Mbps connectivity. Movies, including 3D ones, may be downloaded to tablets and laptops using 5G technology, along with games and medical services. Piconet and Bluetooth will inevitably become obsolete with the arrival of 5G. 5G phones are expected to be similar in size and functionality to tablet PCs [20,21].

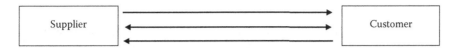

FIGURE 3.3 Flow in Supply Chain.

3.1.7 INTRODUCTION OF SUPPLY CHAIN MANAGEMENT

Supply chains include steps such as acquiring raw materials, transforming them into intermediates, and then completing the finished product. The distribution process then begins with the final products being delivered to customers. Consider the following issues when it comes to the supply chain. This is a change from previous times when all of the supply chain organizations worked in isolation from each other along the supply chain. Organizations have their own goals, which can be at times in conflict with one another. There is a requirement for the integration of these various functions to work effectively. As a result, supply chain management has emerged as a strategy for achieving such integration [22].

3.1.8 FLOWS IN THE SUPPLY CHAIN

Chain reaction–like flow involves the movement of goods from one party to another, namely the supplier and the customer (Figure 3.3). Information is exchanged between the supplier and the customer. Additionally, money moves back and forth between the customer and the supplier [23].

3.1.9 FEATURES OF EFFECTIVE SCM

If there is a need to develop a strong digital supply chain, the following eight characteristics are required in supply chain management software [24]:

A. **Integration across the entire supply chain:** With the help of technology, a digital solution can provide features that span the entire supply chain and integrate multiple parties such as suppliers and OEMs as well as shippers and storage facilities [25]. Everything from enterprise software to legacy systems and third-party applications should be able to communicate with your help desk and email, regardless of the source of information or the platform on which it is running. This can solve the problem of a network's inability to communicate, allowing data to flow more freely throughout the chain. A single source of truth for all stakeholders is created by integrating disparate systems, people, and processes. In addition to saving money by eliminating the need to maintain multiple applications, this also prevents features from duplicating each other. This improves production planning, and logistics, and avoids stock-outs or excess inventory, all of which benefit the company. The use of APIs, data connectors, and SDKs makes it possible to integrate with core systems, thus taking advantage of the infrastructure that is already there. Operations managers, for example, benefit from the ability to create orders & bill customers from a single location. It reduces the likelihood of errors due to omissions or misunderstandings in the chain of command. To meet the needs of different customer groups or product categories, the order management, and billing function has a wide range of customizable options [26].

B. **The ability to work together in real-time:** Avoiding bottlenecks, missing goods, and dissatisfied customers is impossible without access to real-time information. Organizations can respond to supply chain changes in real time thanks to real-time capabilities. Multiple stakeholders should be able to collaborate on a project without having to constantly communicate back and forth or manually update their information.

For example, customers, fleet managers, and truck drivers located in different locations could all see the same order at the same time, increasing collaboration and ensuring transparency throughout the supply chain [27,28]. This can be accomplished in the following ways:

- **Individually tailored dashboards:** A dashboard tailored to the needs of each stakeholder provides them with the information they need at their fingertips. It streamlines governance by providing real-time status updates on all processes.
- **Real-time notifications:** All supply chain activities can be viewed in real-time-thanks to-real-time notifications. Stakeholders are kept up to date on developments, allowing them to respond quickly [29].
- **Portals for self-service:** Many people are involved in a supply chain network, and they all depend on one another to be successful. However, they aren't all interconnected, which can lead to misunderstandings. With the help of a password-protected website, portals make it possible for users to exchange information and plan operations based on what their colleagues are up to.
- **Authentication based on roles:** Only some of the data in your application should be accessible to everyone. Because of this, you choose supply chain management software with roles and permissions that restrict access to the data only needed by suppliers, customers, and other parties involved in the supply chain. Zoho Creator is one option with all of these features. Using this platform, teams, leaders, and companies will be able to work more efficiently. With Zoho Creator, you can plan, monitor, automate, and report on all of your work [30].

C. **The ability to optimize the process:** Shorter product life cycles, reduced paper-based documentation requirements, and tighter links between manufacturing, warehouses, and delivery can all be achieved through automation of the order-to-cash cycle. Businesses can use AI and machine learning to automate a wide range of tedious tasks, not just operational ones. When a product is only approved when it is at its finest, the software may be configured to do so instead of needing human interaction. Organizations may explore more flexible methods of working, better handle high levels of complexity, and call-in human assistance only when required using this strategy. Logistics and transportation firms benefit from cost-effective, high-efficiency optimization techniques. As fuel prices continue to rise and national and regional laws continue to evolve, this must be taken into consideration to avoid delays or uncertainty in the transportation of goods [31].

D. **Insights from data and projections:** Good supply chain management software should have built-in analytics and forecasting capabilities to assist you in evaluating your business, as well as automating day-to-day tasks. It is possible to balance supply and demand discrepancies using advanced software that provides data on both internal (consumption) and external trends. Software that uses artificial intelligence and machine learning identifies risks and volatility on its own, alerting stakeholders. As a result, stakeholders can more effectively plan their procurement and production processes, resulting in lower costs because they don't have to purchase extra raw materials or store excess finished goods in warehouses. Innovative processes like predictive dispatch can be implemented by managers to anticipate future demand, while leaders can use accurate data and insights to better plan future needs and run what-if simulations to mitigate risks. Because of this, analytics may be utilized to enhance future processes and systems by making use of the information now available [32,33].

E. **Customization:** Customizing business rules and using pre-built application components enable organizations to swiftly adjust and go to market with tailored solutions for clients. It is possible to modify certain supply chain systems using Java and Python programming languages. Open architecture also allows enterprises to design their applications to suit their requirements, such as generating various variants of a product to match the demands of different client groups and so increase revenue.

F. **Access to and mobility within the cloud:** To manage, track, and monitor the status of transactions, authorized individuals may access organizations utilizing cloud-based supply chain software from any place and at any time. On the other hand, cloud-based solutions may be put up at a lesser cost, in a shorter amount of time, and with a lower risk than an on-premise system. With the use of a smartphone app, businesses can monitor things like order progress and shipment more quickly. It is possible to send alerts directly to users' mobile phones in real-time, enhancing collaboration between various parties and allowing for immediate action in the event of any issues [34].

G. **Security:** The safety of company data is at the core of every piece of software. As part of the selection process, companies should consider the following factors: encryption of data, virus-scanning, keeping tabs on the network, and an audit trail that is a record of what happened [35].

H. **Scalability:** There is no such thing as a static piece of software. Additionally, a supply chain system should be able to manage the increased volume that comes along when a company expands its product ranges, customer acquisitions, and so on. The overall performance of the system should not be affected by the use of multiple apps and extra channels [36,37].

3.1.10   The Evolution of Supply Chain Management

In the 1960s, this tendency was underlined as a significant area for future productivity improvements due to the fragmentation of the system, and it has continued ever since. It was in the 1970s and 1980s when logistics responsibilities were divided into two independent functions: materials management and physical distribution. Logistics as we know it now was born in the 1990s when globalization led to a functional integration and a unified management viewpoint for the many components of the supply chain [38] (Figure 3.4).

However, it was only with the introduction of SCM that current information and communication technology allowed for a full integration. The integration of information, money, and products movements allows for a wide variety of innovative production and distribution systems. When it comes to managing the supply chain, value capture and competitiveness are the primary goals. Physical distribution and materials management have both evolved in recent years as a result of increasing levels of automation in supply chains. Storage, material handling, and packaging have all seen significant increases in automation as a result of this push toward digitalization in the distribution center. Automated delivery vehicles are one possible outcome of automation. Two early instances of logistical engineering are high-rack storage and package transfer by flat robots. Initially, supply, storage, manufacturing, and distribution were all distinct yet interdependent undertakings with little overlap. Rather than rising expenses, companies were able to meet the need for flexibility by adopting new organizational and managerial practices. Outsourcing and off-shoring allowed many corporations to take advantage of cheaper labor in emerging nations. Consolidation of management activities occurred as production became more dispersed. Economies of scale in the distribution process led to spatial fragmentation as a by-product.

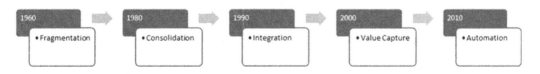

FIGURE 3.4   Evolution of Supply Chain Management.

3.1.11 The Importance of Supply Chain Management

Many businesses already know the importance of supply chain management to their overall performance and the happiness of their consumers. The following points highlight the importance of supply chain management:

A. Increased Focus on Customer Care

As a result, clients have high standards for the products and services they receive. It is expected that products will be available at the right location for customers. When an auto repair company can't fix your vehicle in a day or two because they don't have all the essential components on hand, they lose clients. Orders must arrive on schedule for customers to be. When they buy anything, they want it to be fixed right away. If, for example, clients' furnaces break down in the winter and repairs take days to complete, they are less happy [39,40].

B. Lower the Cost of Operation

- **Decreases purchasing cost** – To avoid having to keep expensive products in store longer than necessary, retailers employ supply chains to expedite delivery of the goods. If a retailer wants to keep their inventory costs to a minimum, they require 60-inch flat-panel plasma HDTVs supplied rapidly.

- **Reduces the cost of production** – As a result of material shortages, manufacturers rely on supply chains to guarantee that supplies are delivered to assembly plants on time. An unanticipated components shipping delay, for example, might force a car assembly facility to shut down, resulting in millions of dollars in lost earnings per day and $20,000 per minute.

- **Reduces the overall cost of the supply chain** – A supply chain manager's job is to create networks that meet customer service goals while also reducing total costs for both manufacturers and retailers. The competitiveness of a corporation is bolstered by the effectiveness of its supply network. For example, Dell's supply chain revolution was founded on building and distributing computers directly to customers. A lack of computer inventory at warehouses and retail locations allowed Dell to save millions of dollars in the long run. Dell avoids stockpiling outdated computers since the industry moves at such a rapid pace [41].

C. To improve your financial situation

- **Profit leverage is enhanced** – Companies prize supply chain managers for their expertise in controlling and reducing the expenses of the supply chain. It might have a negative impact on the company's bottom line. In only five years and 13 billion boxes of cereal, the corporation could save $13 million, or the 2.7 billion boxes of cereal eaten annually in the United States.

- **Decreases supply chain** – Managers help organizations save money by reducing the utilization of big fixed assets like manufacturing facilities, distribution warehouses, and vehicles. To save money, the corporation might instead reconfigure its supply chain so that it could service U.S. consumers from six warehouses instead of ten, rather than constructing four more costly facilities.

3.1.12 The Supply Chain and IoT

SCM has been transformed by the Internet of Things. The location of items, how they are stored, and their arrival times may all be discovered via the Internet.

- **Locate goods and verify their location at any time:** Storage containers, raw materials, and finished goods, for example, can all have Internet of Things (IoT) devices connected to them. An IoT device transmits its location to GPS satellites to monitor the movement of goods [42,43].

- **Track the movement of the goods and when they will arrive:** When items' speed and traffic flow are recorded, it is much easier to forecast how they will move through the supply chain. As a result of preparations made by suppliers, manufacturers, and distribution centers, handling time is reduced and material processing is more efficient [44].
- **Assure that raw materials and finished goods are properly stored:** Food and chemicals, for example, necessitate optimal storage conditions. Keeping an eye on product quality and preventing waste are both made much simpler as a result of this.
- **Slash the time it takes to get things done:** It is possible to track and plan the route of goods via IoT devices to find out where and when they are being held up. Because of this, the supply chain can be made more efficient by considering alternate routes in case of emergencies.
- **Find stuffed animals in a closet:** It is possible to use IoT devices to track goods even after they have been delivered to a distribution center.
- **Immediate delivery of goods upon receipt:** SCM now can verify the exact time that goods have arrived thanks to IoT tracking devices.

3.1.13 INTEGRATING SCM WITH IoT

Customers, suppliers, manufacturers, distributors, and retailers all have a part when it comes to managing the supply chain (SCM). In order to reduce costs, improve service, diminish bullwhip effect, better use assets, and respond to changes in the economic environment, supply chain coordination may be a difficult task to manage. For these reasons, it is essential that all supply chain participants' work together to manufacture, distribute, and support a final product in an efficient and effective manner, both inside and beyond the firm. An example of integration is a transition from managing individual utilitarian procedures to overseeing coordinated chain operations; this may be illustrated by cooperation, coordination, data sharing, trust, associations, and shared innovation. As part of the overall strategic planning process, supply chain management (SCM) must be incorporated into the overall goals and policies of the company, which are based on client requirements and the strengths of the supply chain as a whole [45].

Supply chain performance may be significantly impacted by data sharing. Information exchange has been shown to be critical in integrating the supply chain by researchers as well. Improving channel partner cooperation, and fostering better performance. For a supply network to be successful, it must have extensive information-sharing capabilities, as well as advanced inter- and intra-channel systems. Because SCM is the primary focus of the research, the literature focuses on IoT-based SCM. There have been several recent reviews on the Internet of Things that focus on the many open issues that remain. Few researchers have attempted to link Green IoT and supply chain management [46].

While IoT-based warehouses and supply chains are just beginning to take off, they are already making a significant impact. Since the market is still experimenting with various applications and approaches, there is no definitive answer at this time. Throughout 2015–2019, the global logistics market is expected to expand at an 8.4% CAGR (Source: Technavio). Over the next ten years, the Internet of Things is expected to increase supply and logistics operations by $1.9 trillion. Concerns on both sides of the coin need to be addressed equally. As more and more connected devices enter the market, the issue of IoT security will become more and more prominent. Additionally, the absence of design standards for IoT connection has been noted as a danger. Which one has a stronger influence on the supply chain? It emphasizes further that neither alone can make the supply chain operate. People and technology, on opposing ends of the supply chain, are responsible for moving it towards its objectives. As a consequence, keeping the trust and loyalty of all of your channel partners is crucial to your overall performance in the supply chain. Because of this, they must recognize and understand the new technology's potential and how it will enhance their services. Few academics have looked at how HRM and supply chain management (SCM) are interwoven to increase collaboration and performance. HR management has a substantial influence on SCM's capacity to function smoothly, according to these researches. Effective communication, team management, and continual lifecycle innovation are vital in this dynamic context. Insights into human aspects in these areas are vital for a successful supply chain. the supply chain has become increasingly dependent on the use of technology and the right people.

3.1.14 5G AND IoT: EMERGING TECHNOLOGIES WITH ENDLESS USE CASES

The 5G IoT will allow IoT devices to connect more quickly and reliably, making slowness a thing of the past. As long as you have an open mind, you can imagine a world where smart homes can unlock doors by scanning your face, adjusting lighting and temperature, and managing chronic diseases around the clock without a single accident. It is expected that 5G of wireless technology will deliver more than just a high-speed network connection. As a result, it will help you redefine the network and set a new global wireless speed standard. 5G technologies serve as a bridge to the future, connecting us to it. It's not just another wireless technology; it's the foundation for the 5G revolution, which will transform how technology advances in the future. We've all heard of IoT and how it works. With minimal human intervention, digital machines and objects can exchange real-time data via IoT [47].

IoT depends largely on hardware, connection, data processing, and user interface. As a result of 5G technology, IoT might surge even further (fifth generation). Many IoT devices may be connected owing to the wide range of 5G and IoT technologies. A new mobile ecosystem has been created owing to the Internet of Things and 5G, which will link billions of networks in the next five years and fundamentally alter our environment. Take a look at some numbers to see what IoT enabled by 5G is capable of almost three times as many people will be using 5G smartphones by the end of this year than in 2020, according to Ericsson's latest Mobility Report. Subscriptions to 5G networks are expected to reach a billion by 2022. By 2030, the IoT will have linked more than 50 billion devices. The speed of current LTE networks will be ten times that of 5G. With this boost in speed, IoT devices will be able to exchange data much more quickly than they have in the past. The total number of 5G IoT devices expected to be sold to businesses in 2030 is expected to be 44.8 million. Industry 4.0 applications, also known as smart factories, are expected to account for nearly half of all of these new jobs. 8.4 million units of 5G IoT devices are expected to be sold in smart city use cases, the second-largest share of forecasted sales.

3.1.15 NEW SUPPLY CHAIN CHALLENGES: 5G, IoT, AND BEYOND

There has been a lot of good and bad luck for supply networks in the last year. Every facet of manufacturing and delivery was put to the test by unprecedented demand, driven by a suddenly at-home public's dependence on online shopping. For industries that have been slowly adapting to the digital world, supply chain issues such as international shipping holdups and quarantine delays at food processing plants and docks, as well as logistical issues such as COVID-19 quarantine delays, have exposed their growing pains. Looking ahead, we can't assume that 2020 will be a year of unusually high demand or increased strain on supply networks. This year's surge in e-commerce has been more of a prediction than a response to the pandemic. Global freight demand may more than triple in the next 30 years, according to some outlandish predictions, while short-term forecasts point to continued economic growth, particularly in supply-chain sectors. Supply networks will need to incorporate new technology in the future to boost efficiency and fight possible hurdles. Investment in cutting-edge technologies that serve to shift global economic demands has never been a better moment for the business [48].

3.1.16   OPTIMIZATION OF SHIPPING VIA TELEMETRIC

There have been a lot of digital advancements in ship safety and efficiency during the previous decade. Because of the extensive use of GPS monitoring, it is now possible to follow the chain of custody across the supply chain, from containers to trailers to vehicles to distribution facilities to last-mile deliveries. When mandated by law in 2017, electronic logging devices (ELD) raised the bar for drivers' hours-of-service accountability and oversight (HOS). As a result, the trucking industry has seen another technological boom, this time focused on driver safety and providing fleet managers with additional data to help them find new efficiencies. With ELDs, a driver can use them to find the best route to their destination and reduce the amount of time spent filling out paperwork. ELDs can also record driving habits like speeding, harsh braking, hard turns, and idling to improve safety, fuel management, and equipment utilization.

Unpowered equipment like trailers, containers, and chassis may benefit from the usage of telemetric technologies to enable fleets of all sizes to optimize their tractor-trailer ratios, save costs, and find available equipment more rapidly. As a result, fleets might be transformed by smart technology advancements in the Internet of Things (IoT) and other smart devices. Advances in intelligent data streams may boost cargo throughput while also ensuring safe shipment management for delicate or specialty items. Sensors and cameras on trailers and other assets are increasingly being used by fleets to locate additional capacity, monitor the quality of cargo or freight, check for shifts during travel to detect or avoid spoiling, and more. Shippers' relationships with carriers may be improved by carriers giving more precise arrival times or anomalies that enable receiving docks to be prepared in advance and so minimize the time spent on the dock. Advances in power technologies are helping to facilitate the convergence of new telemetric tools. Supercapacitor batteries and solar power have made it possible for stand-alone tracking devices to operate without being tethered, increasing the reliability and value of sensor and camera systems. A more intelligent supply chain is made possible by all of these technological developments. It's a completely different story when you put these efficiencies into action.

3.1.17   LOGISTICAL SOFTWARE INNOVATION ACHIEVES BREAKTHROUGH

All the hardware breakthroughs in the world won't help much if there are no improvements in wireless technology and logistical software. Older technologies like GPS and Bluetooth depended on the 3G cellular network at that time new sensors and IoT integrations need quicker connections and more data capacity, which is available with the advent of 4G/LTE and 5G networks and the demise of the 3G network by 2022 so that fleet and operations managers can access more data at a faster pace. These faster connections are also being accompanied by improvements in logistics technology, which will aid in the deciphering of all the data. For optimal supply chain co-ordination, fleet managers can keep track of the whereabouts and status of all their cargo, as well as communicate that information directly to shippers and drivers. Correcting supply chain in-efficiencies and faults in real time is now possible thanks to data that can be quickly translated, interactive dashboards that display the status of deliveries, and alarms that can be set up for any potential problems. A decade ago, this improved dependability was only an idea. Now it's a reality, thanks to an increase in industry demand. In 2020, several sectors that had been undergoing steady transformation for decades were put to the test. As demand rises in the next years, supply chain professionals must make use of the transformative potential of the technologies at their disposal to maximize both efficiency and safety.

The objectives of the chapter are:

1. To elaborate research related to supply-chain management, IoT, and 5G along with methodology and limitation.

2. To highlight issues and challenges faced in the implementation of the supply-chain management system that made use of IoT and 5G.
3. To propose an efficient, scalable, and high-performance supply-chain management system for IoT and 5G applications to support real-life survey requirements.
4. To conduct comparative analysis interest, awareness in case of male and female regarding supply-chain management system.

Organization of Chapter: Section 3.2 explains previous research along with methodology, work, and limitations. Section 3.3 explains research methodology. In this section, research in the area of supply-chain management, IoT, and 5G has been considered. Section 3.4 focuses on the proposed work. Section 3.5 highlights general awareness, interest, and experience of the population in retail supply-chain management along with their preferences. Section 3.6 concludes the chapter with a future scope.

3.2 LITERATURE REVIEW

Muthu et al. [1] (2020) expressed that food habits and environmental conditions have made it harder for people of excellent health to thrive in today's world. As a result, if they were to survive, they must raise public knowledge about health issues. Lack of proper medical information, preventable errors, data security threats, incorrect diagnoses, and delayed transmission were just a few of the problems that healthcare systems were dealing.

Atlam et al. [2] (2020) cited IoT as a revolution on the Internet which could link almost any piece of environmental equipment to the Internet so that their data can be exchanged to build new services and apps that enhance quality of life of people. The IoT employed low-cost sensors to address, identify, and locate a broad variety of objects and things from immediate environment. Many benefits come with the Internet of Things (IoT); yet, certain problems exist, especially in terms of privacy and security.

Pavithran [3] (2020) discussed the process of developing a blockchain foundation for the Internet of Things. In addition to cryptocurrency, blockchain was a very promising technology. Among other things, it's still not obvious how this will operate in IoT networks. More research was needed. For the most part, IoT devices and blockchain technology were both built on ledgers. Numerous possibilities were possible with the Internet of Things (IoT).

Mabodi et al. [5] (2020) expressed that connected, universal, and smart node with autonomous interaction was predicted by the Internet of things (IoT) when it comes to services. IoT devices were prime targets for gray whole attacks due to their high computing power, open architecture, and extensive distribution. In a gray hole attack, the attacker pretends to have the fastest route to the target. These packets were never delivered to their intended location. The MTISS-IoT technique, which was based on the AODV routing protocol and aimed to reduce gray hole attacks by leveraging check node information, were developed and was now being tested.

Kaur et al. [6] (2020) presented artificial intelligence (AI), digital twins, IoT, blockchains, and other emerging technologies which have the potential to change the way they think about globalization in the future. Most businesses throughout the world were likely to be affected by digital twin technology, which replicates the physical model for control, remote monitoring, and viewing. By using data collected in real time from various IoT sensors and devices, it served as a live model of a physical system and predicts the future behavior of the corresponding physical counterparts using machine learning/artificial intelligence. A digital twin with IoT capabilities has been studied in terms of its architecture, applications, and problems.

Safara et al. [7] (2020) used sensors and monitors to acquire vast amounts of data about their surroundings. One of the numerous difficulties that IoT systems provided was the transmission of data collected by IoT devices to the cloud via relay nodes. A problem with data transmission includes, but was not limited to, concerns with fault tolerance, security, energy usage, and load balancing. Authors proposed methods to reduce energy consumption and RPL model was used to identify the path to be taken by the data packets in their method in which data was sent to the destination using timing patterns that take into account network traffic, audio, and image data.

Islam et al. [8] (2020) explained the development of an IoT-enabled monitoring system for medical care in the future monitoring systems in hospitals and other health centers that has grown tremendously, and portable healthcare monitoring systems employing growing technologies have become a problem in many countries across the world. Health care consultations can now move from in-person to telemedicine with ease, thanks to the proliferation of IoT technologies. Researchers have developed an Internet of Things (IoT)–enabled healthcare system that can track patients' vital signs and room settings in real time. To collect data from the hospital environment, authors used five sensors, including a heartbeat sensor, a body temperature sensor, a CO sensor, and a CO_2 sensor. For each scenario, the developed technique's error % fell inside a set range.

AI et al. [9] (2020) reported that an increase in botnet assaults has attributed to the growth of susceptible IoT devices in recent months. Due to the lack of basic security measures in IoT devices, botnets carry out DDoS attacks. Because current IoT botnet detection techniques still have certain flaws, such as depending on labeled data, and not being verified with newer botnets, and employing very sophisticated machine learning algorithms, developing new ways to identify hijacked IoT devices was necessary. To detect IoT botnet assaults, the authors highlighted anomaly detection technologies are promising because there was a lot of normal data available. One class support vector machine was observed as a method that can be used to detect anomalies (OCSVM).

Verman and Ranga [10] (2020) presented intrusion detection systems using machine learning for IoT applications. The IoT and its applications were the most prominent topics of research at the time. Aside from the fact that it's easy to use in the real world, IoT's properties also exposed it to cyber threats. One of the most devastating assaults against IoT was DoS. Using machine learning classification techniques to protect IoT from DoS assaults was examined in their research. A thorough investigation was conducted on the best classifiers for developing anomaly-based IDSs. To evaluate the performance of classifiers, metrics and validation techniques were used. Classifiers were tested on popular data sets such as CIDDS-001, UNSWNB15, and NSL-KDD. Classifiers were compared statistically using Friedman and Namely tests and classifiers on IoT-specific hardware were tested using a Raspberry Pi in addition.

Al-Emran et al. [12] (2020) examined that IoT and its applications had been the subject of a wide range of review papers that had attempted to assess and synthesize its use. The IoT was being used in education, but no systematic review study had been conducted yet. Hence, the study's major goal was to highlight the recent advancements in the use of IoT applications in education and to identify potential opportunities and obstacles. About IoT in education, their paper outlined the prospects for adoption, as well as the use of green IoT in educational settings as well as wearable technology in educational settings.

Bansal and Kumar [14] (2020) cited that IOT was playing an increasingly important role in the development of new applications in a wide range of fields, including healthcare, education, smart cities, the home, and agriculture, among others. The IoT ecosystem was surveyed in their article. Everything needed to know about IoT was laid out in detail. Wireless connectivity and the Internet allowed the smart sensors to work together without the involvement of a human to create clever apps that run automatically. An M2M technology was the first stage of the Internet of Things in today's Internet environment. To deal with the massive amounts of data and devices generated by IoT, a wide range of technologies is being brought together which include big data, AI, and machine learning.

Khalid et al. [17] (2020) presented IoT authentication using a decentralized, lightweight blockchain. In IoT, the heterogeneous mechanism was used to create smart, ubiquitous items that were effortlessly connected to the Internet. Smart cities, smart health, and smart communities were just a few of the many areas in which these devices might be put to use. These Internet of Things (IoT) gadgets generate enormous amounts of data that must be protected for reasons of privacy and security. The security of these devices was therefore critical to the system's safety and performance.

Al-Qerem et al. [13] (2020) found that clients might avoid continually connecting with the cloud server through the upstream communication channel. A new version of the optimistic concurrency control protocol was developed as a result of this research. For read-only transactions, an improved partial validation technique was used at the fog node. Updates alone were sent for final confirmation to the cloud. In a fog node, update transactions were partially validated before they were delivered to the cloud, making them more opportune. Using this protocol, applications operating in such contexts used less computing power and communication bandwidth while still obtaining the transactional services they need.

Tseng et al. [18] (2020) expressed the rise of Bitcoin and other cryptocurrencies; the notion of blockchain had received unprecedented interest. One of the most common applications of blockchain was a distributed database. For IoT applications, authors gave preliminary studies on problems and opportunities as databases. While latency was significant for IoT applications, consistency, a feature that determined how the system orders operations over blocks, was critical for application developers. Consistency in blockchain databases wasn't well understood, especially when the network also isn't synced and the system was dynamic—both of which were typical in an IoT setting.

Ye et al. [19] (2020) presented the world's population ages; healthcare systems around the world face increased operational expenses and resource demands. The use of IoT and wearable technology could alleviate some of their stress and provide more efficient healthcare services. Elderly people's well-being can be enhanced while healthcare systems were relieved of some of their burdens and costs were reduced. Data collection methods and gadgets for geriatric healthcare were discussed in their detailed overview of IoT and wearable technology applications and the study sheds light on current IoT/wearable application areas while also pointing to potential future research directions, such as robotics and other forms of integrated technology.

Schiliro et al. [15] (2019) analyzed business processes that rely on both data and expertise. As a result, the Internet of Things (IoT) has evolved as a common platform for the development and integration of data-driven business processes both inside and between enterprises. Computer-enabled technology, such as CCTV cameras, police cars, and drones, may be employed in law enforcement to get a better understanding of the real world.

Ilapakurti and Vuppalapati [16] (2015) presented IoT as the cutting-edge electronic monitoring framework recommended in the study, and it allowed dairies to reduce the financial effect of HS while maximizing return on assets and return on investment (ROI) via operational efficiency improvements. More significantly, a contented cow translates into more prosperous dairy business and, therefore, richer and creamier dairy products for consumers. Dairy IoT may be used both online and offline using the suggested architecture. Prototyping solutions and their applications were discussed in their study along with a few experimental outcomes.

Dhanda et al. [35] (2019) highlighted several different types of assaults that can be used against information in IoT. Confidentiality and integrity were maintained by using cryptographic algorithms. It's difficult to apply the resource-intensive standard cryptographic algorithms because of the small size, restricted computational performance, limited memory, and limited power resources of the devices. As a result, developing IoT security solutions that weren't too burdensome becomes imperative. To address the issue of resource-constrained devices in IoT, comprehensive research on lightweight cryptography was conducted. The goal of the work was to present an in-depth and current survey of the many lightweight cryptographic primitives that were currently accessible.

Gupta et al. [49] (2020) proposed an excellent strategy for producing green communications and regulating network traffic. The rising use of smartphone apps means that edge computing on smartphones may greatly aid network traffic control. The construction of an AI-based data processing approach was a difficult challenge because smartphones have limited computational capability. However, because of the customers' need for cost-effective technology, it cannot be readily penetrated. Use of smartphone end-devices instead of expensive IoT sensors to monitor the health of agricultural vehicles was proposed in the research, which can accomplish both objectives.

Krishnamoorthy et al. [50] (2019) proposed a methodology by which, the system recommends the item by taking up from the shelf and placed in a virtual shopping cart when it was detected. It increases the system's appeal from the standpoints of both sellers and buyers. As soon as the customer exited the shop, the products they had purchased were tallied up and charged to the user's bank account. Using a camera and face recognition technology called Kairos, the business can keep track of who is entering and exiting. In addition, Google's Firebase database serves as the backbone for all of the activities.

Bhattacharaya et al. [36] (2020) aimed to capture the intellectual structure of this subject and research trends from the quantitative and statistical analysis of research papers. The study's conceptual framework was built utilizing methods and techniques from the field of social network analysis, which was also utilized to build conceptual links.

Cao and Wachowicz [37] (2020) played a critical role in facilitating anticipatory learning and IoMT systems were reviewed as part of their study. The entire potential of IoMT systems in future smart cities, in terms of proactive decision-making and decision delivery, was realized via an anticipatory action/feedback loop. Authors also examined the limitations and promise of anticipatory learning in specific systems.

3.3   RESEARCH METHODOLOGY

Research in the area of supply-chain management, IoT, and 5G have been considered. The issues regarding performance, accuracy, and flexibility are resolved by proposing an efficient supply-chain management system based on IoT and the 5G model that is making use of the classifier and filtering mechanism. The propsoed methodology makes use of IoT based 5G model to take a survey of males and females to understand their perspectives on supply-chain management (Figure 3.5).

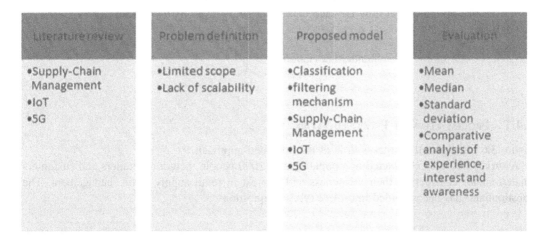

FIGURE 3.5   Research Methodologies.

3.4 PROPOSED WORK

The objectives of the research work is to identify current trends in the unorganized retail supply chain in India. In supply chain management, men's and women's perspectives are being studied. This study was carried out utilizing a 5G IoT network. Unorganized and organized retail supply-chain management has been examined via the development of a questionnaire for both controlled and disorganized supply-chain data. The data was then subjected to statistical analysis, using the formulas of mean, median, and standard deviation.

It is important to know the mean and median values of data in order to make sense of it. The goal of calculating the central tendency is to find a data set based on information about the average value of a group of variables. When the data is dispersed, the standard deviation is helpful in deciphering the measurements. The higher the standard deviation, the more dispersed the data is. The standard deviation, on the other hand, can't be negative at all.

This research is considering the awareness, interest, and experience of retailers and customers regarding supply-chain management. In order to address statistical issues, the following statistics formulae will be most often used:

- **Mean:** The formula for calculating the mean of a data collection is as follows:

$$Mean\ (\bar{x}) = \frac{\Sigma x}{N}$$

- **Median:** For the median, we have two formulae to choose from. We apply the following formula if the data set has an odd number of terms:

$$Median = \frac{n+1}{2}^{th} observation$$

If the data collection has an even number of terms, we may apply the following formula.

$$Median = \frac{\frac{n}{2}^{th} observation + \left(\frac{n}{2}+1\right)^{th} observation}{2}$$

- **Standard Deviation:** It is possible to compute the standard deviation as the square root of variance by looking at how far each data point strays from its own mean.

$$Standard\ deviation(\sigma) = \sqrt{\frac{\Sigma(x_i - \mu)^2}{N}}$$

3.4.1 PROCESS FLOW OF PROPOSED WORK

Figure 3.6 highlights the process flow of the propsoed approach.

A survey has been conducted in a population of 1000 people including retailers and customers (male/female) considering their awareness and interest in retail supply-chain management. The questionnaire has been divided into three types of questions.

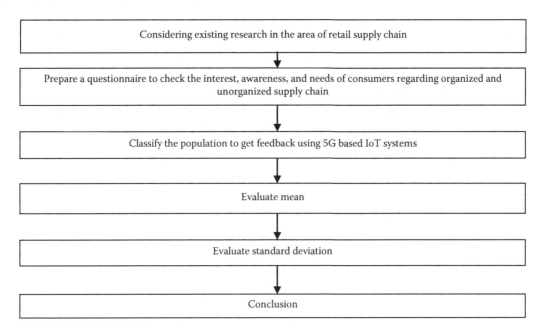

FIGURE 3.6 Process Flow of Proposed Research.

3.4.2 QUESTION CATEGORY

Category 1 questions are yes/no based.
Category 2 questions are ranking based where rank lies from 1 to 5.
Category 3 questions are good/bad/average based.
Category 4 questions are yes/no/cannot say based.

3.4.2.1 Category 1 Questions (General Question Awareness)

1.1. Have you heard about unorganized retail supply-chain management?

   Yes   no

1.2. Have you heard about organized retail supply-chain management?

   Yes   no

3.4.2.2 Category 2 Questions (to Test Interest)

2.1. What would rate organized retail supply?

   1   2   3   4   5

2.3. What would rate unorganized retail supply?

   1   2   3   4   5

3.4.2.3 Category 3 Questions (to Know Their Experience)

3.1. What is your shopping experience in an unorganized supply chain?

   Good   bad   average

3.2. What is your shopping experience in an organized supply chain?

Good   bad   average

3.4.2.4   Category 4 Questions (to Know Their Preference)

4.1. Would you prefer online product buying from an unorganized supply chain?

Yes   No   Cannot-say

4.2. Would you prefer online product buying from an organized supply chain?

Yes   No   Cannot-say

3.4.3   DATA COLLECTION FOR CATEGORY 1

The population of 2000 people has given feedback where 1000 are male and 1000 are female candidates.

Considering Table 3.1, Figure 3.7 has been plotted.

Considering the Table 3.1 mean has been calculated for unorganized and organized and is shown in Table 3.2, presenting almost equal awareness regarding unorganized and organized retail chain management in the case of males (Figure 3.8).

TABLE 3.1
Male Feedback for General Question

	Unorganized	Organized
Knows about Retail Supply-Chain Management	559	566
Don't Know about Retail Supply-Chain Management	541	534

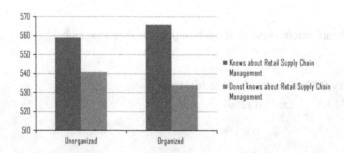

FIGURE 3.7   Male Feedback for General Question.

TABLE 3.2
Mean

Unorganized	Organized
550	550

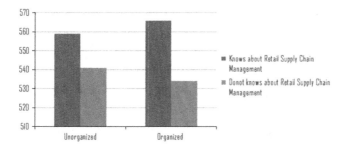

FIGURE 3.8 Mean of Male Feedback for General Question.

In similar way, considering the Table 3.1 median has been calculated for unorganized and organized and shown in Table 3.3, presenting almost equal awareness regarding unorganized and organized retail chain management in the case of males (Figure 3.9).

But the Table 3.1 standard deviation has been calculated for unorganized and organized and shown in Table 3.4, presenting variation in unorganized and organized retail chain management in the case males (Figure 3.10).

TABLE 3.3
Median

Unorganized	Organized
550	550

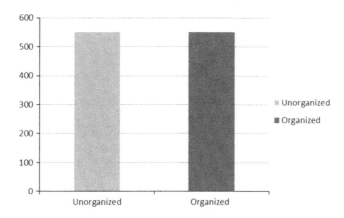

FIGURE 3.9 Median of Male Feedback.

TABLE 3.4
Standard Deviation

Unorganized	Organized
12.72792	22.62742

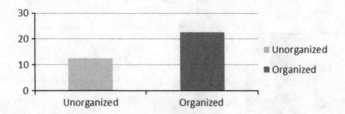

FIGURE 3.10   Standard Deviation of Male Feedback.

Considering Table 3.5, Figure 3.11 has been plotted.

Considering the Table 3.5 mean has been calculated for unorganized and organized and shown in Table 3.6, presenting almost equal awareness regarding unorganized and organized retail chain management in the case of females (Figure 3.12).

In similar way, considering the Table 3.5 median has been calculated for unorganized and organized and shown in Table 3.7, presenting almost equal awareness regarding unorganized and organized retail chain management in the case of females (Figure 3.13).

But the Table 3.5 standard deviation has been calculated for unorganized and organized and shown in Table 3.8, presenting variation in unorganized and organized retail chain management in the case females (Figure 3.14).

TABLE 3.5
Female Feedback General Question

	Unorganized	Organized
Knows about Retail Supply-Chain Management	558	562
Don't Know about Retail Supply-Chain Management	542	538

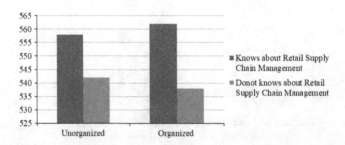

FIGURE 3.11   Female Feedback for General Question.

TABLE 3.6
Mean

Unorganized	Organized
550	550

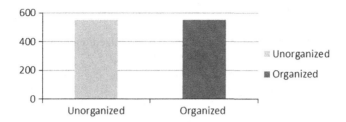

FIGURE 3.12   Mean of Female Feedback.

TABLE 3.7
Median

Unorganized	Organized
550	550

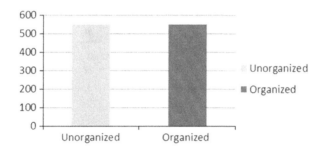

FIGURE 3.13   Median of Female Feedback.

TABLE 3.8
Standard Deviation

Unorganized	Organized
11.31371	16.97056

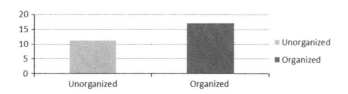

FIGURE 3.14   Standard Deviation of Female Feedback.

3.4.4 DATA COLLECTION FOR CATEGORY 2

The population of 2000 people has given a rating to organized/unorganized retail supply chain management in the general sector (Figure 3.15).

3.4.5 DATA COLLECTION FOR CATEGORY 3

Experience has been considered in this chart, where 0 is presenting bad, 1 is presenting average, and 2 is presenting well (Figure 3.16).

3.4.6 DATA COLLECTION FOR CATEGORY 4

Preference has been considered in this table, where 0 is presenting no, 1 is presenting cannot say, and 2 is presenting yes (Figure 3.17).

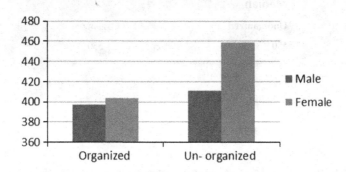

FIGURE 3.15   Rating Chart.

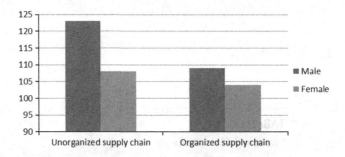

FIGURE 3.16   Experience Chart.

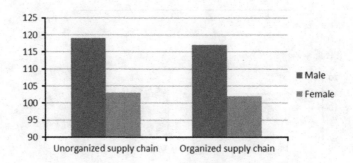

FIGURE 3.17   Preference Chart.

3.5 FINDINGS

The findings of the research is to present general awareness, interest, and experience of the population in retail supply-chain management along with their preferences. Table 3.1 highlights that number of males that know retail supply-chain management are more than that of person who does not know it. Tables 3.2, 3.3, and 3.4 highlights the mean, median, and standard deviation of a data set of male awareness, respectively. Table 3.5 represents that the number of females that know retail supply-chain management are more than that of person who does not know it. Tables 3.6, 3.7, and 3.8 consider mean, median, and standard deviation of data set of male awareness, respectively. Table 3.9 presents that females showed more interest in organized and unorganized retail supply, whereas Table 3.10 highlights that male experience is better than females. Table 3.11 presents the preference of supply chain by males and females.

TABLE 3.9
Summary of Rating

	Organized	Un-Organized
Male	397	411
Female	404	459

TABLE 3.10
Summary of Experience

	Unorganized Supply Chain	Organized Supply Chain
Male	123	109
Female	108	104

TABLE 3.11
Summary of Preference

	Unorganized Supply Chain	Organized Supply Chain
Male	119	117
Female	103	102

3.6 SIGNIFICANCE OF RESEARCH

In order to make sense of data, the mean and median values must be understood. Calculating the central tendency is all about finding a data set based on the average of a collection of variables' values. The standard deviation is useful in interpreting data that is spread. The more widespread the data is, the bigger the standard deviation. A negative standard deviation would be impossible, on the other hand. Present research has considered the mean, median, and standard deviation.

3.7 CONCLUSION AND FUTURE SCOPE

Retail supply-chain management has been a topic of research. The study is focussed on supply-chain management in various industries. This study was based on prior research in the field of retail supply-chain management. Over a population of 2000 people, a survey was done to determine their knowledge, interest, and attitudes in supply-chain management organized and unstructured. Findings conclude that the overall rating, experience, and preference of unorganized is found more than organized. IoT is no longer an exception to this trend. It's changing the way we work. Almost everything from how we use our appliances to how we monitor the environment has altered in the last few decades, according to a new study. Similarly, the industry has a role to play. 4G and 5G technologies can provide smart services such as smart health, smart homes, smart energy, and smart surroundings by analyzing data from the IoT. A need, not a luxury, is the introduction of 5G technology. IoT with 5G technology could be efficiently used to conduct real-time survey. A lack of wireless technology and logistical software means that this data is ineffective and hence worthless. Thus, proposed research would open new doors for further research.

REFERENCES

[1] Muthu, B., Sivaparthipan, C. B., Manogaran, G., Sundarasekar, R., & Kadry, S. (2020). IoT-based wearable sensor for diseases prediction and symptom analysis in the healthcare sector. *Peer-to-Peer Netw. Appl.* 13, 2123–2134. https://doi.org/10.1007/s12083-019-00823-2

[2] Atlam, H. F., & Wills, G. B. (2020). *IoT Security, privacy, safety, and ethics.* Springer International Publishing. 10.1007/978-3-030-18732-3

[3] Pavithran, D. (2020). Towards building a blockchain framework for IoT. *Cluster Computing.* 10.1007/s10586-020-03059-5

[4] Shukla, A., Ahamad, S., Rao, G. N., Al-Asadi, A. J., Gupta, A., & Kumbhkar, M. (2021). Artificial intelligence assisted IoT data intrusion detection. In *2021 4th International Conference on Computing and Communications Technologies (ICCCT)*, pp. 330–335. 10.1109/ICCCT53315.2021.9711795

[5] Mabodi, K., Yusefi, M., Zandiyan, S., & Irankhah, L. (2020). Multi-level trust-based intelligence schema for securing of internet of things (IoT) against security threats using cryptographic authentication. *The Journal of Supercomputing, 0123456789.* 10.1007/s11227-019-03137-5

[6] Kaur, M. J., Mishra, V. P., & Maheshwari, P. (n.d.). *The convergence of digital twin, IoT, and machine learning: Transforming data into action.* Springer International Publishing. 10.1007/978-3-030-18732-3

[7] Safara, F., Souri, A., Baker, T., & Al, I. (2020). PriNergy: A priority-based energy-efficient routing method for IoT systems. *The Journal of Supercomputing, 0123456789.* 10.1007/s11227-020-03147-8

[8] Islam, M., Rahaman, A., & Islam, R. (2020). Development of smart healthcare monitoring system in IoT environment. *SN Computer Science, c.* 10.1007/s42979-020-00195-y

[9] Al, A., Hossam, S., & Ibrahim, F. (2020). An unsupervised intelligent system based on a one-class support vector machine and Grey Wolf optimization for IoT botnet detection. *Journal of Ambient Intelligence and Humanized Computing, Angrishi 2017.* 10.1007/s12652-019-01387-y

[10] Verma, A., & Ranga, V. (2019). Machine learning-based intrusion detection systems for IoT. *Wireless Personal Communications, 0123456789.* 10.1007/s11277-019-06986-8

[11] Garg, M., Gupta, A., Kaushik, D., & Verma, A. (2020). Applying machine learning in IoT to build intelligent system for packet routing system. *Materials Today: Proceedings.* 10.1016/j.matpr.2020.09.539

[12] Al-emran, M., Malik, S. I., & Al-kabi, M. N. (n.d.). *A survey of internet of things (IoT) in education: Opportunities and challenges.* Springer International Publishing. 10.1007/978-3-030-24513-9

[13] Al-qerem, A., Alauthman, M., Almomani, A., & Al-qerem, A. (2019). IoT transaction processing through cooperative concurrency control on the fog–cloud computing environment. *Soft Computing, 0123456789.* 10.1007/s00500-019-04220-y

[14] Bansal, S., & Kumar, D. (2020). IoT ecosystem: A survey on devices, gateways, operating systems, middleware, and communication. *International Journal of Wireless Information Networks, 0123456789.* 10.1007/s10776-020-00483-7

[15] Schiliro, F., Beheshti, A., Ghodratnama, S., Amouzgar, F., Benatallah, B., Yang, J., & Sheng, Q. Z. (n.d.). *iCOP: IoT-enabled policing processes* (Vol. 1). Springer International Publishing. 10.1007/978-3-030-17642-6

[16] Ilapakurti, A., & Vuppalapati, C. (2015). Building an IoT Framework for Connected Dairy, *2015 IEEE First International Conference on Big Data Computing Service and Applications*. pp. 275–285. 10.1109/BigDataService.2015.39

[17] Khalid, U., Asim, M., Rafferty, L., & Asim, M. (2020). A decentralized lightweight blockchain-based authentication mechanism for IoT systems. *Cluster Computing*, 6, 2067–2087. 10.1007/s105 86-020-03058-6

[18] Tseng, L., Yao, X., & Otoum, S. (2020). Blockchain-based database in an IoT environment: Challenges, opportunities, and analysis. *Cluster Computing*, 7, 2151–2165. 10.1007/s10586-020-03138-7

[19] Ye, S., Tun, Y., Madanian, S., & Mirza, F. (2020). Internet of things (IoT) applications for elderly care: A reflective review. *Aging Clinical and Experimental Research, 0123456789*. 10.1007/s40520-020-01545-9

[20] Kaushik, D., & Gupta, A. (2021). Ultra-secure transmissions for 5G-V2X communications. *Materials Today: Proceedings*. 10.1016/j.matpr.2020.12.130.

[21] Palattella, M. R., Dohler, M., Grieco, A., Rizzo, G., Torsner, J., Engel, T., & Ladid, L. (2016). Internet of Things in the 5G era: Enablers, architecture, and business models. *IEEE Journal on Selected Areas in Communications*, 34(3), 510–527. 10.1109/JSAC.2016.2525418

[22] Rachit, Bhatt, S., & Ragiri, P. R. (2021). Security trends in Internet of Things: A survey. *SN Applied Sciences*, 3(1), 1–14. 10.1007/s42452-021-04156-9

[23] Kumar, S., Tiwari, P., & Zymbler, M. (2019). Internet of Things is a revolutionary approach for future technology enhancement: A review. *Journal of Big Data*, 6(1). 10.1186/s40537-019-0268-2

[24] Wang, P., Valerdi, R., Zhou, S., & Li, L. (2015). Introduction: Advances in IoT research and applications. *Information Systems Frontiers*, 17(2), 239–241. 10.1007/s10796-015-9549-2

[25] Mahmoud, R., Yousuf, T., Aloul, F., & Zualkernan, I. (2016). Internet of things (IoT) security: Current status, challenges, and prospective measures. In *2015 10th International Conference for Internet Technology and Secured Transactions, ICITST 2015*, pp. 336–341. 10.1109/ICITST.2015.7412116

[26] Zhang, Q., & Fitzek, F. H. P. (2015). Mission-critical IoT communication in 5G. *Lecture Notes of the Institute for Computer Sciences, Social-Informatics and Telecommunications Engineering, LNICST*, 159, 35–41. 10.1007/978-3-319-27072-2_5

[27] Gupta, A., Kaushik, D., Garg, M., & Verma, A. (2020). Machine learning model for breast cancer prediction. In *2020 Fourth International Conference on I-SMAC (IoT in Social, Mobile, Analytics and Cloud) (I-SMAC)*, pp. 472–477. 10.1109/I-SMAC49090.2020.9243323

[28] Ben-Daya, M., Hassini, E., & Bahroun, Z. (2017). Internet of Things and supply chain management: A literature review. *International Journal of Production Research*, 57(15–16), 1–24.

[29] Verma, A., Gupta, A., Kaushik, D., & Garg, M. (2021). Performance enhancement of IOT based accident detection system by integration of edge detection. *Materials Today: Proceedings*. 10.1016/j.matpr.2021.01.468.

[30] De Vass, T., Shee, H., & Miah, S. J. (2018). The effect of 'Internet of Things' on supply chain integration and performance: An organisational capability perspective. *Australasian Journal of Information Systems*, 22, 1–29.

[31] Bansal, R., Gupta, A., Singh, R. & Nassa, V. K. (2021). Role and impact of digital technologies in E-learning amidst COVID-19 pandemic. In *2021 Fourth International Conference on Computational Intelligence and Communication Technologies (CCICT)*, pp. 194–202. 10.1109/CCICT53244.2021.00046

[32] Gunasekaran, A., Patel, C., & McGaughey, R. E. (2004). A framework for supply chain performance measurement. *International Journal of Production Economics* 87(3), 333–347.

[33] Gupta, A., Singh, R., Nassa, V. K., Bansal, R., Sharma, P., & Koti, K. (2021). Investigating application and challenges of big data analytics with clustering. In *2021 International Conference on Advancements in Electrical, Electronics, Communication, Computing and Automation (ICAECA)*, pp. 1–6. 10.1109/ICAECA52838.2021.9675483.

[34] Burow, K., Hribernik, K., & Thoben, K.-D. (2018). First steps for a 5G-ready service in cloud manufacturing.Paper presented at the*2018 IEEE International Conference on Engineering, Technology, and Innovation (ICE/ITMC)*.

[35] Dhanda, S. S., Singh, B. & Jindal, P. (2020). Lightweight cryptography: A solution to secure IoT. *Wireless Pers Commun*, 112, 1947–1980. 10.1007/s11277-020-07134-3

[36] Bhattacharya, S., Kumar, R., & Singh, S. (2020). Capturing the salient aspects of IoT research: A Social Network Analysis. *Scientometrics*, 125(1), 361–384. 10.1007/s11192-020-03620-4

[37] Cao, H., & Wachowicz, M. (2020). A holistic overview of anticipatory learning for the internet of moving things: Research challenges and opportunities. *ISPRS International Journal of Geo-Information*, 9(4). 10.3390/ijgi9040272

[38] Khan, N. A., Jhanjhi, N. Z., Brohi, S. N., & Nayyar, A. (2020). Emerging use of UAV's: Secure communication protocol issues and challenges. In *Drones in Smart-Cities: Security and Performance*. (pp. 37–55). Elsevier. https://doi.org/10.1016/b978-0-12-819972-5.00003-3

[39] Nayyar, A., Rameshwar, R. U. D. R. A., & Solanki, A. (2020). Internet of Things (IoT) and the digital business environment: A standpoint inclusive cyber space, cyber-crimes, and cybersecurity. *The Evolution of Business in the Cyber Age*, 10, 9780429276484-6.

[40] Krishnamurthi, R., Nayyar, A., & Solanki, A. (2019). Innovation opportunities through internet of things (IoT) for smart cities. *Green and Smart Technologies for Smart Cities*, 261–292.

[41] Rathee, D., Ahuja, K., & Nayyar, A. (2019). Sustainable future IoT services with touch-enabled handheld devices. *Security and Privacy of Electronic Healthcare Records: Concepts, Paradigms and Solutions*, 131, 131–152.

[42] Lee, C. K. M., Yeung, C. L., & Cheng, M. N. (2016). Research on IoT-based cyber-physical system for industrial big data analytics. In *IEEE International Conference on Industrial Engineering and Engineering Management, 2016-January*, pp. 1855–1859. 10.1109/IEEM.2015.7385969

[43] Cheng, J., Chen, W., Tao, F., & Lin, C.-L. (2018). Industrial IoT in 5G environment towards smart manufacturing. *Journal of Industrial Information Integration*, 10, 10–19.

[44] Rai, A., Patnayakuni, R., & Seth, N. (2006). Firm performance impacts of digitally enabled supply chain integration capabilities. *MIS Quarterly*, 30(2), 225–246. 10.2307/25148729

[45] Mahapatra, B., Turuk, A. K., Nayyar, A., & Sahoo, K. S. (2021). Multilevel authentication and key agreement protocol for D2D communication in LTE based C-IoT network. *Microprocessors and Microsystems*, 103720.

[46] Solanki, A., & Nayyar, A. (2019). Green internet of things (G-IoT): ICT technologies, principles, applications, projects, and challenges. In *Handbook of research on big data and the IoT* (pp. 379–405). IGI Global.

[47] Krishnamurthi, R., Kumar, A., Gopinathan, D., Nayyar, A., & Qureshi, B. (2020). An overview of IoT sensor data processing, fusion, and analysis techniques. *Sensors*, 20(21), 6076.

[48] Shee, H., Miah, S. J., Fairfield, L., & Pujawan, N. (2018). The impact of cloud-enabled process integration on supply chain performance and firm sustainability: The moderating role of top management. *Supply Chain Management*, 23(6), 500–517.

[49] Gupta, N., Khosravy, M., Patel, N., Dey, N., Gupta, S., Darbari, H., & Crespo, R. G. (2020). Economic data analytic AI technique on IoT edge devices for health monitoring of agriculture machines. *Applied Intelligence*, 50(11), 3990–4016. 10.1007/s10489-020-01744-x

[50] Krishnamoorthy, A., Vijayarajan, V., & Sapthagiri, R. (2019). Automated shopping experience using real-time IoT. In *Advances in intelligent systems and computing* (Vol. 862). Springer Singapore. 10.1007/978-981-13-3329-3_20

4 Impact of Artificial Intelligence on Agriculture Value Chain Performance: Agritech Perspective

C. Ganeshkumar
Indian Institute of Plantation Management, Bengaluru, India

Jeganthan Gomathi Sankar
BSSS Institute of Advanced Studies, Bhopal, India

Arokiaraj David
Department of Management Studies, St. Francis Institute of Management and Research, Mumbai, India

CONTENTS

4.1 INTRODUCTION

The United Nations' Sustainable Development aims to attain zero hunger by bringing advancements in the agricultural sector by 2030. Moreover, interruptions due to unpredictable weather, greenhouse gas emissions and global water scarcity have raised the following significant concerns (Lakshmi & Corbett, 2020). The transition to a sustainable mode of growing food from conventional agricultural practices can lead to a healthy environment and social and economic equity. Hence, minimizing the negative environmental impacts by accelerating

DOI: 10.1201/9781003264521-4

agricultural productivity, has become a precedence. Agricultural information technology (AIT) is known as the use of information technology within agricultural practices, which in the last 20 years have shown a remarkable progression (Wang & Siau, 2019). To ameliorate agricultural productivity, the direct tool that can be used is the AIT and as an indirect mechanism, it can also be used to make informed decisions for empowering the farmers. The IT integration within the agriculture practices has led to the advent of precision farming, which is a new approach to revolutionize agriculture and that utilizes the data-intensive tools and techniques to inform farming decisions and value-added agriculture by harnessing the vast amount of data (Lakshmi & Corbett, 2020). The implementation of machine learning (ML) and AI in the agricultural sector helps to generate value by addressing sustainability concerns and enhancing crop yields. From the IS research community, precision farming has accumulated limited interest as it is gaining ground in practice and its prior studies is all about the study related to precision agriculture as well as the adoption and diffusion of AIT. The tools of AI and ML provides actionable insights on water conditions, weather and soil which ultimately leads the farmers to exercise discretion on "irrigation, planting, and harvesting."

The applications of AI in agriculture is still developing, so that the challenges, motives, and impacts of technological innovation remains trackless, where these limitations seem to be significant as the traditional or business contexts differs from the agricultural contexts, where IS typically studied. The conditions including both physical and natural in the natural eco-system makes the risk anticipation and the decision-making processes more complex. The effect of technology on the desired outcomes is determined by the atmospheric conditions, dynamic soil, and the weather along with a myriad of biological interactions. By improving the use of AI and motivating to build knowledge, this research helps in analyzing the impacts and objectives of the deployment of AI in agriculture, which addresses sustainability and seeks to understand the improved agricultural outputs. The productivity and performance of an organization are dependent on the availability of resources, based on the resource grounded view of the organization. This is where the organization attains profitability with the capabilities that direct these resources. In particular, dynamic capabilities are the "capacities to reconfigure an organization's resources and routines, by its top decision-makers in an appropriate manner" that have helped to explain the sources of value creation and the business change in organizations. Based on this approach, the agricultural enterprises gain a competitive advantage by protecting, reconfiguring and combining their capabilities and resources that reduce waste by exploiting and enhancing productivity by deploying AI. Likewise, the farmer's intuition about their agricultural lands and the amalgamation of AIT can give unique results to various issues related to weather patterns, land, and water encountered on farms (Lakshmi & Corbett, 2020; David et al., 2022).

This research investigates and looks into how the agricultural organizations operating within various geographical regions as well as the ones operating globally creates value with AI, and in in turn addresses the sustainability concerns. The Centering Resonance Analysis has been conducted to answer this question where the archived secondary data of agricultural organizations in the form of press releases and media reports that are actively deploying or plan to deploy AI is used. The results suggest that, globally, AI is used to increase efficiency and production when it is primarily applied and during the process, the technology also serves to address environmental issues as well as labor shortages. AI deployment is found active in Europe and North America at the regional level and it is forging ahead with its efforts in Africa and Asia.

Organization of Chapter: The chapter is organized as: Section 4.2 elaborates literature review. Section 4.3 outlines research methodology. Section 4.4 highlights the results and discussions. Section 4.5 concludes the chapter with implications.

4.2   REVIEW OF LITERATURE

Sustainable Development Goals (SDGs) have been established by the United Nations (UN) in 2016 for having certain objectives (Sachs, 2012). In total, 17 goals have been developed by the UN to replace the previously developed eight Millennium Development Goals (Sachs, 2012). The newly introduced SDGs integrated economic growth, social inclusion, and environmental protection (Haliscelik & Soytas 2019). The objectives of SDGs are developed to address various issues that are firmly related to each other (Ashraf et al., 2019). There are some goals identified to be relevant to the agriculture industry such as zero hunger, and responsible consumption and production. The aforesaid agriculture-related SDGs invite to introduce alternative solutions for conventional farm management thus difficulties that occurred in the old farm management can be eradicated. The data-intensive intelligent principles can be implemented in the agriculture industry as it has been widely employed in other industries such as manufacturing, transportation, and retail (Liao et al., 2020). High uncertainty results complicated problems in conventional farming operations. Technological development and present information technology innovation provide adequate support for the agriculture industry to improve efficiency, productivity, and profitability (Liao et al., 2020). The industrial transformation has been witnessed in various industries with the help of information technology and attained successful outcomes. The fourth Industrial Revolution, named Industry 4.0, was introduced in the last decade in various industries (Liao et al., 2020). To attain effective operational results and eliminate complications, the agricultural industry also inevitably experienced the transition. Farming experienced digitalized from conventional methods. In this process, data have been gathered by various agricultural platforms to assist farmers by providing planting suggestions, customized recommendations, sensemaking processes, data stemming from the farm directly. Artificial intelligence (AI) can assist the logic of the farmers and human intelligence in operational disruptions of the agricultural industry (Ganeshkumar et al., 2021). Artificial intelligence is having a significant influence in agriculture in accordance with complex problem solving. This can provide effective information on soil management, spray schedule, and congenial conditions and will assist the farmers in effective decision making for agronomy.

A survey conducted by National Sample Survey Organisation (NSSO) identified that nearly 40% of the farmers are feeling of quit farming if they had a choice. This is evidence of the state of agrarian crisis farmers are in (Choudhary & Choudhary, 2013). The agricultural industry is facing an important challenge to feed an increasing population with decreasing natural resource availability such as soil fertility, water scarcity, climatic change, and urbanization. The aforesaid problems can be managed by implementing AI in agriculture to attain productivity, efficiency, and sustainability (Awuor et al., 2013). AI has a great potential application in agriculture by assisting farmers to know about soil quality, time to sow, spray herbicide, and the possibility of pest infestations. To improve efficiency in accordance with soil and crop monitoring, forecasting the weather, agricultural analytics, and supply chain efficiency, AI technologies are being employed (Bhar et al., 2019). Agriculture robotics is one of the most widely implemented AI techniques in agriculture followed by soil management and monitoring. Crop selection, deployment of available resources, crop investment, scouting procurement location, and cost management are a few examples where AI plays a major role in agriculture. Biomimicry is one kind of AgriTech that can help to design innovative methods to assist agricultural problem solving and decision making (Pathak et al., 2019). Farming is highly supported by nature such as insects and birds. One of the evolving concepts of artificial intelligence (AI) is collective intelligence, also known as swarm intelligence, in which problems have been solved by exhibiting noteworthy competencies that are being confronted by conventional methods (Li and Clerc 2019). Swarm intelligence is basically not having continuous centralized control, it is working based on self-organizing behaviours usually as a logical method (Li and Clerc, 2019). Artificial intelligence is defined as a "cognitive process and especially to reasoning (Pomerol, 1997). Human problem-solving skills and behaviour are being employed to solve real-world tasks by AI (Duan et al., 2019). The activities that require human intelligence to perform are done by digital technologies in AI (Bawack et al., 2019). With the help of AI, various applications have been performed in agrarian operations that are commonly known as agriculture technology (AgriTech). AI and machine learning (ML) can be used to understand the farming land condition. Farmers can learn and gain knowledge about their crops and yields with the help of AgriTech, and also AgriTech assists to increase the rate of production (Pham & Stack, 2018). Along with human intelligence, AI can identify the methods to improve farming land condition and change it which in turn influences to achieve sustainability (Yahya, 2018). The rate of production, resource allocation, and proactive action can be achieved by applying AgriTech (Fountas et al. 2015). The application of AgriTech can significantly change everyday farming activities. There are various AgriTech has been employed, drones are the most widely used one. AgriTech drones are used for gathering data to monitor the real-time growth of the crop. It influences production efficiency and effective decision making. It can be used in situations when human interaction is limited and also employed on a remote basis. The sensor usage has also been used in smart farming and smart irrigation; this is known as EI–embedded intelligence (Yin, 2018). The usage of AI in agriculture has registered imperative growth in the last decade. In the United States of America (USA), it is expected to reach $1.1 billion USD by the year 2025. In 2017, it was recorded $240 USD. From the above discussion, AI has been a widely used practice in developed nations and farmers have also gained dividends from this intervention by AI in agriculture; the scope and spread of technology is required for developing and underdeveloped countries too.

4.3 RESEARCH METHODOLOGY

This section discusses in depth the methods used by the researcher to carry out the proposed research. The section provides explanations of the research design used for the analysis. The objective of this study is to examine and define the various characteristics and dynamics of agritech companies in relation to artificial intelligence (AI) value chain performance; research methodology suggested is, therefore, descriptive research design. This research is based on primary as well as secondary data. Secondary data related to the number of agritech firms in Bangalore. Primary data on the AI in value chain performance was collected through the survey approach by administering a proper structured questionnaire to the Bangalore-based agritech firms' executives. A well-structured questionnaire was used to gather primary data, which was presented directly to the executives of the agritech firms based in Bangalore that use AI. The personal interview process was used to collect primary data. The questionnaire consisted of both quantitative and qualitative aspects relating to the AI-based agritech units interviewed and their possible effect on the success of agritech firms in the value chain performance. The sampling units were selected using a simple random sampling method using random numbering for the survey, and used as a sample frame of the study. The sampling units were selected using a simple random sampling method using random numbering for the survey. The sampling technique employed in this research is simple random sampling through the method of random number generation. The IBM-SPSS 21 is the statistical package used. The statistical tools used to evaluate data are as follows: 1) simple mean 2) Chi-square test 3) ANOVA–analysis of variance 4) factor analysis 5) cluster analysis 6) correspondence analysis 7) canonical correlation.

4.4 RESULTS AND DISCUSSION

This section consists of two sub-sections; the first sub-section presents a descriptive analysis of characteristics of the agritech industries. Followed by second sub-section highlighting AI benefits, dimension analysis was analyzed using statistical tools, namely simple mean analysis, simple mean, Chi-square test, independent sample t-test, analysis of variance (ANOVA), factor analysis, cluster analysis, discriminant analysis, correspondence analysis, and canonical correlation. Results are represented in tabular and figurative forms.

TABLE 4.1
Ranking of AI Value Chain Performance

Sl. No.	Value Chain Performance with Company Profile	F-Value	P-Value	Significance Level
1	Value Chain Position	2.365	0.047	Significant
2	Number of Employees	2.339	0.079	Not Significant
3	Nature of Industry	1.002	0.396	Not Significant
4	Type of Business Organization	0.65	0.585	Not Significant
5	Agritech Category	5.348	0.006	Significant
6	Market Coverage	0.843	0.434	Not Significant
7	Number of years in this business	1.034	0.382	Not Significant

4.4.1 RANKING OF AI VALUE CHAIN PERFORMANCE

The performance of AI in the value chain such as reusability, energy and water efficiency, product reliability, traceability, storage and transport condition, promotion, convenience, client services, working conditions, appearance, product safety, customer satisfaction, awareness in retail/super-market, emissions, pesticide use, taste, product lateness, customer response time, customer complaints, backorder, volume flexibility, nutrition, lead time, inventory levels, delivery flexibility, shipping errors, return on investment, shelf life, pesticide/chemical use, profit, production costs, transaction costs, last sales, order fill rate, and transportation costs are ranked based on the mean values assigned to them and the results are exhibited in Table 4.1.

It could be seen from Table 4.1 that more reusability in the AI value chain is appreciated by the agritech units that are sampled. It is ranked the top as it helps the companies in reducing resource usage and projects them to be an environmentally friendly business. The other factors that are ranked following more reuse are less energy and water use. This also shows that the agritech units aspire to reduce resource wastage. The performance factors that are ranked 12th, 13th and 14th are better product safety, better customer satisfaction, and better awareness at retail, respectively. This portrays their concern for their end consumers. More profit is observed to be ranked as 31st, which reveals that the agritech companies are more concerned about the environment and consumer-related factors than profit maximization. All cost-related factors such as less production cost, less transaction cost, less transportation cost are ranked among the last five. This implies that the agritech companies focus less on cost minimization in their value chains. This could be due to their need to satisfy their value chain partners.

4.4.2 ANALYSIS OF AGRITECH INDUSTRY PROFILE AND AI VALUE CHAIN PERFORMANCE

The mean value difference in the AI value chain performance with respect to agritech industry profile is analyzed using the statistical tool ANOVA and the results are represented in Table 4.2.

It can be seen from Table 4.2 that the mean values of categories grouped based on value chain position and agritech category differ significantly. Other profile characteristics do not differ significantly. The value chain of companies differs when their offerings differ. This could be a reason for the significant variation in the mean value rating in companies categorized under agritech category and value chain position. The mean value of the various categories is classified based on the value chain position, shown in Table 4.3.

Table 4.3 shows that the mean value of categories in the value chain position falls between 3.0758 and 3.7435. Also, as per Table 4.2, it is found that the difference in mean values is significant. AI-related technologies are used by the production stage to perform tasks related to production and operation. Hence, they feel a moderate to a high level of impact in the performance

TABLE 4.2

ANOVA for Agritech Profile and AI Value Chain Performance

Sl. No.	Value Chain Performance with Company Profile	F-Value	P-Value	Significance Level
1	Value Chain Position	2.365	0.047	Significant
2	Number of Employees	2.339	0.079	Not Significant
3	Nature of Industry	1.002	0.396	Not Significant
4	Type of Business Organization	0.650	0.585	Not Significant
5	Agritech Category	5.348	0.006	Significant
6	Market Coverage	0.843	0.434	Not Significant
7	Number of years in this business	1.034	0.382	Not Significant

TABLE 4.3

Mean Values for Categories under Value Chain Position

Sl. No.	Value Chain Position	In Number	Mean	Mean
1	Processing/Value Addition Stage	11	3.0758	
2	Distributor/Retailer Stage	17	3.3856	
3	Combination	8	3.4931	
4	Agri-input/Supplier Stage	12		3.662
5	Supporting Services Stage	24		3.7373
6	Production/Farmer Stage	17		3.7435

after implementation of AI in their system than other categories. The mean value of the various categories is classified based on the number of employees are shown in Table 4.4.

It can be observed from Table 4.4 that the mean value of companies employing more than 300 staff is higher. Nonetheless, it cannot be stated that companies employing more than 300 employees only feel more impact on the value chain performance since the difference in mean values is insignificant as observed from Table 4.2. All agritech companies regardless of their employee count feel that the AI-related technologies impact the performance of the value chain at a moderate to a high level. Employees are slowly trying to assimilate the technology since AI is a recent implementation in the agritech industry. The mean value of the various categories is classified based on the nature of the industry, shown in Table 4.5.

It is seen from Table 4.5 that the mean values of categories split based on the nature of the industry falls between 3.427 and 3.861. However, based on the observations from Table 4.2, it

TABLE 4.4

Mean Values for Categories under Number of Employees

Sl. No.	Number of Employees	In Number	Mean	Mean
1	101–300	15	3.1944	
2	31–100	25	3.5544	
3	Less than 30	31		3.621
4	More than 300	18		3.7546

TABLE 4.5

Mean Values for Categories under Nature of Industry

Sl. No.	Nature of Industry	In Number	Mean
1	Micro Scale	05	3.4278
2	Medium Scale	45	3.4864
3	Small Scale	29	3.5852
4	Large Scale	10	3.8611

could be understood that the difference in mean values among these categories is not significant. Therefore, all companies irrespective of their nature feel an impact on the AI value chain performance. The value chain is a hot spot for any company, as it connects the manufacturer with all the channel partners and with enhanced technology usage; the performance of the value chain is likely to improve. The mean value of the various categories is classified based on the type of business organization, shown in Table 4.6.

It is seen from Table 4.6 that the mean value of the companies categorized based on the type of business organization falls between 3.1389 and 3.5887. Also, as observed from Table 4.2, it is found that the difference in mean values among the categories is insignificant. It can be inferred that all the companies regardless of their type of ownership feel a moderate to the high impact of AI-related technologies in the value chain performance. The ownership of a company may change but the value chain performance may still remain unchanged. The mean value of the various categories is classified based on the agritech category, shown in Table 4.7.

It is observed from Table 4.7 that the mean value rating of the companies that serve the customers with a combination of products and services is higher (3.7554) than the other categories. Table 4.2 shows that there is a significant difference in mean values. It could therefore be understood that the impact of AI in the value chain performance of companies offering a combination of products and services is higher than companies that offer only products or services. The value

TABLE 4.6

Mean Values of Categories under Type of Business Organization

Sl. No.	Type of Business Organization	In Number	Mean
1	Sole Proprietor	04	3.1389
2	Public Limited	04	3.4444
3	Partnership	09	3.5432
4	Private Limited	72	3.5887

TABLE 4.7

Mean Values for Agritech Category

Sl. No.	Agritech Category	In Number	Mean	Mean
1	Product	28	3.254	
2	Service	25		3.6122
3	Both	36		3.7554

TABLE 4.8
Mean Values for Categories under Market Coverage

Sl. No.	Market Coverage	In Number	Mean
1	Domestic Market	76	3.5212
2	International Market	01	3.9167
3	Both	12	3.7569

TABLE 4.9
Mean Values for Number of Years in the Agritech Business

Sl. No.	No of Years in This Business	In Number	Mean
1	More than 10 years	13	3.4487
2	More than 3 years–5 years	36	3.4576
3	More than 5 years–10 years	24	3.6319
4	Less than 3 years	16	3.7587

chain of companies that offer different sets of benefits to the customers is different from each other and hence it could be obvious that the value chain performance also differs, including the AI technologies which assist the value chain. The mean value of the various categories classified based on the market coverage is shown in Table 4.8.

It is seen from Table 4.8 that the mean values for companies grouped based on their market coverage stand between 3.5212 and 3.7569. Also, Table 4.2 shows that there exists no significant difference in mean values among these categories. All the companies regardless of which market they serve experience moderate to the high level of impact on the AI value chain performance. The market operations of the companies serving the markets would be similar in many aspects. Hence, all of them experience moderate to high impact of the AI technologies on their value chain performance. The mean value of the various categories is classified based on the number of years in the business, shown in Table 4.9.

Table 4.9 shows that the mean value rating of the companies categorized based on their number of years in the agritech business is found to lie between 3.4487 and 3.7587. Also, results from Table 4.2 reveal that there is no significant difference in mean values among these categories. Therefore, it can be stated that the companies irrespective of their business experience feel moderate to the high level of impact of the AI-related technologies on the value chain performance. Since AI-related technologies are naïve in the Indian market and companies are in the process of adopting it in their value chain, all companies irrespective of their years of experience find moderate to high levels of impact on their value chain performance.

4.4.3 ASSOCIATION BETWEEN VALUE CHAIN POSITION AND AI DIMENSIONS

To make a strategic, tactical, and operational decision in the firm, it is necessary for any firm to understand the nature of AI dimensions with respect to the profile of agritech companies. In this section, the characteristics of AI dimensions are analyzed through the Chi-square test, and correspondence analysis. The Chi-square values, along with their level of significance, are shown in Table 4.10.

To understand the characteristics of these AI dimensions, the association among the value chain position of the agritech company variables are analyzed. The Chi-square test is applied to test the

TABLE 4.10

Chi-Square Test for Profile of Agritech Company and AI Dimensions

Sl. No.	Value Chain Position	Chi-Square Value	P Value	Significance Level
1	Value Chain Performance	24.67	0.214	Not Significant

significance of associations. The Chi-square values and significant value reveal that AI problems, AI future, value chain performance, and organizational performance have no significant association with the value chain position of the agritech company variables, while there is a significant association between AI benefits and value chain position of the agritech company variables. The forthcoming paragraphs shall throw light on a detailed analysis of the nature of the relationship among the value chain position variables and AI dimensions.

4.4.4 AI VALUE CHAIN PERFORMANCE

Table 4.10 shows that the Chi-square value for the association between AI value chain performance and value chain position is 24.67 and the significance value is 0.214. This implies that there is no significant association between the variables. Figure 4.1 shows the association between variables.

Figure 4.1 shows that the value chain performance is very high in the supporting service stage and moderate in the Agri input/supplier stage. As observed there is no significant association between the variables. Hence, it cannot be stated that the performance of the value chain is very high because of its value chain position. The higher value chain performance of supporting service stages could be due to varied reasons such as the nature of service they offer or due to the inherent nature of the value chain.

4.4.5 VALUE CHAIN PERFORMANCE DIMENSION ANALYSIS

The value chain performance of agritech companies is studied with the aid of variables namely, reusability, energy and water efficiency, product reliability, traceability, storage and transport condition,

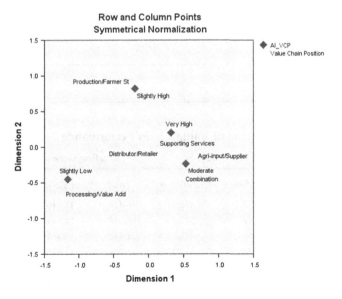

FIGURE 4.1 AI Value Chain Performance – Correspondence Diagram.

promotion, convenience, client services, working conditions, appearance, product safety, customer satisfaction, awareness in retail/supermarket, and better traceability. Each of these variables and their nature of relevance with value chain performance are discussed in detail in the following sections.

4.4.5.1 Factorization of Value Chain Performance

Factor analysis has been used to compress the number of elements or variables into a minimal number of measurable factors. Results of Kaiser–Meyer–Olkin (KMO) and Bartlett's test are shown in Table 4.11.

It reveals that the KMO value is 0.809 which shows that the factor analysis is useful to this set of data. For Bartlett's Sphericity test, the value of Chi-square is 3367 and the value of significance is 0.00. The variance and description of the Eigenvalue for each factor (performance in the value chain) is shown in Table 4.12.

Factors with eigenvalue in excess of one are taken as reduced factors. Such factors are now taking on the role of real factors in further research. From the above table, it can be observed that factor analysis derived six factors from 36 variables pertaining to the performance of the value chain. The six variables derived alone account for 74.37% of the overall variance which is quite important. The results of value chain performance factor loadings and the names assigned to each of those factors are summarized in Table 4.13.

Out of the 36 variables studied, nine variables are accommodated in the first factor, eight variables are accommodated in the second factor, seven variables are accommodated in the third factor, five variables are accommodated in the fourth factor and five variables are accommodated in the fifth factor and the remaining two other variables are placed in the sixth factor. The first factor was labeled "sustainability-related factor," while the second factor was labeled "auality performance-related factor." The third, fourth, fifth and sixth factors were named as "production performance-related factor," "operations performance-related factor,"

TABLE 4.11
KMO and Bartlett's Test

Kaiser–Meyer–Olkin Measure of Sampling Adequacy	0.809	
Bartlett's Test of Sphericity	Approx. Chi-Square	3367
	DF	630
	P-Value	0.000

TABLE 4.12
Variance Explained by Factor of Value Chain Performance

Component	Initial Eigenvalues			Rotation Sums of Squared Loadings		
	Total	% of Variance	Cumulative %	Total	% of Variance	Cumulative %
1	15.209	42.247	42.247	6.414	17.816	17.816
2	4.362	12.118	54.365	5.058	14.049	31.866
3	2.691	7.475	61.840	4.848	13.467	45.333
4	1.777	4.937	66.777	4.261	11.835	57.168
5	1.470	4.082	70.859	3.630	10.084	67.252
6	1.264	3.511	74.370	2.562	7.118	74.370

TABLE 4.13

Factor Loading of Supply Chain Performance

Value Chain Performance Items	Sustainability-Related Factor	Quality Related Factor	Production-Rrelated Factor	Operations Related Factor	Sales & Distribution Related Factor	Retail Related Factor
More Re-use	0.857					
Less Energy use	0.846					
Less Water Use	0.836					
Less Pesticide use	0.818					
Less Emissions	0.812					
Good Promotion	0.622					
Working Conditions	0.616					
Better Client Services	0.614					
Pesticide/ Chemical use	0.594					
Salubrity/Nutrition		0.841				
Less Shelf Life		0.819				
Good Taste		0.701				
Better Product Safety		0.659				
Product Reliability		0.611				
Good Appearance		0.597				
Better Convenience		0.559				
Storage Condition		0.510				
Less Transaction Costs			0.848			
Less Production Costs			0.835			
Transportation Costs			0.810			
Return on Investment			0.721			
Optimal Inventory			0.630			
More Profit			0.597			
Customer Satisfaction			0.546			
Better Lead time				0.785		
Less Shipping Errors				0.780		
Customer Complaints				0.772		
Customer Response				0.706		
Less Product lateness				0.685		
Less Last Sales					0.777	
Better Backorder					0.694	
Delivery Flexibility					0.692	
Better Order Fill Rate					0.686	
Volume Flexibility					0.545	
Awareness in retail						0.607
Better Traceability						0.526

"sales & distribution performance-related factor," and "retail performance-related factor," respectively. Sustainability-related factors hold crucial positions as these components' alone account for 17.81% of the overall variance. This shows that all manufacturing units' primary value chain performance is sustainability and manufacturing units differ from each other mainly based on this characteristic.

4.4.5.1.1 Sustainability-Related Factors

The performance of any business unit may be affected by the sustainability measures it adapts in its day-to-day functioning. These sustainability measures including reusability and efficient usage of resources may prove to be cost-saving in the long run for the company. The sustainability goals also project the company as an environmentally friendly unit in the minds of the customers. Since all sustainability-related variables are grouped under this factor; it is labeled as "sustainability-related factor."

4.4.5.1.2 Quality Performance-Related Factors

The quality of the product/service offered by any firm determines its stability in the market. If the quality of the offering is not satisfying the customer, the product will experience a fall sooner. Thereby, the quality of the product affects the performance of the company. The statements which come under quality-related aspects are grouped and the factor is labeled as "quality performance-related factor."

4.4.5.1.3 Production Performance-Related Factors

The production involves the conversion of any form of resource into a useful good with minimal wastage. This directly affects the cost incurred by the organization. Optimizing the use of resources may help in reducing the production cost and other related costs. The statements which pertain to the production process are clubbed under one factor which is labeled as "production performance-related factor."

4.4.5.1.4 Operation Performance-Related Factors

The operations of any business firm focus on performing the functions related to the different departments of the organization efficiently. Performing operations on time and with the right quantity of resources will enhance the reputation of the company among the stakeholders. The variables pertaining to the operations of the agritech companies are grouped under a single factor and are labeled as "operation performance-related factors."

4.4.5.1.5 Sales and Distribution Performance-Related Factors

The sales and distribution functions ensure that the product/service reaches the customer on time. The performance is decided based on the flexibility and promptness of the system. A flexible sales and distribution network rewards the company with an increased volume of sales and enhanced market coverage. The statements related to the sales and distribution are grouped and are brought under a single factor which is labeled as "sales and distribution related factors."

4.4.5.1.6 Retail Performance-Related Factors

The manufacturer's responsibility does not end when the product leaves the manufacturing unit. The manufacturer must take good care of the retail functions thereby helping his retail partner in selling his goods. This involves keeping the final consumers informed about the nature of the product and the origin of the good. The consumer must know the manufacturer of the good he is consuming. This results in better customer satisfaction and hence impacts the performance of the business. The statements related to this are grouped under one factor and are labeled as "retail performance-related factors."

4.4.5.2 Ranking of Value Chain Performance Factors

By using factor analysis, the 36 value chain performance variables are grouped into six factors, and these factors are labeled as sustainability performance-related factor, quality performance-related factor, production performance-related factor, operations performance-related factor, sales & distribution performance-related factor, and retail performance-related factor based on

TABLE 4.14
Strength of Value Chain Performance

Value Chain Performance	Mean	Standard Deviation	Rank
Sustainability Performance-related factor	3.7111	.86107	I
Retail Performance-related factor	3.6901	.87338	II
Quality Performance-related factor	3.6389	.73234	III
Operations Performance-related factor	3.4978	.85709	IV
Sales & distribution Performance-related factor	3.4089	.83567	V
Production Performance-related factor	3.4079	.82290	VI

variables loaded under each factor. Mean values for these value chain performance factors are displayed in Table 4.14.

Of the six value chain performance factors the agritech companies surveyed have conveyed a stronger opinion about the sustainability-related factor. The mean value is observed as 3.7111 from Table 4.14, which shows that this performance factor is considered to be the most important factor for manufacturing units. Sustainability-related factors such as reusability, less emissions and optimal resource usage does not only help the organization but also contributes to keeping the environment safe. Consumers nowadays prefer to use the goods manufactured responsibly. Understanding the importance this is gaining across the globe, the agritech companies have started to consider this factor crucial. They also try to incorporate sustainability measures in their practices.

4.4.5.3 Segmentation of Value Chain Performance

The agritech companies have been clubbed based on the similarities among these six factors namely sustainability performance-related factor, quality performance-related factor, production performance-related factor, operations performance-related factor, sales & distribution performance-related factor, and retail performance-related factor. Cluster analysis has been used to segment the agritech units based on their value chain performance levels related to the mentioned factors. The final clusters are displayed in Table 4.15.

From Table 4.15, it can be noted that the agritech units were divided into three clusters based on the mean value of their performance scores. The first cluster is known as a "very high-performance

TABLE 4.15
Final Cluster Centers

Value Chain Performance	Cluster		
	1	2	3
Sustainability Performance-related factor	5.00(I)	2.67(II)	2.33(III)
Quality Performance-related factor	5.00(I)	3.13(II)	3.00(III)
Production Performance-related factor	5.00(I)	3.57(II)	2.00(III)
Operations Performance-related factor	5.00(I)	3.60(II)	2.00(III)
Sales & distribution Performance-related factor	5.00(I)	3.60(II)	2.00(III)
Retail Performance-related factor	5.00(I)	4.00(II)	2.00(III)
Average	5.00	3.43	2.22

TABLE 4.16
ANOVA Result

	Mean Square	df	Mean Square	df	F	Sig
Sustainability Performance-related factor	19.97	2	.321	87	62.19	.000
Quality Performance-related factor	11.23	2	.290	87	38.72	.000
Production Performance-related factor	12.46	2	.406	87	30.68	.000
Operations Performance-related factor	15.27	2	.400	87	38.16	.000
Sales & distribution Performance-related factor	15.29	2	.363	87	42.16	.000
Retail Performance-related factor	18.80	2	.326	87	57.62	.000

group" as compared to the other two clusters, the mean value of the value chain performance in this category is high. The second group was classified as a "high-performance group" because in the middle of the five-point scale, they have a moderate mean score. For the third group, the mean performance factor value is moderate in the five-point scale, and this category can be referred to as the "moderate performance category in the supply chain." The companies under the first cluster are the ones performing better than those in the other two clusters. The reason could be attributed to any factor that favours their growth such as their experience in the field and their efforts to satisfy the customer. Table 4.16 displays ANOVA values for the three clusters pertaining to the performance factors in the value chain.

From Table 4.16, it is seen that there exists no significant difference in mean values of the performance factors with respect to the clusters. Hence it can be concluded that all of the performance factors in the value chain play a significant role in bifurcating the agritech units into three classes. All the performance factors considered are very crucial for the functioning of any organization and they are often found to be interdependent. A brief overview is given below regarding the three value chain performance groups such as very high-value chain performance units, high-value chain performance units, and moderate value chain performance units.

4.4.5.3.1 Very High-Value Chain Performance

This cluster's degree of value chain performance is very high with respect to all the variables relevant to the value chain performance factors. Also, the mean value of this group stands at five on the five-point scale making it be grouped under the "very high" category. Within this class, lie all the agritech companies that rated the highest in terms of all the value chain performances. The group of very high-performance units comprises 50% of the 90 agritech units surveyed (45 units).

4.4.5.3.2 High-Value Chain Performance Units

The second cluster of agritech units is known as "High-Value Chain Production Units" with respect to supply chain performance factors. This group has a high level of performance in the value chain on all performance variables considered, as this group's mean value in terms of performance in the value chain glides near three in the high five-point scale. The high-performance group accounts for 27% of the 90 agritech units surveyed (24 units).

4.4.5.3.3 Moderate Value Chain Performance Units

The mean score values are 2.22 in respect of this segment's supply chain performance factors. This group can be classified as a "moderate value chain performance group" because the mean glides around two on the five-point scale. This category has a moderate range of performance-related to sustainability, efficiency, production, service, sales and distribution, and retail. About 21% of the agritech units surveyed fall under this category. The number of agritech firms under each cluster are presented in Table 4.17.

TABLE 4.17
Number of Cases in Each Cluster

Cluster	No. of Agritech	Percentage
Very High-value Chain Performance	45	50%
High-value Chain Performance	24	27%
Moderate Value Chain Performance	21	23%
Total	90	100%

Table 4.17 shows that the very high-value chain performance groups constitute a big chunk ie., about half of the companies studied for the research. This could be due to the opportunities for business development and the congenial environment in the study area. The high and moderate value chain performance groups together constitute the remaining half.

4.4.5.4   Characteristics of Value Chain Performance

In the previous segment, the value chain performance has been divided into three groups namely moderate value chain performance units, high-value chain performance units and very high-value chain performance units based on the six supply chain performance factors. It is evident that the overall performance of the companies displaying good performance in the value chain would be higher. In this section, the characteristics of value chain performance segments are analyzed using statistical tools such as the chi-square test added with correspondence analysis followed by analysis of variance (ANOVA) and canonical correlation. Chi-square values to find the association between the profile characteristics and the performance factors and their significance values are portrayed in Table 4.18.

In order to study the association between the profile of the agritech companies and the value chain performance factors, the data related to the same are subjected to a chi-square test. It can be seen from Table 4.18, that the companies categorized based on the profile characteristics such as value chain position and agritech category show significant association with the value chain performance factors.

4.4.5.5   Relationship between Value Chain Performance and Profile of Manufacturing Industries

The Chi-square test shows that there is a significant association between the value chain performance factors with the profile characteristics of the agritech enterprises namely, value chain

TABLE 4.18
Chi-Square Test for Profile of Manufacturing Industries

Sl. No.	Variable	Chi-Square Value	P-Value	Significance Level
1.	Value Chain Position	**25.339**	**0.005**	**Significant**
2.	Number of Employees	8.485	0.205	Not Significant
3.	Nature of Industry	4.305	0.635	Not Significant
4.	Type of Business Organization	4.816	0.568	Not Significant
5.	Agritech Category	**10.46**	**0.033**	**Significant**
6.	Market Coverage	5.872	0.209	Not Significant
7.	Number of years in this business	4.040	0.671	Not Significant

position and agritech category. The next section shall illustrate the significant relationship between agritech units categorized based on their profile and their value chain performance factors.

4.4.5.5.1 Value Chain Position

The Chi-square value is 25.339 and the significant value is 0.005 (Table 4.18). This indicates that there exists a significant association between the agritech firms categorized based on their value chain position and their value chain performance. The companies surveyed have different value chain positions and their performance has varied based on this. This could be due to the inherent nature of each value chain in the agritech business. Some value chains require a lot of input from the company to perform well and some do not. The value chains requiring more focus should be given the attention to perform well, failing which, their performances decrease. The association between the value chain position of the agritech enterprises and value chain performance factors are shown in Figure 4.2.

It can be inferred from the above figure that those agritech units grouped based on their value chain positions like supporting service stage and production/farmer stage are associated with the "very high-value chain performance units," while those agritech units belonging to processing/ value addition are associated with "high-value chain performance units" and those units belonging to agri input/ supplier are associated with "moderate value chain performance units." The supporting services stage involves less risk than other value chain positions and is less prone to the influence of other external factors. The production stage may flourish based on the quality of resources spent on it. The higher the quality of the product, the greater will be their overall performance. Good quality products create their own space in the market and with technologies like AI, the agritech companies in the production stage seldom make defective products. Hence, the value chain performances of companies in this value chain performance are very high. The significance in the mean value differences among the value chain positions of the agritech enterprises and value chain performance factor are displayed in Table 4.19.

It can be observed from the above table that significant differences exist in production-related factors and operation related factors of the agritech units categorized based on the value chain positions. The other performance factors are alike for the agritech companies irrespective of their value chain positions. This could be due to the fact that all the agritech companies are unanimously working towards sustainability and other factors and they differentiate their product in the minds of the customer using their production and operation strategies. The mean values of the different value chain positions for the sustainability-related factors are displayed in Table 4.20.

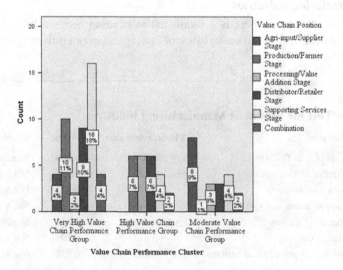

FIGURE 4.2   Value Chain Position and Performance – Cluster Bar Chart.

TABLE 4.19
ANOVA for Value Chain Position and Value Chain Performance

Value Chain Performance	F-Value	P-Value
Sustainability Performance-related factor	1.082	0.377
Quality Performance-related factor	1.855	0.111
Production Performance-related factor	3.761	0.004
Operations Performance-related factor	4.283	0.002
Sales & distribution Performance-related factor	1.879	0.107
Retail Performance-related factor	0.944	0.457

TABLE 4.20
Mean Values for Sustainability Performance Factors

Sustainability Performance-Related Factor	In Number	Mean
Processing/Value Addition Stage	11	3.4141
Distributor/Retailer Stage	18	3.4630
Combination	08	3.4722
Production/Farmer Stage	17	3.7778
Supporting Services Stage	24	3.8287
Agri-input/Supplier Stage	12	4.0278

It was observed from Table 4.20 that the mean values fall between 3.4141 as lowest and 4.0278 as the highest which indicates that the performance falls between moderate to high. Also, according to Table 4.18, there is no significant difference in mean values of the value chain positions. All agritech companies exhibit a moderate to the high level of sustainability performance. This may be due to the gradual shift that is happening in Indian agriculture which now focuses on ensuring sustainability while working on the growth of the sector. Policies such as the National Mission for Sustainable Agriculture (NMSA) focus on promoting climate change adaptation measures and thereby increasing agricultural productivity. The mean values of the different value chain positions for the quality related factors are displayed in Table 4.21.

Table 4.21 shows that the mean values lie between 3.295 and 4.020 which indicates that the quality performance of the value chain positions is moderate to high. Also, from Table 4.18, it is observed that no significant difference exists in the mean values of different value chain positions

TABLE 4.21
Mean Values for Quality Performance Factors

Quality Performance-Related Factor	In Number	Mean	Mean
Processing/Value Addition Stage	11	3.2955	
Combination	08	3.4375	
Distributor/Retailer Stage	18	3.4583	
Supporting Services Stage	24	3.6563	
Production/Farmer Stage	17	3.8529	
Agri-input/Supplier Stage	12		4.0208

with respect to the quality performance factors. The companies irrespective of their value chain positions strive hard to maintain the quality of the product. The focus of the customer has shifted from getting more quantity to more quality products. Even the start-ups are nowadays supported by government-sponsored incubators where they are mentored on developing quality products through standard operating procedures. The mean values of the different value chain positions for the production-related factors are displayed in Table 4.22.

It is observed from Table 4.22, that the highest mean value among the value chain positions is 3.8036, which is of the supporting service stage and the lowest value is 2.8831, which is of the processing/value addition stage. Table 4.18 shows that there exists a significant difference in these mean values. Hence, different companies exhibit different levels of production performance with respect to their value chain position. The production performance of the supporting service stage is almost high whereas that of the processing stage is near moderate. The agritech companies in different value chain positions work with different technologies including AI, robotics, IoT, and farm management software. The production performance is also largely dependent on the technology used. The technologies developed so far may be more conducive for the supporting service stage than to the processing stage. The mean values of the different value chain positions for the operation related factors are represented in Table 4.23.

It can be observed from Table 4.23 that the mean value of the supporting service stage is 3.7917, which is the highest among all. Since Table 4.18 illustrates that the mean values have a significant difference with respect to the operations performance, it can be stated that the value chain position with the highest mean value is higher. The operational performance of the processing stage ranges between low to moderate. All the companies are not alike in the way they operate. The operation performance of the agritech companies based on their value chain positions differ from each other since there could be differences in their way of functioning. The mean

TABLE 4.22
Mean Values for Production Performance Factors

Production Performance-Related Factor	In Number	Mean	Mean
Processing/Value Addition Stage	11	2.8831	
Agri-input/Supplier Stage	12	2.9762	
Distributor/Retailer Stage	18	3.1984	
Combination	08	3.5179	
Production/Farmer Stage	17		3.6639
Supporting Services Stage	24		3.8036

TABLE 4.23
Mean Values for Operation Performance Factors

Operations Performance-Related Factor	In Number	Mean	Mean
Processing/Value Addition Stage	11	2.5273	
Combination	08		3.3750
Distributor/Retailer Stage	18		3.5000
Agri-input/Supplier Stage	12		3.6333
Production/Farmer Stage	17		3.6706
Supporting Services Stage	24		3.7917

TABLE 4.24

Mean Values for Sales and Distribution Performance Factors

Sales & Distribution Performance-Related Factor	In Number	Mean	Mean
Processing/Value Addition Stage	11	2.8000	
Agri-input/Supplier Stage	12	3.3333	
Distributor/Retailer Stage	18	3.3667	
Supporting Services Stage	24	3.4750	
Combination	08		3.5750
Production/Farmer Stage	17		3.7294

TABLE 4.25

Mean Values for Retail Performance Factors

Retail Performance-Related Factor	In Number	Mean
Processing/Value Addition Stage	11	3.4091
Distributor/Retailer Stage	18	3.5000
Production/Farmer Stage	17	3.6471
Combination	08	3.8125
Agri-input/Supplier Stage	12	3.8750
Supporting Services Stage	24	3.9375

values of the different value chain positions for the sales and distribution related factors are represented in Table 4.24.

The table shows that the mean value of the value chain positions with respect to their sales and distribution performance factors lies between 2.8000 and 3.7294. However, Table 4.18 illustrates that the difference in these mean values is insignificant. The sales and distribution performance irrespective of the value chain position of the agritech company ranges from moderate to high. To ensure proper distribution through all the channels, the agritech companies are innovating to create many markets linkage models. The agritech companies are working towards developing a successful model like price forecast system for important commodities and information dissemination platforms, irrespective of their value chain positions. The mean values of the different value chain positions for the retail-related factors are represented in Table 4.25.

It is observed from Table 4.25, that the mean values of the value chain positions lie between 3.4091 for the processing stage and 3.9375 for the supporting service stage. This does not imply that the supporting service stage companies outperform the processing stage in retail performance as Table 4.18 indicates that there is no significant difference in mean values. All companies regardless of their value chain position are trying to outperform their competitors in the retail where the product reaches the final consumer. With government initiatives like eNAM, it is high time for the private agritech players to work towards creating innovative retail strategies. The programs like Agriculture Grand Challenge also focus on encouraging the agri-based units to develop solutions for retailing of the farm produce and offer them handholding support.

4.4.5.5.2 Agritech Category

From Table 4.18, it was noted that the Chi-square value is 10.46 and the significant value is 0.033 which clearly indicates the presence of significant association between agritech enterprises

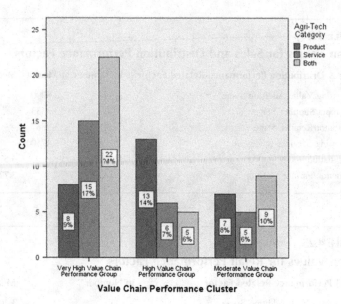

FIGURE 4.3   Agritech Category and Performance – Clustered Bar Chart.

categorized based on their agritech category and their value chain performance factors. The performance levels of the companies are associated with their offering such product/service/ combination. The cost, selling price, and usage rate differ among categories and hence the performance might be influenced. The association between agritech category of the units and the value chain performance factors are shown in Figure 4.3.

It can be observed from Figure 4.3, that those agritech units offering a combination of product and service to the market are associated with the "very high-value chain performance" while those units offering only product are associated with the "high-value chain performance" and those manufacturing units offering only service are associated with the "low supply chain performance units." The very high performance of the agritech companies offering both product and service could be due to their extended scope of operation. On the other hand, the companies offering only services fall into moderate performance groups. Agriculture is a highly volatile sector and the farmers tend to cut down on the services during difficult times. This could be a reason for their moderate performance. The relationship between agritech category and value chain performance factor has been illustrated in Table 4.26.

It can be observed from Table 4.26 that only production, operations, and sales and distribution performance factors show significant differences with respect to their category. But other factors such as sustainability, quality and retail performance do not show a significant difference in mean

TABLE 4.26

ANOVA for Agritech Category and Value Chain Performance

Value Chain Performance	F-Value	P-Value
Sustainability Performance-related factor	2.387	0.098
Quality Performance-related factor	1.344	0.266
Production Performance-related factor	3.606	0.031
Operations Performance-related factor	9.164	0.000
Sales & distribution Performance-related factor	4.959	0.009
Retail Performance-related factor	2.611	0.079

TABLE 4.27
Mean Values for Sustainability Performance Factors

Sustainability Performance-Related Factor	In Number	Mean	Mean
Product	28	3.4246	
Service	26	3.6880	3.6880
Both	36		3.8981

values. The agritech companies serve a homogeneous set of audience and hence they are in a need to outperform their competitors by varying mainly their strategies in production, operation and sales and distribution. The mean values in respect to the sustainability performance factor of agritech firms classified based on agritech category have been displayed in Table 4.27.

Table 4.27 shows that the mean value of the companies grouped based on their agritech category falls in the range between 3.4246 and 3.8981 which implies that the performance is moderate to high. Also, Table 4.26 shows that there exists no significant difference among these mean values. It can therefore be stated that irrespective of the agritech category all companies are performing at a moderate to high pace towards sustainability. The companies are developing sustainable products that do not harm the environment. They try to emit fewer pollutants thereby safeguarding the population to which they serve. The mean values in respect to the quality performance factor of agritech firms classified based on agritech category have been displayed in Table 4.28.

It is observed from Table 4.28 that the mean values of the agritech companies classified based on their category fall between 3.4732 and 3.7743 ie., moderate to high. Also, there is no significant difference in mean values as observed from Table 4.26. The agritech companies regardless of their categories work toward offering quality products/services to the consumer. This could be due to the realization of the current purchase trend of the consumers by these companies. Quality products contribute to the long-term growth and profitability of the company. Also, stringent quality standards brought forth by the government has made the agritech companies ensure the quality of their offering. The mean values in respect of the production performance factor of agritech firms classified based on agritech category have been displayed in Table 4.29.

TABLE 4.28
Mean Values for Quality Performance Factors

Quality Performance-Related Factor	In Number	Mean
Product	28	3.4732
Service	26	3.6298
Both	36	3.7743

TABLE 4.29
Mean Values for Quality Performance Factors

Production Performance-Related Factor	In Number	Mean	Mean
Product	28	3.0765	
Both	36		3.5159
Service	26		3.6154

It is observed from Table 4.29, that the service category shows the highest mean value of 3.6154, which implies that it performs better than the other categories of companies with respect to production. Also, from Table 4.26 it is seen that there exists a significant difference in mean values among these categories. The process of providing a service to the consumer is meant to be production for a service company. The agritech service companies require less investment than the other two categories. To help the agritech companies that require more funds, schemes like ACABC were launched by the government to assist the agripreneurs both technically and financially. The companies in the product category could make use of such schemes to improve their performance in production. The mean values in respect to the operation performance factor of agritech firms classified based on agritech category have been presented in Table 4.30.

Table 4.30 shows that the mean value of the companies serving the market with both product and service stands higher at 3.8556. Also, the differences in mean values of these categories are significant as observed from Table 4.26. It implies that the operation performance of the agritech companies offering both product and service is higher compared to other categories. The companies offering only product categories are performing moderately in terms of operations. Consumers find it easier to value products than services. Errors occur more frequently in product manufacturing than in service rendering. Hence, the operation performance of the agritech companies offering products might have been low when compared to companies offering a combination of benefits. The mean values in respect to the sales and distribution performance factor of agritech firms classified based on agritech category have been presented in Table 4.31.

It is found through observation from Table 4.31 that the mean value of the companies offering a combination of product and service is higher (3.6556). Table 4.26 illustrates that the difference in mean values among these different categories are significant. It can therefore be stated that the companies offering both product and service outperform companies offering only product or service in terms of sales and distribution. This difference could be due to the different marketing strategies that are adapted based on the nature of the product or service. Tangible goods require a lot of efforts for selling and distributing product. The government has implemented transport schemes for agricultural exports and reimburses freight costs through DBT as an act to boost agribusiness firms and to ease the burden of sales and distribution overseas.

TABLE 4.30

Mean Values for Operations Performance Factors

Operations Performance-Related Factor	In Number	Mean	Mean
Product	28	3.0071	
Service	26		3.5308
Both	36		3.8556

TABLE 4.31

Mean Values for Sales and Distribution Performance Factors

Sales & Distribution Performance-Related Factor	In Number	Mean	Mean
Product	28	3.0286	
Service	26		3.4769
Both	36		3.6556

TABLE 4.32

Mean Values for Retail Performance Factors

Retail Performance-Related Factor	In Number	Mean	Mean
Product	28	3.4107	
Service	26	3.8077	3.8077
Both	36		3.8750

The mean values in respect to the retail performance factor of agritech firms classified based on agritech category have been presented in Table 4.32.

Table 4.32 shows that the mean values of the categories fall between 3.4107 and 3.8750, which is between moderate to high. Also, from Table 4.26, it was found that there is no significant difference in the mean values. It implies that the retail performance of the companies irrespective of their agritech category is moderate to high. Retail is the point where the final product reaches the consumer. Hence, all companies strive hard to compete for the attention of the consumer in retail. This could be a reason for the moderate to the high performance of the companies belonging to different categories.

4.4.5.5.3 Canonical Correlation between Value Chain Performance and Agritech Industries Profile Variables

Canonical correlation is a tool used to predict the shared relationship between two or more variables. This study establishes the relationship between two variables and also discusses the overall relationship between two or more variables. These sections deal with the canonical correlation between two sets of variables. The first set of variables is value chain performance factors namely, sustainability performance factors, quality performance factors, production performance factors, operations performance factors, sales and distribution performance factors and retail performance factors while the second set of variables consists of profile characteristics of agritech enterprises namely, value chain positions and agritech category. Canonical correlations in respect of supply chain performance with regard to different profile characteristics of manufacturing units are displayed in Table 4.33.

Table 4.33 demonstrates the results of the canonical correlation conducted on the basis of the two data sets. The coefficients of canonical correlation for these two factors are 0.4521 and 0.3766, respectively. Certain results are shown in the table above including the df1 value of 12, the df2 value of 164, the f value of 2.8727, Wilk's λ value of 0.682792 and the p-value of 0.001, which is less than 0.05, indicate that there exists a significant relationship between the two data sets. In order to determine the overall relationship between these two data sets, Wilk's (λ) value should be subtracted from one. The r^2 value from the three sets of canonical functions is 0.3172. It implies that 31% of the variation in the value chain performance is explained by the two profile characteristics. Therefore, a strong positive correlation exists between the two data sets, which are the six value chain performance factors and the two profile characters such as the value chain position and category of agritech enterprises. The value chain position of the agritech company may affect the overall value chain performance since it determines the companies target audience and ways to serve them. The agritech category determines the nature and form of the benefit that would reach the customer. Possessing a good service component in the value chain may add value to the product. The Integrated Scheme for Agricultural Marketing aims to promote companies falling into different agricultural value chain positions to market their produce by extending financial support.

TABLE 4.33

Canonical Correlation for Supply Chain Performance

Linear Combinations for Canonical Correlations   Number of obs = 90

	Coef.	Std. Err	t	P> |t|	[95% Conf. Interval]	
u1						
Sustainabi~y	.3236593	.3764387	0.86	0.392	−.4243164	1.071635
Quality	−.5894719	.4442609	−1.33	0.188	−1.472209	.2932651
Production	−.3750011	.3812552	−0.98	0.328	−1.132547	.382545
Operations	1.096746	.3676719	2.98	0.004	.3661898	1.827302
Salesdistr~n	.6158473	.4046846	1.52	0.132	−.1882523	1.419947
Retail	−.116429	.3708737	−0.31	0.754	−.8533471	.6204891
V1						
ValueChain~n	−.0531948	.1350815	−0.39	0.695	−.3215989	.2152093
AgritechCa~y	1.198244	.2543129	4.71	0.000	.6929301	1.703559
U2						
Sustainabi~y	.3387309	.469404	0.72	0.472	−.593965	1.271427
Quality	1.100233	.5539757	1.99	0.050	−.0005051	2.200971
Production	−1.006411	.4754101	−2.12	0.037	−1.951041	−.0617811
Operations	−.193179	.4584722	−0.42	0.675	−1.104153	.7177955
Salesdistr~n	.3185421	.5046255	0.63	0.529	−.6841381	1.321222
Retail	−.4022431	.4624647	−0.87	0.387	−1.321151	.5166643
V2						
ValueChain~n	−.6364014	.1684413	−3.78	0.000	−.9710907	−.3017122
AgritechCa~y	.0987574	.317118	0.31	0.756	−.5313493	.7288642

(Standard errors estimated conditionally)

Canonical correlations: **0.4521 0.3766**

Tests of significance of all canonical correlations

	Statistic	df1	df2	F	Prob>F
Wilks' lambda	.682792	12	164	2.8727	0.0013 e
Pillai's trace	.346191	12	166	2.8957	0.0012 a
Lawley-Hotelling trace	.422129	12	162	2.8494	0.0014 a
Roy's largest root	.256906	6	83	3.5539	0.0035 u

e = exact, a = approximate, u = upper bound on F

4.5   CONCLUSIONS AND IMPLICATIONS

Findings derived from analyses based on AI value chain performance dimensions are presented here; more re-use has been ranked as the topmost AI value chain performance. The ability to reuse any good has a lot of advantages which includes cost benefits, environment-friendly identity and reduced wastage. This could be one of the reasons why more reuse is seen as a major contributor to the value chain performance by the agritech companies. It is observed that more profit is ranked as 31st, which reveals that the agritech companies are more concerned about environmental and consumer-related factors than maximizing profit.

ANOVA results to find a significant difference between the agritech industry profile and the AI value chain performance shows that companies categorized based on value chain position and agritech category show a significant difference in mean values. The mean values of companies categorized based on number of employees, nature of the industry, type of business organization, market coverage and the number of years in the business stand between 3.1944 and 3.7546 (moderate), 3.4278 and 3.8611 (moderate), 3.1389 and 3.5887 (moderate), 3.5212 and 3.7569 (moderate), and 3.4487 and 3.7587 (moderate), respectively. This shows that the performance variation is not due to the influence of any of the above-mentioned profile characteristics. The mean values of companies categorized based on their value chain position differ significantly. The supporting service stage has the highest mean value among all the stages. The AI value chain performance of the supporting service stage is higher than other stages. The supporting service stage usually is a less risk business dealing with farms. Hence, these companies make better use of AI in their value chain than companies in other value chain positions. The mean value of groups under the agritech category differ significantly which implies that the companies which offer both product and service enjoy better AI value chain performance than companies that offer only product or service. The companies which offer both products and service are less susceptible to the volatile nature of the market. These companies can therefore utilize their resources in strengthening their internal operations rather than spending much on promotional activities. Hence, the AI value chain performance of such companies stands higher comparatively.

Chi-square tests to find the association between value chain position and AI value chain performance reveal that there is no significant association between their mean values. This implies that the AI value chain performance is not associated with their value chain position. The enhanced performance could be due to other factors as well since the mean value association is not significant as per the test results.

Correspondence analysis, to find the level of association between the value chain position variables and AI value chain performance shows that the value chain performance is very high in the supporting service stage and moderate in the agri input/supplier stage. However, since the difference in mean values is not significant according to the results of the chi-square test, it cannot be concluded that the value chain performance of the supporting stage is higher because of its position in the value chain.

Value chain performance variables are factored into five factors namely sustainability performance-related factor, quality performance-related factor, production performance-related factor, operations performance-related factor, sales & distribution performance-related factor, and retail performance-related factor. The agritech units are grouped into three namely very high-value chain performance units, high-value chain performance units and moderate value chain performance units.

Chi-square test to find the association between the profile of the agritech companies and the value chain performance factors reveal that the companies categorized based on value chain position and agritech category show significant association. Other profile characteristics did not show significant association with respect to the value chain performance.

Using cluster bar charts, the level of association between the profile characteristics and the value chain position is found. It was inferred that the supporting stage is associated with "very high-value chain performance units", the processing/value addition stage is associated with "high-value chain performance units," and the agri input/supplier stage is associated with "moderate value chain performance units." ANOVA to test the significance in mean value difference between the value chain position and value chain performance shows that significant differences exist with respect to the production-related factors and operation-related factors of the agritech units categorized based on the value chain positions. Other factors like sustainability performance do not differ significantly. This could be due to the fact that all the agritech companies have realized the need for product differentiation and are adapting various processes to ensure the same. The mean values of the value chain position for the sustainability performance factors, quality performance factors, sales and distribution performance factors, and retail performance factors stand between 3.4141 and 4.0278 (moderate to high), 3.2955 and 4.0208 (moderate to high), 2.8000 and 3.7294 (low to moderate), and 3.4091 and 3.9375 (moderate), respectively. The mean value of the supporting service stage with respect to the production performance-related factors stands at 3.8036 and the processing stage stands at 2.8831. There is a significant difference in mean values as per the ANOVA test results which implies that the performance-related factors of the supporting stage are better than the other stages. Similarly, the mean value of the supporting stage with respect to the operation-related factors is higher than other value chain positions.

Using cluster bar charts, the association between the value chain performance and the agritech category is found. It is found that those agritech units offering a combination of product and service are associated with the "very high-value chain performance", those units offering only product are associated with the "high-value chain performance" and those agritech units offering only service are associated with the "low supply chain performance units." The relationship between agritech category and value chain performance was found using ANOVA. The results suggest that production, operations and sales and distribution performance factors show significant differences with respect to their category. The agritech companies try to outperform their competitors in terms of production, operation and sales and distribution. The mean values of agritech category for the sustainability performance factors, quality performance factors and retail performance factors stand between 3.4246 and 3.8981 (moderate), 3.4732 and 3.7743 (moderate), and 3.4107 and 3.8750, respectively. The mean value of the service category with respect to the production performance factors is 3.6154. And since the mean values differ significantly for the production factors, it could be said that the service category outperforms the other two categories in production. The mean value of companies offering both product and service with respect to the operation performance factors and sales and distribution performance factors are 3.8556 and 3.6556 which is higher than the other two categories in their respective performance factors. This implies that companies offering both product and service excel in their operation procedures and sales and distribution process.

Canonical correlation is used to measure the relationship of sharing between two sets of variables. The first set of variables are value chain performance factors, namely sustainability performance factors, quality performance factors, production performance factors, operation performance factors, sales and distribution performance factors, and retail performance factors and the second set of variables is profile of agritech industry variables namely value chain position and agritech category. The findings show that 31% of the variation in the value chain performance is attributed to the two profile characteristics such as value chain position and agritech category.

The Bangalore agritech firms' value chain and organizational performance were studied using the dimensions of AI benefits, AI problems, AI future, AI value chain performance, and organizational performance. These measurements were evaluated using different variables, and these variables differ depending on the profile of the agritech units. This study attempted to analyze the differences between the agritech companies with different positions in the value chain in the Indian context. Based on their profile characteristics, the agritech companies were categorized into groups in order to truly comprehend their corresponding nature and characteristics. Understanding the characteristics of agritech units in terms of their value chain would be of great benefit to practitioners and policymakers.

Policymakers can construct appropriate and effective agricultural policies to encourage more investment into the growing agritech sector, whereas agritech professionals can devise and change their strategies that meet the ever-changing needs of the Indian agritech sector.

The most relevant part of this study is the creation of a conceptual framework after carrying out all the tests required to determine its reliability and validity on the basis of the data collected from agritech companies. Such a conceptual framework could be used by researchers as a baseline for future studies. This research has found that value chain performance influences the organizational performance of the agritech firms strongly, while the value chain performance of the agritech firms is strongly influenced by AI benefits and the AI future of the agritech units. Hence, agritech firms focusing on the AI benefits and AI future can greatly enhance their organizational performance as these factors impact the performance indirectly by their impact on value chain performance. Therefore, managers should focus on AI benefits and AI future in order to enhance the performance of their companies.

The overall contributions from this research work are as follows:

1. The study used many variables to analyze the characteristics of the agritech value chain performance. These variables have been checked to be accurate and reliable in the analysis of agritech enterprise value chain performance. Future scholars and researchers can very well use these variables to undertake extensive analysis on the different factors involved in the performance of the value chain. Business managers could therefore use the various tools proposed by this research to effectively manage the critical value chain issues.

2. Business managers can use the tools suggested by this research work to enhance their knowledge of various value chain components and the integrally-linked relationships of the different dimensions of AI benefits, AI future, AI problems, AI value chain performance, and their probable effect on the organizational performance.

3. This work provides useful inputs for improving academic insight and justifications on theory and proposal, measurement size, methods of addressing research problems, and operational consequences of the value chain for agritech.

4. This study has made an enormous contribution by appropriately developing a theory that integrates various aspects related to the value chain of agritech and the various problems related to the value chain that can significantly affect an agritech enterprise performance.

REFERENCES

Ai, F.-f., Bin, J., Zhang, Z.-m., Huang, J.-h., Wang, J.-b., Liang, Y.-z., ...Yang, Z.-y. (2014). Application of random forests to select premium quality vegetable oils by their fatty acid composition. *Food Chemistry*, 143, 472–478.

Ashraf, M., Ullah, L., Shuvro, M. A., & Salma, U. (2019). Transition from millenniu m development goals (MDGs) to sustainable development goals(SDGs): blueprint of Bangladesh for implementing the sustai nable development goals(SDGs) 2030. *Medicine Today*, 31(1), 46–59.

Awuor, F., Kimeli, K., Rabah, K., & Rambim, D. (2013). ICT solution architecture for agriculture. *In 2013 IST-Africa Conference & Exhibition* (pp. 1–7). IEEE.

Bawack, R., Fosso Wamba, S., & Carillo, K. (2019). Where information systems research meets artificial intelligence practice: towards the development of an AI capability framework.

Bhar, L. M., Ramasubramanian, V., Arora, A., Marwaha, S., & Parsad, R. (2019). Era of Artificial Intelligence: Prospects for Indian Agriculture.

Choudhary, H. R., & Choudhary, A. (2013). Why Indian farmers and rural youth are moving from farming. *Popular Kheti*, 1, 60–66.

David, A., Kumar, C. G., & Paul, P. V. (2022). Blockchain Technology in the Food Supply Chain: Empirical Analysis. *International Journal of Information Systems and Supply Chain Management (IJISSCM)*, 15(3), 1–12.

Duan, Y., Edwards, J. S., & Dwivedi, Y. K. (2019). Artificial intelligence for decision making in the era of Big Data-evolution, challenges and research agenda. *International Journal of Information Management*, 48, 63–71.

Fountas, S., Carli, G., Sorensen, C. G., Tsiropoulos, Z., Cavalaris, C., Vatsanidou, A., ... & Tisserye, B. (2015). Farm management information systems: Current situation and future perspectives. *Computers and Electronics in Agriculture*, 115, 40–50.

Ganeshkumar, C., Jena, S. K., Sivakumar, A., & Nambirajan, T. (2021). Artificial intelligence in agricultural value chain: review and future directions. *Journal of Agribusiness in Developing and Emerging Economies*, 12(5), 143–156.

Halisçelik, E., & Soytas, M. A. (2019). Sustainable development from millennium 2015 to Goals 2030. *Sustainable Development*, 27(4), 545–572.

KPMG and Skills Impact (2019). Agricultural workforce: Digital capability framework. Narrabri, New South Wales: Cotton Research and Development Corporation, 1–92.

Kern, A. G. (2018). *Blockchain Technology: A Technology Acceptance Model (TAM) Analysis*. Universidade Catolica Portuguesa. (Doctoral dissertation).

Khamis, A. (2018). Artificial intelligence and Machine Learning: Promises and Limitations. from Paperpicks https://www.paperpicks.com/artificial-intelligence-machine-learning-promises-and-limitations/

Kobayashi-Solomon, E. (2018). AgTech: A Great Investment for the Future. *Forbes*.

Lakshmi, V., & Corbett, J. (2020). *How Artificial Intelligence Improves Agricultural Productivity and Sustainability: A Global Thematic Analysis*. Paper presented at the Proceedings of the 53rd Hawaii International Conference on System Sciences.

Li, X., & Clerc, M. (2019). Swarm intelligence. In *Handbook of Metaheuristics* (pp. 353–384). Springer, Cham.

Liao, C., Qiu, J., Chen, B., Chen, D., Fu, B., Georgescu, M., ... & Wu, J. (2020). Advancing landscape sustainability science: theoretical foundation and synergies with innovations in methodology, design, and application. *Landscape Ecology*, 35(1), 1–9.

Michael White (2018). Emerging technologies in agriculture: Consumer perceptions around emerging agtech. AgriFutures National Rural – AgriFutures Australia, Publication No. 18/048, (pp. 1–97). Australia.

Mungarwal, A.K., & Mehta, S.K. (2019). *Why Farmers Today Need to Take Up Precision Farming*. India: downtoearth.org.

Pathak, P., Pal, P. R., Shrivastava, M., & Ora, P. (2019). Fifth revolution: Applied AI & human intelligence with cyber physical systems. *International Journal of Engineering and Advanced Technology*, 8(3), 23– 27.

Pham, X., & Stack, M. (2018). How data analytics is transforming agriculture. *Business Horizons*, 61(1), 125–133.

Pomerol, J. C. (1997). Artificial intelligence and human decision making. *European Journal of Operational Research*, 99(1), 3–25.

Radhika, K.T.P. (2018, December 30). Agritech Sprouts Start-Ups. *The Hub. Magazine-Business Today*, 1–11.

Sachs, J. D. (2012). From millennium development goals to sustainable development goals. *The lancet*, 379(9832), 2206–2211.

The Biotechnology and Biological Sciences Research Council (BBSRC) (2017). Research in Agriculture and Food Security Strategic Framework. UK. 1–24.

Van Henten, E. J., Goense, D., & Lokhorst, C. (2009). *Precision agriculture'09*: Wageningen Academic Publishers.

Venugoban, K., & Ramanan, A. (2014). Image classification of paddy field insect pests using gradient-based features. *International Journal of Machine Learning and Computing*, 4(1), 1.

Vitón, R. C., Ana; Lopes Teixeira, Tomas. (2019). Agtech: Agtech Innovation Map in Latin America and the Caribbean. from Inter-American Development Bank https://publications.iadb.org/en/agtech-agtech-innovation-map-latin-america-and-caribbean

Von Grebmer, J., Bernstein, R., Mukerji, F., Patterson, M., Wiemers, R., N¡ Ch,illeachair, C., Foley, S., Gitter, K., Ekstrom, H., & Fritschel, K. (2019). Global Hunger Index by Severity, Map in 2019 Global Hunger Index: The Challenge of Hunger and Climate Change. Retrieved from https://policycommons. net/artifacts/2079864/global-hunger-index-by-severity-map-in-2019-global-hunger-index/2835162/ on 25 Nov 2022. CID: 20.500.12592/khqwmg.

Waheed, T., Bonnell, R., Prasher, S. O., & Paulet, E. (2006). Measuring performance in precision agriculture: CART—A decision tree approach. *Agricultural Water Management*, 84(1-2), 173–185.

Wang, W., & Siau, K. (2019). Artificial intelligence, machine learning, automation, robotics, future of workand future of humanity: A review and research agenda. Journal of Database Management (JDM), 30(1), 61–79.

Wilson, H. J., Daugherty, P.R., & Davenport, C. (2019). The future of AI will be about less data, not more. *Harvard Business Review*, 14(1).

Yahya, N. (2018). Agricultural 4.0: Its implementation toward future sustainability. In *Green Urea* (pp. 125–145). Springe, Singapore.

Yin, J., Wang, Y., & Gilbertson, L. M. (2018). Opportunities to advance sustainable design of nano-enabled agriculture identified through a literature review. *Environmental Science: Nano*, 5(1), 11–26.

5 Applications of Artificial Intelligence of Things in Green Supply Chain Management: Challenges and Future Directions

Arshi Naim and Hamed Alqahtani
Department of Information Systems, King Khalid University, Abha, Saudi Arabia

Sadaf Fatima
Department of Business Administration, Aligarh Muslim University, Aligarh, India

Mohammad Faiz Khan
Security Forces Hospital, Ministry of Interiors, Abha, Saudi Arabia

CONTENTS

DOI: 10.1201/9781003264521-5

5.1   INTRODUCTION

The concept of "green smart business" [1] has expanded in recent years as a new form of sustainable development and represents a model that incorporates all alternative approaches to improving the quality and performance of a business in order to better interact between production and service space and customers [2]. The modern business environment depends on data application and this new way of doing business has created issues, challenges but created new opportunities too. New information sources provide opportunities for new applications [3] to improve the quality of business activities and their relevance to modern life [4]. The IoT, as a social human technology, leads to dramatic environmental and urban technological changes in complexity and diversity [4]. IoT enables the integration of digital and physical structures and provides a completely new class of applications and services that should be used with respect to the stability of the environment [5]. This reveals the importance of concepts such as the green IoT [4,5]. On the green IoT [4,6], sensors, devices, applications, and services are portrayed in terms of energy efficiency [6]. In the domain of green and sustainable smart businesses, increasing the volume of data generation is beyond imagination, and the vast amount of information available in different areas is of great value [7].

5.1.1   CONCEPT OF SUPPLY CHAIN

The supply chain is an interconnection of organizations, activities, resources, people, and information for transforming natural resources and raw materials into a finished product for delivery to the end customer [8]. Among major drivers of the transformation of traditional supply chain are hyper-segmentation, localization of source and produce of products, manufacturing, rising customer expectations and end-to-end visibility to companies, suppliers, and customers [9]. These factors triggered large technology organizations to initiate a collaboration impacting the whole process of the supply chain with increased automation in hybrid architecture [10].

IoT devices are used for daily tasks and data production therefore has increased at a global level. In this scenario, AI has emerged as a branch to aid IoT in its working and providing effective connectivity and automation in many BPM [10]. Business industries are using AIoT at all the levels and experience effective results and increase in profit. HST is also an integral part of business sector where data production and its application is very important for success. Health professionals have identified that medical sectors cannot work without computing technologies and AIoT is the most important amongst them.

5.1.2   CONCEPT OF ARTIFICIAL INTELLIGENCE

An AI system is a computer system that makes a decision or performs a task that a human being is capable of performing [11]. At present, artificial intelligence is a form of advanced analysis that relies on machine learning, optimization, and in-depth learning. Connections enhance the smart elements in products and devices by externalizing their capabilities. This makes monitoring, controlling, and optimization conditions possible [12]. Connected objects themselves do not promote learning but pave the way. Many IoT applications rely on sending data to a cloud system or data center, as well as analyzing and modeling data and applying these perspectives. They ultimately give the firms results and possibly return modified logic to the same devices.

Interconnected devices in order to enhance learning and collective intelligence, and to take advantage of objects' artificial intelligence capabilities, they must understand the value of the information provided to them and used in informal and automated networks [13]. Artificial intelligence of things (AIoT) provides a unique opportunity to enhance learning and personalization at the same time [14]. These AI systems can work well with other AI systems. This is a unique advantage of AIoT; therefore, HST utilized this in its effective flow of medical supplies (MS), production of medicines, their storage, and other logistics [15].

5.1.3 Concept of Internet of Things (IoT)

A thing in the Internet of Things can be a person with a life that has built-in sensors to alert the driver when tire pressure is low or any other natural or man-made object that can be assigned an Internet protocol (IP) address [16] and is able to transfer data over a network.

Many organizations are using IoT in their operations, understanding customers' requirements, and improving customer services, aiding in complex decision-making process that eventually brings good profit to the firms [17].

5.1.3.1 How Does IoT Work?

The working of IoT is a web-enabled system [18] where the devices are connected with the processors, sensors, and communication hardware. These devices then collect, send, and act on data they acquire from their surroundings [19]. IoT devices share the sensor data they collect by connecting to an IoT gateway or other edge device where data is either sent to the cloud to be analyzed or analyzed locally [20]. These smart devices [21] are able to send and receive data to other devices and get the information for the particular cause and results.

IoT offers many benefits to the firms and many benefits are industry-specific, and some are applicable across multiple industries [22]. Some of the common benefits of IoT enable businesses are shown in the Figure 5.1.

IoT promotes firms to reorganize the methods the firm's approach in their businesses and gives them the tools to improve their business strategies [23].

Most commonly, IoT is used in sectors such as manufacturing, transportation, and utility organizations, making use of sensors and other IoT devices [24]. The application is not restricted to these areas only but also includes agriculture, home firms, HST, etc. Leading organizations have clearly explained the advantages of IoT in their working in different departments such as finance, marketing, sales, and advertising [24,25]. Another important concept is that AI has made possible for machines to work and think like human and can contribute in decision making processes. Figure 5.2 shows the various reasons for the importance of AI [26].

AIoT can help to save various areas for all the firms and especially for the HST; some of advantages are given in Table 5.1.

5.1.3.2 Applications of AIoT

AIoT helps in making decisions on buying new machinery and also if there is a requirement of buying at this time or to work with the existing equipment [28]. This helps the firms to save costs and to contribute in saving the environment by reducing the consumption of hardware and software devices [29]. AIoT helps the firms to prevent high-cost investment and encourages repair or to replace only some parts of the big machines instead of changing the entire machine [30]. AIoT aids

FIGURE 5.1 Benefits of IoT in Businesses [22].

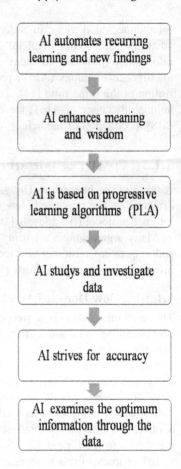

FIGURE 5.2   Reasons for the Application of AI [26].

TABLE 5.1
Application of AIoT in HST [27]

Precautionary maintenance
Management and optimum application of for the best use of time, machine, and money
Low costs of energy
Management of working infrastructure

---

the firm to identify if machines are working properly in a short time or in the moment and when they are required to change completely. Figure 5.3 shows general the benefits of IoT in SCM [30].

There are many challenges also for the applications of IoT devices. Figure 5.4 explains the most important challenges of its applications [31].

5.1.4   Supply Chain Management

Supply chain management (SCM) [32] is the handling of the flow of goods and services from the raw manufacturing of the product through to the consumption by the consumer. This process requires an organization to have a network of suppliers to move the product through each stage [32].

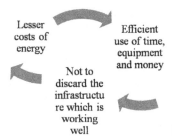

Lesser costs of energy

Efficient use of time, equipment and money

Not to discard the infrastructu re which is working well

FIGURE 5.3   IoT General Benefits in SCM [31].

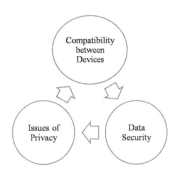

Compatibility between Devices

Issues of Privacy

Data Security

FIGURE 5.4   Challenges of IoT in Its Applications [31].

Effective supply chain management [33] improves the financial position of an organization by delivering value linked to the organization's corporate strategy. Supply chain management plays a significant role in customer satisfaction through the delivery of products and services [34]. Good supply chain management is critical at reducing operating costs from procurement activities, through operations and logistics functions, and throughout the whole supply chain [35]. The scale of profitability for large organizations is relative to the management of an organization's supply chain [36]. Figure 5.5 shows the six components of SCM in the business process management (BPM).

The future of SCM [38] will be different for each organization; however, having a constant awareness to add value and reduce waste will serve an organization well when reviewing their

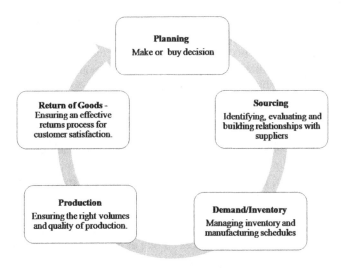

Planning
Make or buy decision

Sourcing
Identifying, evaluating and building relationships with suppliers

Return of Goods -
Ensuring an effective returns process for customer satisfaction.

Production
Ensuring the right volumes and quality of production.

Demand/Inventory
Managing inventory and manufacturing schedules

FIGURE 5.5   Six Components of SCM in BPM [37].

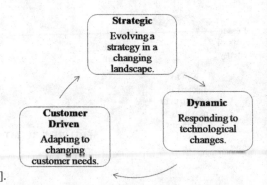

FIGURE 5.6　Future of SCM for a Specific Firm [38].

TABLE 5.2

Definitions of GSCM in BPM [41]

GSCM Definition in Different Areas of BPM

It includes a set of environmental practices that encourage improvements to the environmental practices of two or more organizations within the same supply chain.	It is the process of incorporating environmental concerns into supply chain management including product design, material sourcing and selection, manufacturing, delivery of final products, and the management of product's end-of-life.
It can be achieved by considering environmental issues at the purchasing, product design and development, production, transportation, packaging, storage, disposal, and end of product life cycle management stages.	It is the integration of environmental concerns in the inter-organizational practices of supply chain management.

---

operational activities. Supply chains of the future can be characterized by the three major traits shown in Figure 5.6.

Baz and Iddik (2022) defines GSCM's scope as ranging from reactive monitoring of general environmental management programs to more proactive practices implemented through various Rs (Reduce, Re-use, Rework, Refurbish, Reclaim, Recycle, Remanufacture, Reverse logistics, etc.) [39]. From an entrepreneurial perspective, entrepreneurial GSCM is a new approach to environmental management executed by green entrepreneurs across whole supply chains instead of thinking in terms of individual non-environmental firms [40]. This new holistic view can integrate individuals, companies, and supply chains of different entrepreneurs from various countries together in an environmentally friendly way. Table 5.2 gives a broader perspective on definition of GSCM [41].

5.1.5 Practices of Green Supply Chain Management (GSCM)

The practice of GSCM incorporates sustainability concepts into traditional supply chain management [42]. The goal is to help industries reduce their carbon emissions and minimize waste while maximizing profit [39]. Every area of the supply chain has room for green improvements from manufacturing and purchasing to distribution, warehousing, and movement of MS for HST [43].

GSCM practices offer substantial benefits for the environment. For one, striving for supply chain sustainability by using less energy reduces carbon dioxide (CO_2) [44] emissions and other air pollutants. Produced by activities ranging from industrial work to operating vehicles, CO_2 is one of the main greenhouse gases responsible for global warming [45].

TABLE 5.3

Best Practices of GSCM in Reducing the Cost and Preserving the Environment [47]

Rethinking materials	Using recycled materials during the manufacturing process can save money.
Optimizing transport	Streamlining the transportation logistics provides many ways to save money and reduce the carbon footprint.
Reusing waste or by-products	Reusing the wastes or discard or even unlock new revenue streams from recyclable materials.
Redesigning processes	Considering new ways and technology to streamline the manufacturing processes in order to reduce the cost and carbon emission and enhance the efficiency.
Reducing business risk	Reduce the unsustainable practices.
Cutting back on packaging	Switching to streamlined packaging uses fewer materials and can lead to lower costs.

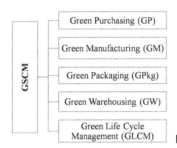

FIGURE 5.7 GSCM Methods to Preserve the Environment [48].

A fundamental principle of GSCM [46] is reducing waste and overall energy use. Needless to say, this can lead to big cost savings. Some of the ways of using GSCM practices to improve environmental performance and reduce costs are given in Table 5.3.

GSCM uses ethical and environmentally sound practices at every stage, with the goal of reducing air, water, and waste pollution. Figure 5.7 shows some methods of following the concept of GSCM.

GP means finding suppliers with environmentally sustainable products and services is just as important as greening the firm's own operations [49].

GM focuses on using fewer nonrenewable natural resources, reducing pollution and waste, and keeping emissions to a minimum, among other green practices. The most critical component of going green at this stage is reducing energy use [50].

GPkg considers every phase of a package's life cycle. That includes everything from how the firm's supplier sources materials to how consumers dispose of the packaging [51].

GW focuses on ensuring warehouses run more efficiently, reducing waste and energy use. Installing installation, using alternative energy sources like hydro and wind power, and adding windows to maximize natural light are just a few ways to improve your facility [52].

GLCM explains the green product design to be considered for the complete life cycle of the item in the BPM [53].

Green supply chain management and sustainable supply chain management [54] share many features in common, but the two fields are not interchangeable. Whereas GSCM practices have the goal of improving environmental health, sustainable supply chains focus on reducing their impact across many areas of life to ensure industry can continue to operate into the future. Naturally, environmental concerns factor into sustainability [55]. The firms also need to consider economic issues like managing sustainable growth [56]. Factors affecting GSCM practices are given in the Table 5.4.

TABLE 5.4

GSCM Best Practices at the Management Level [57]

Leadership Commitment	Managers and executives have to develop a unified approach to creating a green supply chain strategy.
Technology	A variety of software applications and advanced technology support green supply chain management at various steps of the process. These could range from warehouse management systems (WMS) that boost warehouse efficiency to new manufacturing technologies that use less energy to make products or reduce the quantities of hazardous materials involved in the manufacturing process.
Brand Image and Firm	When going green has positive implications for a firm's brand and culture supports these changes, introducing new green practices throughout the firm's supply chain tends to go efficiently.
Cost	Depending on the business and industry, developing a green supply chain may initially seem cost prohibitive. The expense of investing in overhauling your infrastructure and equipment may be daunting, which means the investments can lead to long-term savings.
Knowledge	Participants' expertise stands to have a big impact on supply chain performance at every stage. Involving green architects, consultants, and other green supply chain experts can help companies make the best use of green resources, implement sustainable solutions, and optimize results.

TABLE 5.5

Competitive Benefits of GSCM [58]

Increasing Cost Efficiency

Reducing Waste

Meeting Consumer Demand for Green Products

GSCM practices are critical to the health of globally and the continued sustainability of industry. Table 5.5 shows competitive benefits of GSCM for economic development.

There are many technology developments that are impacting GSCM where the advent of supply chain technology innovations make it easier to achieve green results by optimizing efficiency at all phases of GLCM. Table 5.6 shows the technological applications impacting the GSCM.

GSCM is a phenomenon that has emerged as a result of the development of sustainable and smart business and information technology trends. Sustainable and green supply chains are an innovative phenomenon that uses information technology to improve the quality of activities in operating areas [60]. In this regard, (IoT) is one of the most important components of technology infrastructure for smart. For this purpose, in this research, a framework for implementing a green IoT-based supply chain is presented [61]. This framework illustrates the direct relationship between data generation and how it interacts with the HST is affected by environmental sustainability and outlines a clear pathway for sustainable and green decision making in the supply chain for medical care. This framework has been endorsed by medical experts and BPM expert in the supply chain field as well as health care working for effective implementation of the green supply chain with an emphasis on technology in pharmaceutical firms.

The IoT connects hardware, sensors, and devices with different technologies to each other and to the Internet by receiving, controlling, and analyzing information [62]. The use of low-cost sensors, thanks to scientific advances and low-cost connectivity, has expanded the use of the

TABLE 5.6
Impact of AIoT in GSCM [59]

The Internet of Things (IoT)	IoT enables organizations to monitor equipment, inventory, and energy use in real time. Organizations gain greater awareness of energy waste, overstocking, and other missteps. That leads to a clearer picture of what needs to change throughout the supply chain.
Digitization of the Supply Chain	Improved digital tools, such as smarter WMS, make it possible to increase supply chain efficiency and automate processes—from stock reordering to optimizing warehouse picking paths. This helps supply chain leaders increase accuracy, avoid rush orders that require expedited shipping solutions like air freight, and prevent over-ordering. In turn, these improvements help reduce waste and energy use.
Artificial Intelligence (AI)	AI helps automate processes, boost efficiency, and prevent human error. These capabilities are useful throughout the supply chain, from streamlining manufacturing processes to using data to forecast product demand to analyzing delivery routes and planning the easiest and best methods. AI leads to less wasted effort and resources..
Robotics	Robots hold great potential for streamlining operations throughout the supply chain—particularly when it comes to logistics.
Materials Engineering	In recent years, advances in materials engineering have led to greener and more efficient manufacturing and product packaging.

Internet of Things [63]. The combination of IoT-enabled devices and sensors with machine learning creates a shared world, where innovation and optimal results are the always beneficial. IoT and AI have changed many industries and businesses and HST is one of the important sectors. Patient's related data is collected through IoT and turn it into usable and valuable data to boost HST for maintaining the GSCM for medical care supplies. If the data is used for specialized tasks such as predicting events, and component failure, HST needs to be aware of the type of data and its processing. It is difficult to use traditional tools to analyze data that is constantly changing and abundant [64]. Artificial intelligence is a powerful and effective tool in such cases. These various advances in information technology capabilities have changed the face of the HST more rapidly than in the past decade [65]. Adoption of IoT technology along with artificial intelligence and its efficient implementation can improve the cooperation between supply chain members through the rapid transfer and distribution of accurate information and the use of information systems and increase the efficiency of the supply chain in HST. Medical suppliers can use the IoT, AI, and advanced methods of analysis to determine the risk of refrigerator failure, the health of a consignment of frozen goods, by monitoring location, weather conditions, environmental conditions, traffic patterns, and so on [66].

Due to the complexity of the supply chain and in order to better manage HST, new technologies are considered as a potential factor in improving the performance of their supply chain [67]. The most accurate use of integrated information devices such as IoT of objects in this part of the management of the HST is important. Coverage of this information accurately and in an instant medical facilitates matters and makes the process progress more transparent. Figure 5.8 shows the AIoT in the GSCM in HST [69].

5.1.6 Application of AIoT in Sustainable Development

AIoT in sustainable development is an environmental sustainability that emphasizes reducing the use of natural resources and non-renewable energy, preventing the loss of energy resources, reducing waste production and emphasizing the reuse and recycling of waste, using recyclable

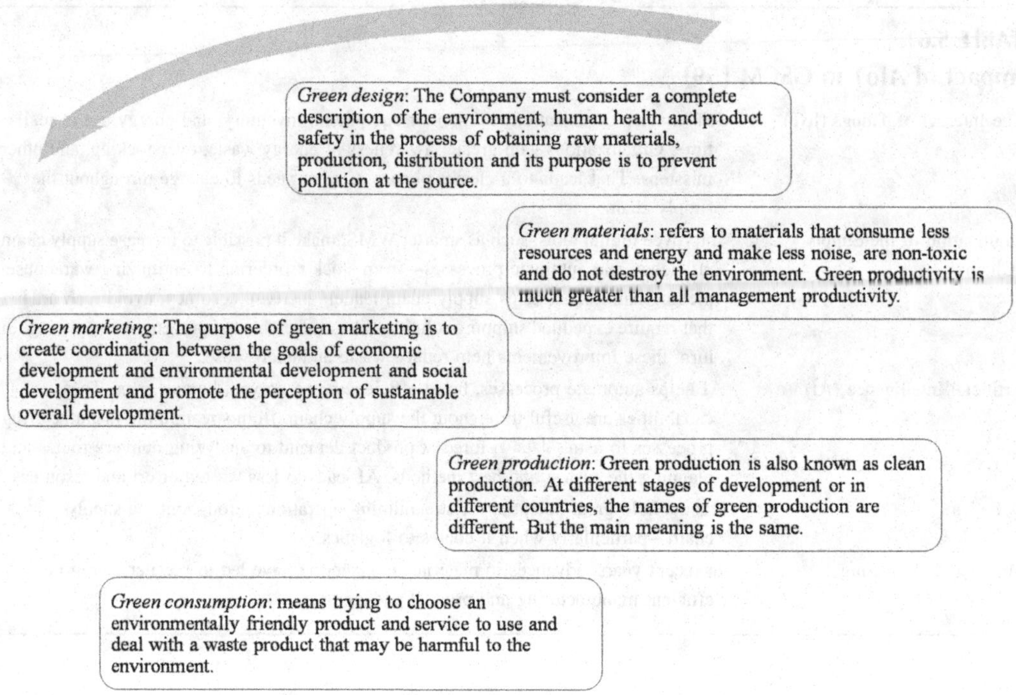

Green design: The Company must consider a complete description of the environment, human health and product safety in the process of obtaining raw materials, production, distribution and its purpose is to prevent pollution at the source.

Green materials: refers to materials that consume less resources and energy and make less noise, are non-toxic and do not destroy the environment. Green productivity is much greater than all management productivity.

Green marketing: The purpose of green marketing is to create coordination between the goals of economic development and environmental development and social development and promote the perception of sustainable overall development.

Green production: Green production is also known as clean production. At different stages of development or in different countries, the names of green production are different. But the main meaning is the same.

Green consumption: means trying to choose an environmentally friendly product and service to use and deal with a waste product that may be harmful to the environment.

FIGURE 5.8 AIoT in GSCM in HST [68].

materials, and reducing pollution in HST. Implementing AIoT to improve the environment is one of the areas that will have a great impact on preserving the environment and reducing pollutants caused by HST. IoT technology provides an opportunity to increase energy and productivity through green energy and renewable energy in HST.

The objectives of this chapter are as follows:

1. To study the background literature for the related studies and to analyze the application of AIoT in the GSCM for HST
2. To understand the advantages as well as challenges of application of AIoT in GSCM in general and specifically for HST through the vision of healthcare professionals
3. Analysis with respect to the results of the applications of AIoT in GSCM for HST for the relevance of AIoT.

Organization of the Chapter: The chapter is organized as: Section 5.2 outlines the literature review and the contribution of this research. In Section 5.3, the general scenario of research methods is explained and that are applied in the study and also describes the research process specifically in the HST. Section 5.4 gives the detailed explanation on results and discussion. Section 5.5 outlines a brief analysis of overall study. Section 5.6 concludes the chapter with future scope.

5.2 LITERATURE REVIEW

IoT was coined by the Kevin Ashton in 1999. IoT gives opportunity to each and every living as well as non-living things like machines to have unique digital identity. All around the globe Internet helps in connecting people making the place a small global village. With IoT, each and every thing can remain well connected, organized, managed, and controlled. All of this is achieved by using apps available on smartphones, computers, and tablets [70]. Moreover, IoT is a newer concept in the IT world that enables people to exchange data using communication networks such as the Internet [71]. The usage of technologies also plays significant role for the development and growth of system that is smarter and intelligent [72]. Today, all the growing businesses pay attention to this fact as with the help of IoT approach, the inter-operability between objects with humans and between objects increases which give rise to the development of new services [73]. Countless fields, for instance e-commerce, e- health, cloud-based production, etc. have used IoT for their growth and development. With the use of IoT, all of these businesses have been transformed. IoT is among the top methods that help to create the big data. Useful and competitive models are created which optimize all these businesses [74]. Thus, IoT usage and acceptance is of great importance and has numerous benefits. By using IoT, business gets the benefits like improved operational processes, value-creation, cost reduction benefit, and minimizing the overall risks involved as IoT ensures flexibility and reliability [75]. The IoT is made on applications' backdrop, which gives empowerment for the technology. These technologies include wireless sensors, radio frequency identification (RFID), smart technologies, as well as nanotechnologies. These apps enable user real-time monitoring and give control of the changes that might happen in the physical state of the associated objects [76]. Many researches have been conducted that point outs the uses of IoT in different sectors and the healthcare industry is also among them [77]. All the technologies used in healthcare sector can be tracked and monitored continuously using IoT technologies; for example, identity recognition, communication capabilities on the IoT, etc. [78]. The use of the IoT in a supply chain system has also been studied from time to time. IoT suggestions several resolutions for monitoring, tracking, and managing supply chain tasks [79]. IoT technology can collect all the process and allocate the data that is related to the chain [80]. The IoT is also used for safer production in mines. IoT technologies detect occurrence of any mine accident. It then provides necessary warnings so that mine accidents can be avoided [81]. IoT is also important in the transportation and logistics sector. Any logistics and transport company can observe the movement of items from its source to respective destination when there are more RFID or sensors in the physical items [80]. Various other studies addressed the linkage between IoT and smart cities or environment [82]. For this particular reason, the environmental applications and green technologies are of great importance in the current scenario.

Michigan State University Industrial Research Association introduced the term GSCM in 1996. In fact, for environmental protection GSCM is a totally new management model. GSCM from GLCM perspective includes all the stages of i) raw materials, ii) product design and manufacturing, iii) product sales and transportation, iv) product use, and v) product recycling for the medical products in the healthcare system. To get optimal usage of resources/energy the HST can cut negative environmental impacts by using green technology and supply chain management. GSCM is the procedure that considers environmental issues throughout its supply chain. Green supply chain management incorporates SCM with environmental requirements at each and every stage like product designing, while selecting a supply chain [83]. The sustainable supply chain management and green supply chain management are two different topics but in the supply chain literature these concepts were used interchangeably. Sustainable supply chain management comprises dimensions of economic, social and environmental sustainability [84]. Thus, the concept of sustainable SCM is a broader concept compared to GSCM and green supply chain management is a sub-part of sustainable SCM. Earlier, the product life cycle (PLC) incorporated the practices from design phase to last phase i.e., consumption. While today, with the environmental-management methodology, it includes the methods of raw-material preparation, designing, creation, usage and recycling, reuse that forms a closed loop of material flow which decreases resource consumption as well as reduces the harmful effects on environment [85].

Green IoT is a notion of sustainable smart company that has evolved as a result of major global developments, including the diffusion of sustainability as well as the growth of ICT. Using IoT technology, this word has become a more powerful idea. IoT entails the large utilization of anticipated network and number-nodes in the future. As a result, there is a need to lower the resources needed to enable all network parts, as well as the energy used by their operation. Today, one of the most critical issues in IoT research is maintaining the appropriate energy consumption rate [86]. As a result, Green IoT (G-IoT) constantly require to reduce energy usage and create a contemporary environment. To that purpose, all of the essential technologies involved with G-IoT must be considered. Among these technologies are: green sensing networks, green tags, and green Internet technologies [87]. These technologies are incorporated in the IoT's PLC and assist to support it. In the context of green Internet technology, both software and hardware should be considered, with the hardware solution producing devices that consume less energy while maintaining performance. The G-IoT has several applications, including green smart factories, green smart cities, green smart logistics, and green smart healthcare.

Since the environmental issues has been connected with the economy, and nations have concluded that environmental protection may boost production, many techniques to realizing these technologies have been pursued, one of which is the most recent, the GSCM strategy. The purpose of GSCM is to eliminate or reduce waste, which serves as a significant innovation in assisting organizations in developing strategies to accomplish common economic and market goals by lowering environmental hazards and improving environmental efficiency [88]. Laws and regulations requiring environmental compliance are the primary motivators for green supply chain adoption. All supply chain interactions can give significant information for analysis and decision making [89]. IoT, being one of the most prominent foundations of big data creation, plays a vital role. HST increases its competitive advantage by refining the environmental role of IoT; by following the environmental standards and laws, by enhancing patient's awareness and reducing the negative environmental impacts on its products as well as services. Since the supply chain is the most important HST unit that covers big range of HST procedures from medical supplier relationships to supply, and from manufacturing practices to making them accessible to HST.

5.3　RESEARCH METHODOLOGY

Qualitative research is defined as a market research method that focuses on obtaining data through open-ended and conversational communication. The qualitative research methods allow for in-depth and further probing and questioning of respondents based on their responses, where the interviewer/researcher also tries to understand their motivation and feelings [90]. Qualitative research methods are designed in a manner that helps reveal the behavior and perception of a target audience with reference to a particular topic.

There are many types of qualitative research methods; Table 5.7 gives the list and brief description on that.

The results of qualitative methods are more descriptive and the inferences can be drawn quite easily from the data that is obtained. Qualitative research includes data collection as a first step and then the analysis of data.

Qualitative data collection allows collecting data that is non-numeric and helps us to explore how decisions are made and provide us with detailed insight [90].

For the purpose of this study, we used qualitative methods such as conducted workshops and an interview for health professionals and stakeholders with open-ended questionnaires. The present research is applied in terms of research (considering its general purpose, which is to identify, discover, and explain the indicators and effective components in the AIoT-based GSCM).

5.3.1　PROPOSED WORK

IoT is an innovative technology for all industries; it has also demonstrated its potential in processes such as supply chain. Combining this technology with artificial intelligence technology helps

TABLE 5.7
Qualitative Research Types and Description [90]

Types	Description
1. One-on-one interview:	Conducting in-depth interviews is one of the most common qualitative research methods. It is a personal interview that is carried out with one respondent at a time.
2,. Focus groups:	A focus group is also one of the commonly used qualitative research methods, used in data collection. A focus group usually includes a limited number of respondents (6–10) from within the target market.
3. Ethnographic research:	Ethnographic research is the most in-depth observational method that studies people in their naturally occurring environment.
4.. Case study research:	It is used for explaining an organization or an entity based on the particular situation in the firm.
5.. Record keeping:	This is similar to going to a library where one can go over books and other reference material to collect relevant data that can likely be used in the research.
6. Qualitative observation:	It is a process of research that uses subjective methodologies to gather systematic information or data.

management, forecasting and monitoring applications to improve the operational efficiency of health care by the distribution and increase transparency in their decisions. Figure 5.9 shows the flow of tasks that we followed for the study of AIoT in GSCM for HST.

The AIoT works on three levels to automate and complete the tasks for planning, doing, checking, and adjusting to offer optimum results. The three levels include IoT, big data, and AI. Figure 5.10 shows the basic workings of AIoT for any general scenario. The same theory we have applied in our study is to find how AIoT automates the GSCM for HST.

AIoT for any sector works under four layers; any system collects data from the environment for the completion of objective sensors of IoT and sensors of AI facilitate in building the communication to connect to the network of networks. Main IoT systems use its applications, aligning with external AI to build AIoT for facilitating the goals of any sector. Figure 5.11 shows the AIoT architecture for any sector and can be applied for the GSCM as presented in our study.

In this chapter, we have shared the experience of three actors of HST: doctors, stakeholders, and pharmacists. We conducted a workshop on explaining the benefits of AIoT in GSCM and in other HST operations. We have also explained the key performance indicators for HST in knowing how AIoT can contribute in cost-effective GSCM. Interview was conducted separately for all these actors in HST and they shared candid experience about the applications of AIoT in GSCM in HST and what advantages and challenges they encountered in the process.

This research proposes to apply AIoT in HST at all levels of BPM and especially at GSCM to know how the environment can be preserved by the application of AIoT in HST. Also, the research explains the general scenario of AIoT in HST for green design, production, and material. AIoT uses its sensor technologies in meeting the objectives of GSCM in HST. Figure 5.12 shows the GSCM areas that are facilitated by AIoT.

AIoT facilitates the tracking and monitoring of MS and creates more transparency in the communication and planning process between health professionals and SCM managers. Figure 5.13 presents the working scenario of AIoT for GSCM for HST.

The study suggests that application of AIoT will solve the complex problems of GSCM in HST and will contribute to improving the services of HST focusing on preserving the nature.

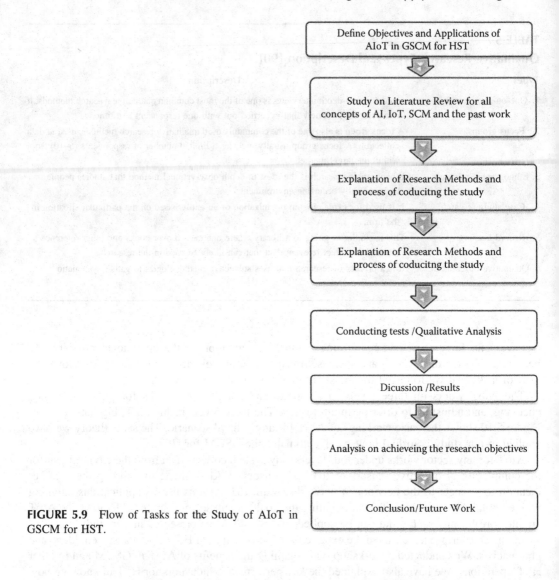

FIGURE 5.9 Flow of Tasks for the Study of AIoT in GSCM for HST.

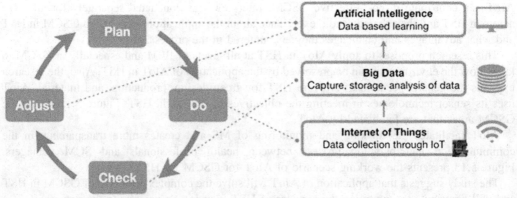

FIGURE 5.10 Working Levels of AIoT for Tasks Completion [91].

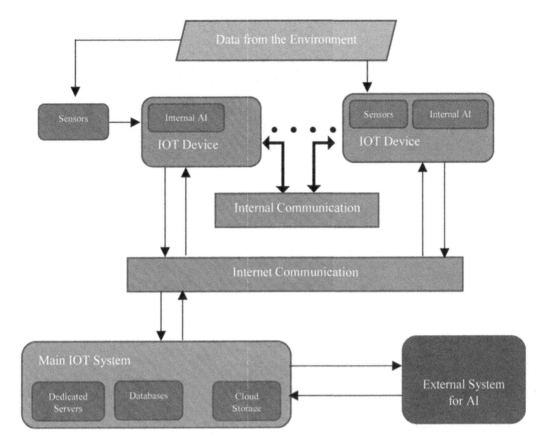

FIGURE 5.11 AIoT Architectural Working.

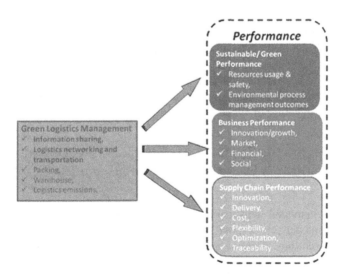

FIGURE 5.12 AIoT Framework for GSCM [91].

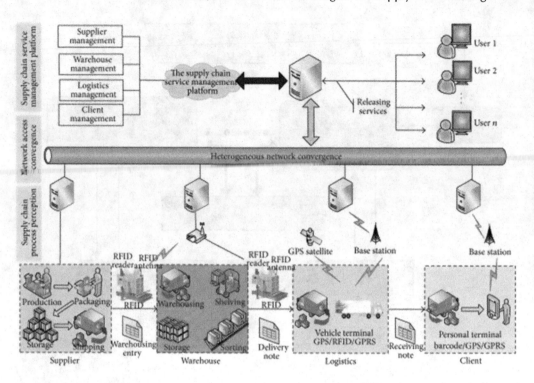

FIGURE 5.13   Working Scenario of AIoT in GSCM for HST [91].

5.4   RESULTS AND DISCUSSION

5.4.1   REAL-TIME MEDICAL CARE SHIPMENT AND LOCATION TRACKING FOR HST AND MEDICAL LOGISTICS

AIoT provides HST workforce that includes the doctors, stakeholders, and other medical staff members with a coherent flow of real-time data about medical products such as medicines, accessories, and equipment with their location and shipping environmental conditions. In this way, HST will be warned if the medicines are transported in the wrong direction and HST can monitor the delivery of medical supplies (MS) on time, shown in Figures 5.14 and 5.15.

The results show that pharmacists have more confidence in AIoT for GSCM for medical supplies, and doctors as well as stakeholders share the good level of confidence based on their experiences for the application of AIoT.

5.4.2   MONITOR THE STORAGE STATUS OF THE MEDICAL SUPPLIES DURING SHIPMENT BY AIoT TO ENSURE IT IS ENVIRONMENTALLY FRIENDLY AND ENHANCE THE APPLICATION OF GSCM

AIoT provides environmental sensors that are applied by the medical professional to track transport conditions and actively respond to changes for the suitability of storing medical supplies. The AIoT provides solutions in the supply chain is displaying information about in-car temperature, pressure, humidity, and other factors that can compromise medical supplies life and the effect on the environment. AIoT facilitates in adjusting the conditions automatically during the SCM, as shown in Figures 5.16 and 5.17.

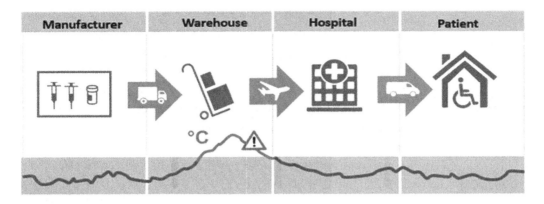

FIGURE 5.14 Real-Time Shipment Location Tracking for Medical Supplies [91].

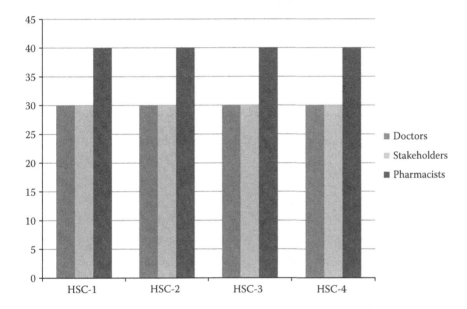

FIGURE 5.15 Effectiveness of Real-Time Shipment Location Tracking for Medical Supplies.

The results show the similar trend; pharmacists have good experiences in AIoT application in monitoring and storing of MS until the MS reach to the respective HST. Doctors and stakeholders show a good response but also claimed that using GSCM increases the cost at some levels.

5.4.3 APPLICATION OF AIoT FOR PREDICTING THE MS MOVEMENT AND ARRIVAL

Health professionals use AIoT devices and data analysis systems to improve decision quality and increase the accuracy of delivery forecasts for MS. By the application of real-time tracking, HST is able to monitor MS during shipment and forecast delivery, as well as predict and reduce risks associated with delays and to maintain the conditions required for the shelf life of MS. AIoT helps in identifying the conditions required for MS in inventory. The integration of AIoT-based supply chain management systems is one of the most important technological trends in inventory that make sure that the devices used are echo-friendly. Its benefits range from increasing the efficiency of inventory processes to better inventory management for MS. Also, AIoT ensures the health and

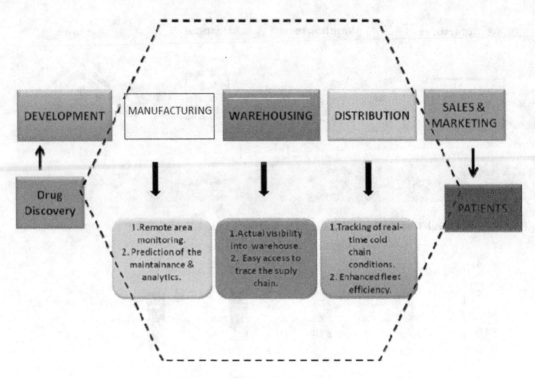

FIGURE 5.16   Monitor the Storage Status of the Medical Supplies by AIoT [92].

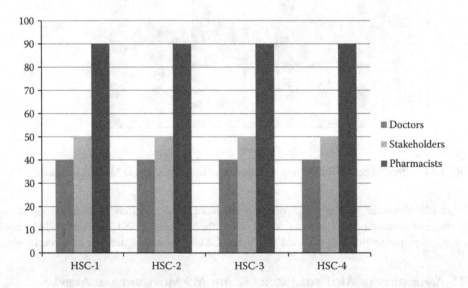

FIGURE 5.17   Monitor the Storage Status of the Medical Supplies During Shipment by AIoT.

safety of the health employees who are working in SCM because some MS needs extra care and can cause damage if not dealt with carefully. AIoT enables integrated workflow and performance that would otherwise be impossible to achieve for MS. AIoT automates the inventory control for MS with artificial intelligence (Figure 5.18).

The results show the same responses based on their experiences and all three actors of the research agree on the application and benefits of AIoT in the GSCM of MS and on the methods of storing MS in inventories.

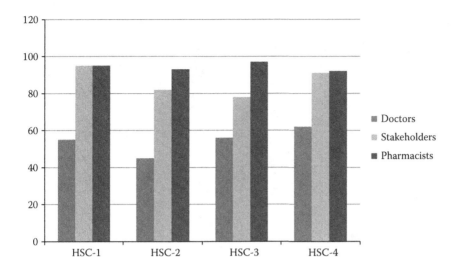

FIGURE 5.18   Movement and Control of MS in Inventory by AIoT.

5.4.4   IMPROVING POTENTIAL PLANNING IN HST BY THE APPLICATION OF AIoT

The IoT and data analytics help SCM plan for preventing delays and potential accidents that can be caused due to misinformation, as shown in Figure 5.19. AIoT controls all the data needed to prepare potentially flexible applications and reach the cause of existing delays. This technology

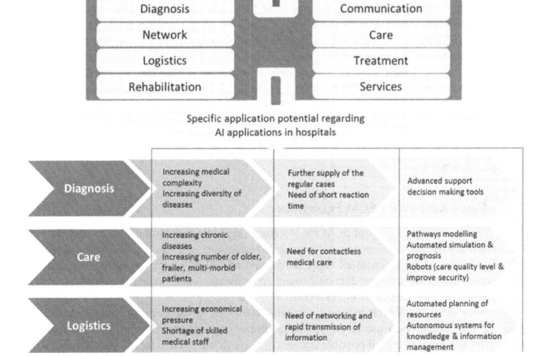

FIGURE 5.19   Potential Application of AIoT in HST [92].

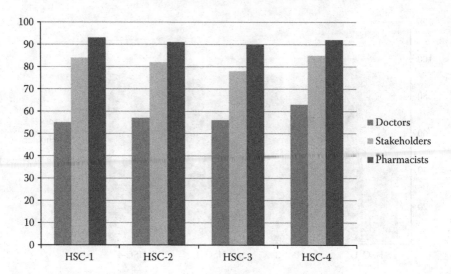

FIGURE 5.20   Potential Application of AIoT in HST to Enhance GSCM.

immediately provides the necessary warnings to HST professionals and these professionals can accordingly with the SCM managers apply AIoT to know the reasons and how to resolve the delays and prevent accidents.

The process is not complex in improving the potential planning in HST; all actors have to play their roles in the management of MS and to control delays as well as accidents. When the flow of MS is carried out SCM will apply AIoT to know number of operations and also how many operations have to work simultaneously. With the help of health professionals these SCM managers will apply AIoT to simplify the processing of the MS. AIoT helps in maintaining the GSCM by alarming the actors of SCM if the process of SCM exceeds the emission limitations and also helps health professionals and SCM managers to comply with the environmental standards. The process of GSCM by AIoT is carried out by the sensors in SCM. These sensors track the MS movement from inventory to pharmacies, clinics, or hospitals. During the process of flow of MS, theses sensors of AIoT calculate the energy used and how to reduce this energy consumption. GSCM uses AIoT and accordingly apply the sensors to reduce the resources such as electricity, cold storage power consumption, water, and other resources by using green strategies in HST (Figure 5.20).

The results show that AIoT is facilitated in maintaining the GSCM and all three actors share their experiences in how AIoT contributed in preserving the environment and enhanced the working of GSCM.

5.4.5   Green AIoT-Based SCM Framework

GSCM focuses on all processes of SCM for MS from supply to distribution with an idea preserve the energy and increase in efficiency and provide all environmental solutions or options based on AIoT. The results of these processes cover all domain of GSCM which use different features of AIoT such as sensor technology, data processing applications, and sustainable computing models in completing various operations in HST (Figure 5.21).

The results show that doctors are mostly convinced for GSCM for all the operations in HST and willing to take initiative to use any measure for preserving the environment. Stakeholders and pharmacists also agree to use all possible features of AIoT to use in HST for the purpose of GSCM.

HST can use AIoT in many areas of their operations and health professional have identified these areas. By the help of AIoT they are able to know the positive impact of AIoT in GSCM [93].

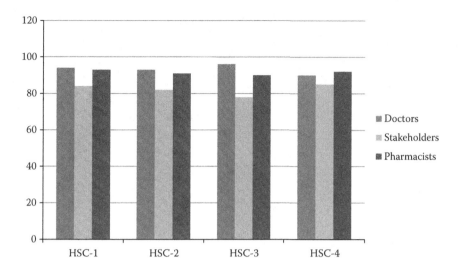

FIGURE 5.21   Effective GSCM by the Features of AIoT.

TABLE 5.8
KPIs of AIoT in GSCM in HST

KPI	GSCM	Target
Green Purchasing	Supply	Achieved
Green Manufacturing	Production	Achieved
Green Design	Production	Achieved
Green Transportation	Distribution	Achieved
Environmental Management	Supply-production- Distribution	Achieved
Operational Performance	Supply-production- Distribution	Achieved
Cold Storage	Supply-production- Distribution	Achieved

The results show that there are six major key performance indicators measured by AIoT for GSCM in different levels in GSCM in HST, shown in Table 5.8.

5.4.6   AIoT and Green Design in HST

The HST should consider a complete description of the environment; human health and product safety in the process of obtaining, production, and distribution of MS and the objective should be to perverse environment and achieve GSCM, as shown in Figure 5.22. This can be achieved by reducing energy consumption and emission of hazardous material (Figure 5.23).

The results show that doctors and pharmacists have more contribution and focus on using green design and are encouraged to apply AIoT to know how this can be enhanced, whereas stakeholders were found slightly less motivated due to the increase in cost and sometimes in availability of green options.

5.4.7   AIoT in Green Materials in HST

The material here is related to use of environmentally friendly options at all the levels of BPM by HST. AIoT has already identified how KPIs can be measured to know if HST will make a profit

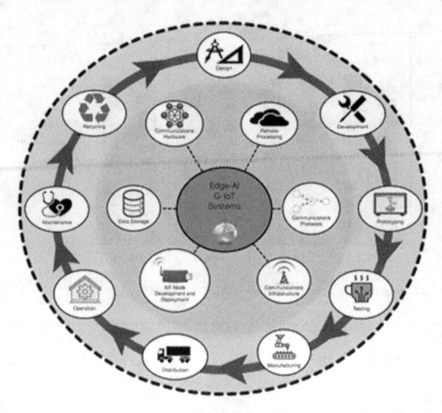

FIGURE 5.22 AIoT in Green Design for GSCM [92].

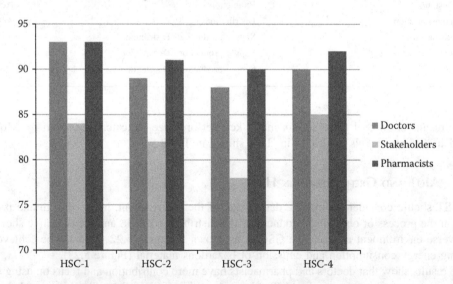

FIGURE 5.23 AIoT in Green Design in HST.

by using green material or not and results have shown that in long term all HST will achieve good profit and more sustainability if they apply green concepts (Figure 5.24).

The results show that doctors and pharmacists have more contribution and focus on using green material and are encouraged to apply AIoT to know how this can be enhanced, whereas stakeholders were found slightly less motivated due to the increase in cost and sometimes of availability of green options.

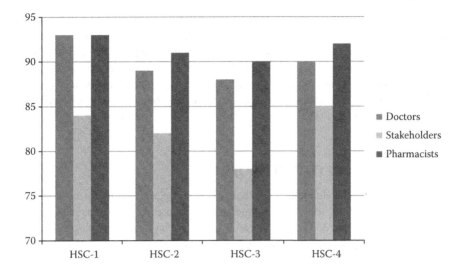

FIGURE 5.24   AIoT in Green Materials in HST.

5.4.8   AIoT in Green Production in HST

Green production is referred to apply all eco-friendly production processes where the consumption of energy is reduced at inventory. AIoT has helped HST to identify the cost-effective options to meet this green target (Figure 5.25).

The results show that doctors and pharmacists have more contribution and focus on using green production and are encouraged to apply AIoT to know how this can be enhanced whereas stakeholders were found slightly less motivated due to the increase in cost and sometimes in availability of green options.

5.5   ANALYSIS

The application of AIoT is encouraged by all sectors and HST is using it for various benefits and maintains logistics while preserving the environment is one of the most important objectives of

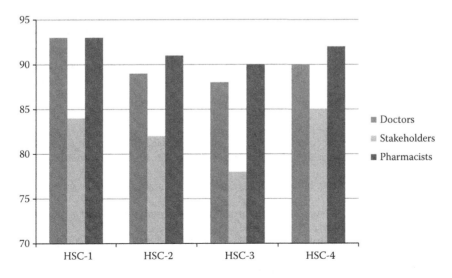

FIGURE 5.25   AIoT in Green Production in HST.

applying AIoT by HST. The three important entities of HST understand the advantages of AIoT in GSCM but stakeholders express the challenges of AIoT in its applications due to short term cost increment. Overall analysis shows the major role of AIoT in applying real-time medical care shipment and location tracking for HST and medical logistics, monitoring the storage status of the medical supplies during shipment by AIoT to ensure it is environmentally friendly and enhancing the application of GSCM, application of AIoT for predicting the MS movement and arrival, and improving potential planning in HST and application of AIoT for green design, green material, and green production in HST.

5.6 CONCLUSION AND FUTURE SCOPE

AIoT is a new form of large-scale application that, due to its operational performance, has received considerable attention from all sectors including the healthcare sector. There is high expectation for environmental gains from AIoT; this research has presented the role of AIoT in GSCM for HST. The purpose of this paper is to identify and explain the advantages of AIoT at all the levels of HST in GSCM. AIoT-enabled sensors are for environmental sustainability and data processing for green design, production, and material for HST and also for GSCM computing models in BPM in HST. The GSCM in HST is aided by AIoT that shows the effective relationship between data and its application in GSCM. The AIoT-based GSCM in HST identifies optimum options for the MS availability, maintain its inventory and other issues in BPM.

This research identified the major benefits of AIoT in GSCM in HST and the main users of this application in the GSCM explained the benefits more holistically. Health professionals confirm that AIoT, which is the combination of AI and IoT, provides smart solutions and can provide good connections between devices and MS and enhance the capabilities of the smart elements to work for GSCM. AIoT improves not only material flow systems, but also global positioning and automatic load detection for HST. It also increases energy efficiency and thus reduces energy consumption for GSCM. AIoT technology is flexible and may have benefits for providing GSCM in the HST by automating the process, reducing the consumption level, and following green design, production, and material at all level of SCM. This chapter shows the positive impact of AIoT in GSCM for HST and concluded that health professionals are willing to apply AIoT for GSCM and other BPM to identify cost-effective options that can reduce harmful emission, reduce energy consumption, and eventually help them to treat patients successfully.

AIoT applications are increasing and in the future HST may apply AIoT in deciding the composition of MS along with maintaining the green logistics. Using new AIoT technologies and integrating them with other technologies such as sensors, it will be possible to achieve continuous communication of MS in the GSCM. This will give instant and accurate results for easy monitoring for the flow of MS and GSCM managers will be able to work on errors by receiving automated reports.

REFERENCES

[1] Nozari, H., Fallah, M., & Szmelter-Jarosz, A. (2021). A conceptual framework of green smart IoT-based supply chain management. *International Journal of Research in Industrial Engineering*, 10(1), 22–34.

[2] Ballina, F. J. (2022). Smart business: The element of delay in the future of smart tourism. *Journal of Tourism Futures*, 8(1), 37–54.

[3] Naim, A. (2021). Green business process management. *International Journal of Innovative Analyses and Emerging Technology*, 1(6), 125–134.

[4] Arshad, R., Zahoor, S., Shah, M. A., Wahid, A., & Yu, H. (2017). Green IoT: An investigation on energy saving practices for 2020 and beyond. *IEEE Access*, 5, 15667–15681.

[5] Liu, X., & Ansari, N. (2019). Toward green IoT: Energy solutions and key challenges. *IEEE Communications Magazine*, 57(3), 104–110.

[6] Kumar, A., Payal, M., Dixit, P., & Chatterjee, J. M. (2020). Framework for realization of green smart cities through the Internet of Things (IoT). *Trends in Cloud-Based IoT*, 85–111.

[7] Naim, A., Alahmari, F., & Rahim, A. (2021). Role of artificial intelligence in market development and vehicular communication. *Smart Antennas: Recent Trends in Design and Applications*, 2, 28.

[8] Nandyala, C. S., & Kim, H. K. (2016). Green IoT agriculture and healthcare application (GAHA). *International Journal of Smart Home*, 10(4), 289–300.

[9] Zanizdra, M. Y. (2018). "Green smart" industry conceptual provisions. *Economy of Industry*, 1(81), 61–85.

[10] Khan, N., Naim, A., Hussain, M. R., Naveed, Q. N., Ahmad, N., & Qamar, S. (2019, May). The 51 v's of big data: Survey, technologies, characteristics, opportunities, issues and challenges. In *Proceedings of the International Conference on Omni-Layer Intelligent Systems* (pp. 19–24).

[11] Hamet, P., & Tremblay, J. (2017). Artificial intelligence in medicine. *Metabolism*, 69, S36–S40.

[12] Friess, P. (2016). *Digitising the industry-Internet of Things connecting the physical, digital and virtual worlds*. River Publishers.

[13] Lampropoulos, G., Siakas, K., & Anastasiadis, T. (2019). Internet of Things in the context of industry 4.0: An overview. *International Journal of Entrepreneurial Knowledge*, 7(1), 4–19.

[14] Aheleroff, S., Xu, X., Lu, Y., Aristizabal, M., Velásquez, J. P., Joa, B., & Valencia, Y. (2020). IoT-enabled smart appliances under industry 4.0: A case study. *Advanced Engineering Informatics*, 43, 101043.

[15] Abedin, S. F., Alam, M. G. R., Haw, R., & Hong, C. S. (2015, January). A system model for energy efficient green-IoT network. In *2015 International Conference on Information Networking (ICOIN)* (pp. 177–182). IEEE.

[16] Ruan, J., Wang, Y., Chan, F. T. S., Hu, X., Zhao, M., Zhu, F., … & Lin, F. (2019). A life cycle framework of green IoT-based agriculture and its finance, operation, and management issues. *IEEE Communications Magazine*, 57(3), 90–96.

[17] Bhbosale, S., Pujari, V., & Multani, Z. (2020). Advantages and disadvantages of artificial intellegence. *Aayushi International Interdisciplinary Research Journal*, 227–230.

[18] Khanzode, K. C. A., & Sarode, R. D. (2020). Advantages and disadvantages of artificial intelligence and machine learning: A literature review. *International Journal of Library & Information Science (IJLIS)*, 9(1), 3.

[19] Mohammad, S. M. (2020). Artificial intelligence in information technology. Available at SSRN 3625444, *Artificial Intelligence*, 7(6).

[20] Alfarsi, G., Tawafak, R. M., ElDow, A., Malik, S. I., Jabbar, J., Al Sideiri, A., & Mathew, R. (2021). General view about an artificial intelligence technology in education domain.

[21] Nahr, J. G., Nozari, H., & Sadeghi, M. E. (2021). Green supply chain based on artificial intelligence of things (AIoT). *International Journal of Innovation in Management, Economics and Social Sciences*, 1(2), 56–63.

[22] De Vass, T., Shee, H., & Miah, S. (2021). IoT in supply chain management: Opportunities and challenges for businesses in early Industry 4.0 context. *Operations and Supply Chain Management: An International Journal*, 14(2), 148–161.

[23] Yaser, T., Sadeghi, S. K., Amiri, R., & Aghajani, H. (2021). Green supply chain network design under uncertainty conditions with the mathematical model and solving it with a NSGA II algorithm. *International Journal of Innovation in Management, Economics and Social Sciences*, 1(3), 58–81.

[24] Sadeghi, M. E., & Jafari, H. (2021). Investigating the dimensions, components and key indicators of supply chain management based on digital technologies. *International Journal of Innovation in Management, Economics and Social Sciences*, 1(3), 82–87.

[25] Shi, H., & Li, Q. (2022). Edge computing and the Internet of Things on agricultural green productivity. *The Journal of Supercomputing*, 1–23.

[26] Reddy, S., Allan, S., Coghlan, S., & Cooper, P. (2020). A governance model for the application of AI in health care. *Journal of the American Medical Informatics Association*, 27(3), 491–497.

[27] Haroun, A. F., Le, X., Gao, S., Dong, B., He, T., Zhang, Z., … & Lee, C. (2021). Progress in micro/nano sensors and nanoenergy for future AIoT-based smart home applications. *Nano Express*.

[28] Eskandari, R., & Feili, H. R. (2021). Designing and solving location-routing-allocation problems in a sustainable blood supply chain network of blood transport in uncertainty conditions. *International Journal of Innovation in Management, Economics and Social Sciences*, 1(4), 32–49.

[29] Dong, B., Shi, Q., Yang, Y., Wen, F., Zhang, Z., & Lee, C. (2021). Technology evolution from self-powered sensors to AIoT enabled smart homes. *Nano Energy*, 79, 105414.

[30] Ragazzi, M., & Ghidini, F. (2017). Environmental sustainability of universities: Critical analysis of a green ranking. *Energy Procedia*, 119, 111–120.

[31] Birkel, H. S., & Hartmann, E. (2019). Impact of IoT challenges and risks for SCM. *Supply Chain Management: An International Journal*.

[32] Haddud, A., DeSouza, A., Khare, A., & Lee, H. (2017). Examining potential benefits and challenges associated with the Internet of Things integration in supply chains. *Journal of Manufacturing Technology Management*, 34, 39–08.

[33] Le, A. N. H., Nguyen, T. T., & Cheng, J. M. S. (2021). Enhancing sustainable supply chain management performance through alliance portfolio diversity: The mediating effect of sustainability collaboration. *International Journal of Operations & Production Management*, 41(10), 1593–1614.

[34] Tu, M. (2018). An exploratory study of Internet of Things (IoT) adoption intention in logistics and supply chain management: A mixed research approach. *The International Journal of Logistics Management*, 29(1), 131–151.

[35] Ali, Y., Saad, T. B., Sabir, M., Muhammad, N., Salman, A., & Zeb, K. (2020). Integration of green supply chain management practices in construction supply chain of CPEC. *Management of Environmental Quality: An International Journal*, 31(1), 185–200.

[36] Chalmeta, R., & Santos-deLeón, N. J. (2020). Sustainable supply chain in the era of industry 4.0 and big data: A systematic analysis of literature and research. *Sustainability*, 12(10), 4108.

[37] Andersson, R., & Pardillo-Baez, Y. (2020). The Six Sigma framework improves the awareness and management of supply-chain risk. *The TQM Journal*, 32(5), 1021–1037.

[38] Ageron, B., Bentahar, O., & Gunasekaran, A. (2020, July). Digital supply chain: Challenges and future directions. *Supply Chain Forum: An International Journal*, 21(3), 133–138.

[39] Esmaeilian, B., Sarkis, J., Lewis, K., & Behdad, S. (2020). Blockchain for the future of sustainable supply chain management in Industry 4.0. *Resources, Conservation and Recycling*, 163, 105064.

[40] Lahane, S., Kant, R., & Shankar, R. (2020). Circular supply chain management: A state-of-art review and future opportunities. *Journal of Cleaner Production*, 258, 120859.

[41] El Baz, J., & Iddik, S. (2022). Green supply chain management and organizational culture: A bibliometric analysis based on Scopus data (2001–2020). *International Journal of Organizational Analysis*, 30(1), 156–179.

[42] Jemai, J., Do Chung, B., & Sarkar, B. (2020). Environmental effect for a complex green supply-chain management to control waste: A sustainable approach. *Journal of Cleaner Production*, 277, 122919.

[43] Rahman, T., Ali, S. M., Moktadir, M. A., & Kusi-Sarpong, S. (2020). Evaluating barriers to implementing green supply chain management: An example from an emerging economy. *Production Planning & Control*, 31(8), 673–698.

[44] Le, T. (2020). The effect of green supply chain management practices on sustainability performance in Vietnamese construction materials manufacturing enterprises. *Uncertain Supply Chain Management*, 8(1), 43–54.

[45] Ta, V. L., Bui, H. N., Canh, C. D., Dang, T. D., & Do, A. D. (2020). Green supply chain management practice of FDI companies in Vietnam. *The Journal of Asian Finance, Economics, and Business*, 7(10), 1025–1034.

[46] De Carvalho, L. S., Stefanelli, N. O., Viana, L. C., Vasconcelos, D. D. S. C., & Oliveira, B. G. (2020). Green supply chain management and innovation: A modern review. *Management of Environmental Quality: An International Journal*.

[47] Mahmood, W. H. W., & Azlan, U. A. A. (2020). QFD approach in determining the best practices for green supply chain management in composite technology manufacturing industries. *Malaysian Journal on Composites Science & Manufacturing*, 1(1), 45–56.

[48] Bag, S., Gupta, S., Kumar, S., & Sivarajah, U. (2020). Role of technological dimensions of green supply chain management practices on firm performance. *Journal of Enterprise Information Management*, 34(1), 1–27.

[49] Notteboom, T., van der Lugt, L., van Saase, N., Sel, S., & Neyens, K. (2020). The role of seaports in green supply chain management: Initiatives, attitudes, and perspectives in Rotterdam, Antwerp, North Sea Port, and Zeebrugge. *Sustainability*, 12(4), 1688.

[50] Deif, A. M. (2011). A system model for green manufacturing. *Journal of Cleaner Production*, 19(14), 1553–1559.

[51] Balon, V. (2020). Green supply chain management: Pressures, practices, and performance—An integrative literature review. *Business Strategy & Development*, 3(2), 226–244.

[52] Hasanova, H., & Romanovs, A. (2020, October). Best practices of technology management for sustainable digital supply chain. In *2020 61st International Scientific Conference on Information Technology and Management Science of Riga Technical University (ITMS)* (pp. 1–6). IEEE.

[53] Ilyas, S., Hu, Z., & Wiwattanakornwong, K. (2020). Unleashing the role of top management and government support in green supply chain management and sustainable development goals. *Environmental Science and Pollution Research*, 27(8), 8210–8223.

[54] Singh, J., Singh, H., & Kumar, A. (2020). Impact of lean practices on organizational sustainability through green supply chain management—An empirical investigation. *International Journal of Lean Six Sigma*, 11(6), 1035–1068.

[55] Micheli, G. J., Cagno, E., Mustillo, G., & Trianni, A. (2020). Green supply chain management drivers, practices and performance: A comprehensive study on the moderators. *Journal of Cleaner Production*, 259, 121024.

[56] Adila, A. S., Husam, A., & Husi, G. (2018, April). Towards the self-powered Internet of Things (IoT) by energy harvesting: Trends and technologies for green IoT. In *2nd International Symposium on Small-Scale Intelligent Manufacturing Systems (SIMS)* (pp. 1–5). IEEE.

[57] Wang, C., Zhang, Q., & Zhang, W. (2020). Corporate social responsibility, green supply chain management and firm performance: The moderating role of big-data analytics capability. *Research in Transportation Business & Management*, 37, 100557.

[58] Khan, N., Sajak, A. A. B., Alam, M., & Mazliham, M. S. (2021, May). Analysis of green IoT. *Journal of Physics: Conference Series*, 1874(1), 012012.

[59] Habib, M. A., Bao, Y., & Ilmudeen, A. (2020). The impact of green entrepreneurial orientation, market orientation and green supply chain management practices on sustainable firm performance. *Cogent Business & Management*, 7(1), 1743616.

[60] Liu, S., Eweje, G., He, Q., & Lin, Z. (2020). Turning motivation into action: A strategic orientation model for green supply chain management. *Business Strategy and the Environment*, 29(7), 2908–2918.

[61] Zhang, G., & Zhao, Z. (2012). Green packaging management of logistics enterprises. *Physics Procedia*, 24, 900–905.

[62] AlKhidir, T., & Zailani, S. (2009). Going green in supply chain towards environmental sustainability. *Global Journal of Environmental Research*, 3(3), 246–251.

[63] Yin, W., & Ran, W. (2021). Theoretical exploration of supply chain viability utilizing blockchain technology. *Sustainability*, 13(15), 8231.

[64] Nahaei, V. S., Novin, M. H., & Khaligh, M. A. (2021). Review and prioritization of investment projects in the waste management organization of Tabriz Municipality with a rough sets theory approach. *International Journal of Innovation in Management, Economics and Social Sciences*, 1(3), 46–57.

[65] Kaur, G., Tomar, P., & Singh, P. (2018). Design of cloud-based green IoT architecture for smart cities. In *Internet of Things and Big Data Analytics Toward Next-Generation Intelligence* (pp. 315–333). Cham: Springer.

[66] Joshi, Y., & Rahman, Z. (2015). Factors affecting green purchase behaviour and future research directions. *International Strategic Management Review*, 3(1-2), 128–143.

[67] Chan, R. Y. (2001). Determinants of Chinese consumers' green purchase behavior. *Psychology & Marketing*, 18(4), 389–413.

[68] Dornfeld, D. A. (Ed.). (2012). *Green manufacturing: Fundamentals and applications*. Springer Science & Business Media.

[69] Suresh, P., Daniel, J. V., Parthasarathy, V., & Aswathy, R. H. (2014, November). A state of the art review on the Internet of Things (IoT) history, technology and fields of deployment. In *International Conference on Science Engineering and Management Research (ICSEMR)* (pp. 1–8). IEEE.

[70] Lueth, K. L. (2015). IoT basics: Getting started with the Internet of Things. White paper, 1–9.

[71] Kim, H., & Hwang, S. K. (2018). Past, present, and future of IoT. *Electronics and Telecommunications Trends*, 33(2), 1–9.

[72] Sharma, N., Shamkuwar, M., & Singh, I. (2019). The history, present and future with IoT. In *Internet of Things and Big Data Analytics for Smart Generation* (pp. 27–51). Cham: Springer.

[73] Savadjiev, P., Chong, J., Dohan, A., Vakalopoulou, M., Reinhold, C., Paragios, N., & Gallix, B. (2019). Demystification of AI-driven medical image interpretation: Past, present and future. *European Radiology*, 29(3), 1616–1624.

[74] Xu, F., Uszkoreit, H., Du, Y., Fan, W., Zhao, D., & Zhu, J. (2019, October). Explainable AI: A brief survey on history, research areas, approaches and challenges. In *CCF International Conference on Natural Language Processing and Chinese Computing* (pp. 563–574). Cham: Springer.

[75] Elliott, A. (2019). *The culture of AI: Everyday life and the digital revolution*. Routledge.

[76] Zimon, D., Tyan, J., & Sroufe, R. (2020). Drivers of sustainable supply chain management: Practices to alignment with un sustainable development goals. *International Journal for Quality Research*, 14(1), 219–236.

[77] Do, A., Nguyen, Q., Nguyen, D., Le, Q., & Trinh, D. (2020). Green supply chain management practices and destination image: Evidence from Vietnam tourism industry. *Uncertain Supply Chain Management*, 8(2), 371–378.

[78] Agyabeng-Mensah, Y., Afum, E., Acquah, I. S. K., Dacosta, E., Baah, C., & Ahenkorah, E. (2021). The role of green logistics management practices, supply chain traceability and logistics ecocentricity in sustainability performance. *The International Journal of Logistics Management*, 32(2), 538–566.

[79] Sahoo, S., & Vijayvargy, L. (2020). Green supply chain management practices and its impact on organizational performance: Evidence from Indian manufacturers. *Journal of Manufacturing Technology Management*, 32(4), 862–886.

[80] Jazairy, A., & von Haartman, R. (2020). Analysing the institutional pressures on shippers and logistics service providers to implement green supply chain management practices. *International Journal of Logistics Research and Applications*, 23(1), 44–84.

[81] Acquah, I. S. K., Agyabeng-Mensah, Y., & Afum, E. (2020). Examining the link among green human resource management practices, green supply chain management practices and performance. *Benchmarking: An International Journal*, 28(1), 267–290.

[82] Farida, N., Handayani, N. U., & Wibowo, M. A. (2019, August). Developing indicators of green construction of green supply chain management in construction industry: A literature review. In *IOP Conference Series: Materials Science and Engineering* (Vol. 598, No. 1, p. 012021). IOP Publishing.

[83] Wibowo, M. A., Handayani, N. U., Farida, N., & Nurdiana, A. (2019). Developing indicators of green initiation and green design of green supply chain management in construction industry. In *E3S Web of Conferences* (Vol. 115, p. 02006). EDP Sciences.

[84] Meixell, M. J., & Luoma, P. (2015). Stakeholder pressure in sustainable supply chain management: A systematic review. *International Journal of Physical Distribution & Logistics Management*, 45(1/2), 69–89.

[85] Wee, H. M., Lee, M. C., Jonas, C. P., & Wang, C. E. (2011). Optimal replenishment policy for a deteriorating green product: Life cycle costing analysis. *International Journal of Production Economics*, 133(2), 603–611.

[86] Lemke, F., & Luzio, J. P. P. (2014). Exploring green consumers' mind-set toward green product design and life cycle assessment: The case of skeptical Brazilian and Portuguese green consumers. *Journal of Industrial Ecology*, 18(5), 619–630.

[87] Özkan, O., Akyürek, Ç. E., & Toygar, Ş. A. (2016). Green supply chain method in healthcare institutions. In *Chaos, Complexity and Leadership 2014* (pp. 285–293). Cham: Springer.

[88] Campos, L. M., & Vazquez-Brust, D. A. (2016). Lean and green synergies in supply chain management. *Supply Chain Management: An International Journal*, 21(5), 627–641.

[89] McCarthy, L., & Marshall, D. (2015). How does it pay to be green and good? The impact of environmental and social supply chain practices on operational and competitive outcomes. In *New Perspectives on Corporate Social Responsibility* (pp. 341–370). Wiesbaden: Springer Gabler.

[90] Darlington, Y., & Scott, D. (2020). *Qualitative research in practice stories from the field*. Routledge.

[91] Bryan, N., & Srinivasan, M. M. (2014). Real-time order tracking for supply systems with multiple transportation stages. *European Journal of Operational Research*, 236(2), 548–560.

[92] Chen, S. W., Gu, X. W., Wang, J. J., & Zhu, H. S. (2021). AIoT Used for COVID-19 Pandemic Prevention and Control. Contrast Media & Molecular Imaging, 2021.

[93] Tapia-Ubeda, F. J., Muga, J. A. I., & Polanco-Lahoz, D. A. (2021). Greening Factor Framework Integrating Sustainability, Green Supply Chain Management, and Circular Economy: The Chilean Case. *Sustainability*, 13(24), 1–33.

6 Effects on Supply Chain Management Due to COVID-19

Shipra Gupta
School of Management, Graphic Era Hill University Dehradun, Uttarakhand, India

Vijay Kumar
Physics Department, Graphic Era Hill University Dehradun, Uttarakhand, India

CONTENTS

6.1 INTRODUCTION

There are various sectors and countries that have witnessed the growth and economic development of the pandemic. Supply chain management is also affected by this pandemic. Generally, supply chain management means to cover all the activities from start to finish such as raw material, supply of goods, data collection, finance, and processing, etc. Supply Chain Management (SCM) covers all the activities from manufacturing of goods, satisfying its demand to supply, managing all the flowing processes to till delivery of the produced goods. The executives of the SCM are directly connected with the manufacturer to consumers. The main purpose of the SCM is to achieve the targets of the organization efficiently and effectively so that the product will be received in high quantity with high quality at minimum cost after those goods have been supplied to customers. Generally, supply chain management and logistics both are used in the same terminology. But there is a significant difference in both the terms. Supply chain management has a wider scope as compared to logistics management. Logistics terms include inventory management (from arranging to supply of demands), warehousing management and transportation management. A successful supply chain operation involves integrated and collaborative work with its channel partners as suppliers, retailers, wholesalers, customers, and distributors. During the complete process of production supply chain, managers also ensures that the waste should be minimum and maximum utilization of the resources at minimum

DOI: 10.1201/9781003264521-6

cost in an efficient manner. The managers keep track of demand, with the collaboration of suppliers and integrated services of goods and services by distributors and wholesalers. Finally, in the whole process, these managers performed effective communication with the collaboration of all the channel partners including customers (Bartoszko et al. (2020)).

During this pandemic time, everyone has to face many problems. Economy compacts, supply chain disruptions, delays in shipment, and many industrial challenges were faced during this pandemic timing. Due to this pandemic and to overcome this situation, the governments of many countries have decided to go through a shutdown. Due to this shutdown, a great variation has come in different products and their demands. Many products were in huge demand and out of stock. But many products were in less demand and buffered stock. This happened due to demand volatility and supply disruptions; demanded products had high prices (Brodin (2020)).

In this timing gateway, many barriers have been faced in the implementation of supply chain management in various industries and a lot of problems were faced by suppliers also. Short-term objectives and price-oriented approaches in the different sectors adversely affected the supply of quality products and relationships with customers (Chan et al. (2015)).

There are various sectors and countries that have witnessed the growth and economic development of the pandemic. Supply chain management is also affected by this pandemic. (Cheng et al. (2020)). Generally, supply chain management means to cover all the activities from start to finish such as raw material, supply of goods, data collection, finance, and processing, etc. Supply Chain Management covers all the activities from manufacturing of goods, satisfying by its demand to supply, managing all the flowing processes to till delivery of the produced goods. The executives of the SCM are directly connected with the manufacturer to consumers (Lu et al. (2020)). Generally, supply chain management and logistics both terms are used in the same terminology. But there is a significant difference in both the terms. Supply chain management has a wider scope compared to logistics management. Logistics terms include inventory management (from arranging to supply of demands), warehousing management and transportation management. The main purpose of the SCM is to achieve the targets of the organization efficiently and effectively so that the product will be received in high quantity with high quality at minimum cost. A successful supply chain operation involves integrated and collaborative work with its channel partners as suppliers, retailers, wholesalers, customers, and distributors (Lai et al. (2020)).

Supply chain management is an immense and productive field. Right now, in the present competitive scenario, those companies not investing in supply chain management, in the end, pay a large amount for this Fu et al. (2020).

Unusual consumer behavior during the pandemic cast unforeseen challenges for all businesses. The majority of customers from different strata of society were stressed about this, not just businesses. Some consumers were hoarding eatables and other necessities, while some of the lower classes were not making their ends meet because of the loss of employment. Some vendors had a good infrastructure for maintaining inventory, buffer, and transportation facilities, while some could not cope with this worldwide panic. (Dona et al. (2020), Wang et al. (2020), Xia et al. (2020), and Yeo et al. (2020)). E-commerce has comparatively less accurate and more virtual touch points that minimize contact and help reduce contagion in such conditions. Hence, e-commerce can help cover up the havoc caused to the supply chain. Since 2020, the number of customers to businesses has declined due to the pandemic. Existing customers get the same materials at higher prices because businesses are getting their raw materials at higher rates (Vander (2007)). They are also expected to be compliant with the covid protocols, requiring time, money, and effort. This also has had a substantial increment in their costs. Resilience is the key to a more stable supply chain (Gidengil et al. (2012)). Because the international borders remained shut for a maximum period, many companies preferred localization over globalization. The service (tertiary) sector has emerged more than the primary or secondary sector. Many brands now realize that what is a product can be turned into a service. Customers in India are also supported with it being convenient for them due to the decreasing cash on hand (Graham (2015), and Oboho et al. (2015)).

Ripple effects have been observed in the supply chain more frequently, which only added to the uncertainty. Hence, the previously projected patterns also stand disrupted Yvonne et al. (2016). The methodology of just-in-time for lean supply chain has returned. Omnichannel retailing has been proven to make the trade shorter by reducing the time and contact since the customer and vendor interact online to some extent as a part of the sale and purchase process. However, some contact remains existent, which can be minimized following the protocols (Habibzadeh et al. (2020)).

The chapter inspects the effect of the coronavirus pandemic on global exchange and distinguishes web-based business and m-trade as its worldwide patterns. A logical and strategic methodology to the investigation of the vector and particulars of the advancement of web-based business and m-trade is framed, given the absence of important factual data to survey the condition of web-based business in a nation and etymological elements of definitions (Jeng (2020)). The impact of web-based business and m-trade on the web shopping and deals industry has been set up. The worldwide circulation of web sources utilized by buyers, worldwide deals in m-business as a level of web-based business, and the effect of the COVID pandemic on Internet-based traffic and changes in exchanges by industry are examined (Kim et al. (2016)). The change of web-based business to m-trade was uncovered because of the expansion in the portion of portable exchanges taking into account the advancement of business sectors in nations zeroed in fundamentally on versatile gadgets and their dynamic use (Lee et al. (2020), Zhu et al. (2020), Zimmerman and Curtis (2020), and Zhang et al. (2020)). Because of the post-pandemic figure in the patterns of global exchange and e-business needs, a promising expansion in world deals in web-based business has been set up. The principal patterns of additional advancement of web-based business and m-trade in the field of global exchange are recognized, for example, huge information, personalization, email showcasing, change of e-business to web-based business, requesting administrations on the web and change of retail to on the web, electronic public acquirement, omnichannel and multichannel, and socially arranged business further developed work with the local area and the requirement for proficient coordinated operations (Lu et al. (2020)).

Nobody questions that life after the pandemic will at this point not be equivalent to what it was previously, because the entire world is living by the new guidelines for the subsequent year in succession. During the coronavirus pandemic, nearly all business processes in the field of global exchange were impacted, specifically in the business housing markets all over the planet (Liu et al. (2020), Sun et al. (2020), and Shen et al. (2020)).

6.1.1 Factors Affecting Supply Chain Management

According to the feedback from various businesspeople, entrepreneurs, and start-up people, the factors affecting supply chain management (Su et al. (2020), Velavan and Meyer (2020), Van et al. (2020), and Stower (2020)) are as follows:

1. Deficiency in the supply of goods
2. Lack of finance
3. High prices
4. Lack of import-export
5. Lack of raw material
6. High unemployment
7. Restrictions in transportation
8. Insufficient medical facilities
9. Lack of medical equipment and technology(pharmacy)

10. Bad impact on property dealing businesses, electrical and electronics items,
11. High demand for grocery items.
12. High demand for Wi-Fi connections
13. High demand for electronic gadgets (mobile, laptop, digital pad, iPad, iPhone, smart-phone, earphone, headphones etc.)
14. Less demand for communication services center
15. Issues:
    a. Health issues
    b. Eye side issues
    c. Backache pain issues
    d. Marital disputes issues
    e. Mental pressure
    f. Increase in the cases of divorce
16. Low-capacity productivity
17. Decrease in the quality of product
18. Decrease in GDP rate
19. Decrease in the revenue created by tourism
20. Lack of ignorance of online learning technology

6.1.2 Challenges after Pandemic

The following challenges were faced post-pandemic:

- After the pandemic, the demand and supply of the products were continuously fluctuating.
- Prices of the raw materials were also fluctuating day by day.
- Transportation has also become delayed.
- The manufacturing capacity and capability of the product become reduced.
- The risk of manufacturing the products by the manufacturers was increased.
- There was a lacking of disruptions plans.
- Contingent plans were not available.
- In maximum products, there was a lack of profit.
- Due to lack of labor, quantity in production was affected.

Supply chain management is an immense and productive field. Right now, in the present scenario, the market has become very competitive. Risk and cost management have many types of complexities.
Many companies failed due to many fundamental factors:

- The costs of the products were rising by the supply chain.
- After the pandemic, the customers demand an improvement in delivery rapidity and satisfaction.
- In the supply chain, there is an unpredictable risk such as market variation, business disputes, etc.

Objectives of the Chapter:

- To highlight the impact of COVID-19 on supply chain management in different categories such as businesspeople, small-scale industries, and new start-ups, and customers' satisfaction.
- To elaborate the affecting factors of supply chain management.
- To understand the challenges after pandemics and suggestions for improvement in supply chain management.
- To estimate the market scenario after pandemics.

Organization of the Chapter: The chapter is organized as: Section 6.2 elaborates the literature review, Section 6.3 expresses materials and methods. Section 6.4 shows observations of the study, Section 6.4 interprets the results and discussion, and Section 6.5 concludes the chapter with future scope.

6.2 LITERATURE REVIEW

Supply chain management plays an important role in economy. Various studies represent the impact of the coronavirus on supply chain management. From among these studies, a few are discussed given below:

Jabbour et al. (2011) measured a framework for supply chain management practices. Their aimed to validate and reliability testing on 107 Brazilian companies and to check the SCM practices implementation scale on collected data. Factor reliability testing has been applied for the measurement of data validity and adequacy.

Salami et al. (2016) conducted a questionnaire survey. For the implementation of a successful supply chain management, 21 barriers are mentioned in this manuscript and a Likert scale ranking has been used by the respondents. The survey had been done through questionnaires and face to face interviews with 107 Turkish construction contractors. Factor analysis had been implied in this study for the analysis of the data and after analysis seven problem-related factors are found. After the study of these finalized factors, it had been concluded that the main problem faced by small to medium size Turkish contractors was a lack of resources such as an expert consultant and hi-tech information technology. The findings of this study were very low suppliers' trust and demand from clients for contractors.

Russell et al. (2020) identified the scope of supply chain management. They also covered the increased uncertainties and factors in a dynamic market like changes in the strategies of supply chain and socio factors. As per the current scenario during the pandemic and after the pandemic, the effects of these factors were highly impacted global economic growth and development. A conceptual framework regarding different factors of logistics had been developed in this manuscript.

Ding et al. (2021) collected statistics from heterogeneous agencies sports activities and surgical. Out of 500 questionnaires, 262 had been filled by the top and middle-level managers from small- and medium-scale enterprises of Pakistan. The result suggested that technological elements had a widespread fantastic dating with B2B e-commerce in sports and surgical SMEs. But aggressive stress as an environmental aspect differed in sports activities and surgical SMEs to apply B2B e-commerce.

Ma et al. (2020) defined the feasibility, sustainability, resilience, and agility of supply chain management. Supply chain management had the viability ability to sustain and maintains itself in this changing and challenging era. Long-term impact cannot affect viable supply chain management as it may be re-plan and redesign its structure. This study also frame worked to the viable supply chain within the ecosystem. As per the conclusion of this study viable supply chain model would be very helpful for making decisions regarding the future and recovering from any pandemic situation. This model Ivanov (2020), had a multi-purpose structural design. It could be used from supply and demand establishment to control the transitions between mechanisms to model.

As per different decision factors, Hussain et al. (2021) identified that impact of the logistics port capacity and port services demand. It had been also cleared in this study that port capacity was not only a combination of sea, land, and platform side but also a combination of other different complicated components. Logistic port and capacity must be flexible; also suggested in this study. This flexibility might be different from different perspectives like ports, countries, types of freight and continents, etc. As per this study, it had been also suggested that various countries have different budget capacities. So, it would be better to minimize the cost at a flexible capacity. In this manuscript, the author identified the pandemic COVID-19 impacts on B2B e-commerce on elements of the technological, organizational, and environmental (TOE) model. The research framework changed into developed primarily based on the diffusion of innovation concept and the TOE version.

Hayakawa et al. (2021) expressed that the pandemic situation had changed the entire business activities scenario forever. Now all businesses used digital solutions for their business-to-business activities. During the pandemic, only such kinds of industries were survived those adapted their business to online mode. In most countries, all the industry's growth was badly affected by COVID-19. E-Commerce provides a crucial role for improvement in the business environment and overall performance of the business. In this study, the authors had also demonstrated that COVID-19 had severely affected all the sectors in its earlier phase. But there was no likeliness of this pandemic creating any further hurdles in the amelioration of the social economy as well as the e-commerce logistics. Logistics fee has become a vital issue for e-commerce enterprises influencing the customer experience and operational impact. Based on this, e-trade logistics fee management techniques have proposed for the prevention and management of COVID-19 surroundings. The logistics fee designed on the premise of the evaluation of the influencing elements of e-trade, and the cross-border logistics approach that conforms to the history of COVID-19 prevention and management. It was called to lessen the e-trade logistics fee than the running fee of logistics enterprises. The e-trade logistics fee management approach proposed was powerful in the context of prevention and management of any epidemic outbreak while the general fee was within the budgeted fee range. The outcomes of this study expressed the effectiveness of the e-business logistics fee management approach. With the help of this approach, many e-business agencies had taken the advantage of right monetary earnings. Because of the traits of e-trade enterprises, this study explored the fee management system of e-trade logistics, conducting a theoretical evaluation first, and afterwards making an in-intensity evaluation combined with realistic cases. The outcomes of this show that logistics costs were primarily based on product, an element that includes e-commerce income and maintenance. According to the outcomes on the connection between fee management, budgeting, and worthwhile accounting, it became glaring that existed a widespread and sturdy courting among fee management and e-trade enterprise. Place problem was not a major issue for e-commerce logistic organizations. The major issue for that was managing the fee management. The electrical enterprise's logistics fee was very low as compared to trade enterprises on logistics. Most logistics enterprises were converted into digital trade organisations and the result of this type of trading was becoming fruitful in powerful fee management.

Cross-border selling had brought social and monetary improvement to the United States, which had a vital price and function in fortifying the management and control of logistics fees. E-trade logistics was an important role in the content, material, power distribution, and evaluation of warehouses in other countries. With the help of e-trade logistics, import and export trade could be easily done in an effective cost and performance manner Bai (2020).

Li et al. (2020) defined the movements of clients in the direction of a commodity and in logistics. It was needed to map the affiliation and accounting to bridge the space in fee control, stock valuation and budgetary management. The authors targeted to empirically check out the position of e-commerce on the exchange effects of COVID-19. The findings of this study summarized as follows: A large variety of confirmed instances of deaths in each exporting nation were significantly lower than the global exchange. However, the authors observed that e-commerce has continuous improved in uploading nations and contributed to mitigate the poor effect of COVID-19 on global exchange. With the chance of an extended COVID-19 pandemic, the viable emergence of recent infectious diseases, and the development of virtual transformation, e-commerce was probable to grow in significance in domestic and global transactions. Indeed, e-commerce could play a significant position in contributing to the boom of the worldwide economic system with the aid of using globalization through multiplied exchange. To gain a healthy boom in e-commerce, the status quo of rules-based unfastened, open, stable, and obvious surroundings for accomplishing e-commerce commercial enterprise was important.

While the guidelines on e-commerce had been built as part of many unfastened exchange agreements, which include the comprehensive and progressive Trans-Pacific Partnership settlement and the EU–Japan Economic Partnership Agreement, a global rule or settlement had now no longer been established. In December 2017, 71 WTO individuals declared that they could provoke exploratory paintings closer to destiny WTO negotiations on virtual exchange. Since then, numerous conferences have been held with extra individuals; however, negotiations have now no longer begun. A variety of demanding situations must be treated earlier than negotiations may even start. It was hoped that negotiations start and are concluded quickly so that global EC commercial enterprise would develop rapidly, contributing to the standard monetary boom (Kumar and Mishra (2020)).

Pathak et al. (2020) focused on the function of government. They played a significant function in the socio-financial transformation of society from the implication of the coronavirus pandemic. For the study, a web questionnaire was performed using a non-probability snowball sampling technique; the researcher had accrued responses from 100 respondents. Various components were covered in this questionnaire regarding socio-financial repute, boundaries in enhancing the profits stage, GDP stage, intake stage, and investment stage of the humans and whether or not the government is substantial in enhancing the requirements of the humans submit the pandemic. All classes of folks who were suffering from this pandemic surveyed through the use of Google paperwork and the volume of development after the pandemic of their social and financial repute was studied. A small enterprise had been made to recognize the pressure and tension of the humans in the course of this pandemic and additionally how the government would assist with the transformation of society Paul et al. (2021). This could be the time to reset. In this study, the authors wanted to recognize the worldwide effect of COVID-19 on supply chain control, a hard and fast questionnaire was designed which addressed four pertinent topics that specialize in product categories consisting of export products, import products, and optional product and emergency product. The questionnaire became centered on supply chain corporations and became disbursed online in businesses with the use of diverse social media platforms. The authors expressed in their study the hurdles generated by COVID-19 and its solutions. This pandemic had created problems in each era in the manufacturing sector, transportation, clothing and social distancing etc. All the firms wanted to remove this problem and formulate a proper technique to heal it. All the firms had to need their survival and supply chains. A study of Bangladeshi ready-made garments firms had been taken for the reference study.

Ivanov et al. (2020) findings could be helpful for decision makers in growing strategic regulations to triumph over the healing demanding situations in the post-COVID-19 era. The examiner furnished a complete set of healing demanding situations to be used in a whole lot of studies and making plans contexts. Further, it analysed the interrelationships among healing demanding situations to discover cause-impact businesses in of the demanding situations. The findings could assist decision makers set their priorities allocating sources inside the healing system to optimize the outcome.

Ivanov et al. (2021) intertwined supply network (ISN) as an entirety of interconnected supply chains (SC) which, of their integrity stable the supply of society and markets with items and offerings. Now intertwined supply chain networks have opened to the dynamic structural based structure. This network would work as buyer-provider connectivity with the help of supply chain networks. The intertwined supply chain networks assured society of a long period of survival. To outbreak of the pandemic had given this new perspective of intertwined network structure. The authors revealed theories a virtual supply chain twin – an automated version that had shown community states for any given second in actual time. COVID-19 has affected the supply chain, but to recover from the post-pandemic, the author desires virtual twins by mapping supply chain networks to make sure. The result of this observation made contributions to the studies and exercise of supply chain danger control through improving predictive and reactive selections to make use of the benefits of supply chain visualization, ancient disturbance statistics analysis, and actual-time disruption statistics and make sure cease-to-cease visibility and commercial enterprise continuity in worldwide companies. A mixture of version-pushed and statistics-pushed decision-making help had become a seen studies fashion inside the ultimate years. The exceptional version-primarily based decision-making help strongly relies upon statistics, its perfectness, entireness, validity, continuity, and well-timed availability. These statistics necessities were of unique significance in SC danger control for predicting disruptions and reacting to them. Industries 4.0 in the fashionable and virtual era particularly supply upward pushed to statistics analytics packages to obtain a brand new exceptional of decision-making help while coping with intense disruptions. The mixture of simulation, optimisation, and statistics analytics constitutes a virtual twin: a unique data-driven framework for coping with disruption dangers inside the SC.

Reardon et al. (2021) and Magableh (2021) explained the impact of COVID-19 on food industries in developing countries. Firms were belonging to e-commerce to procure customers and small trades and food businesses. Supply providers were associated with the food industries for assistance in supply. The unexpected pandemic had badly affected all the areas like innovations, e-trade businesses, shipping, etc. A theoretical framework of the organizational techniques was followed. Now all the food industries were following up with the well-developed countries' policies and infrastructural situations to overcome such type of pandemic scenario. Now companies were developing very fast monitoring techniques, innovative techniques and maintaining the relationship with mediators who would assist the food industries to adopt supply chains. The author explained the impact of pandemic COVID-19 coronavirus on the supply chain. This pandemic had affected all over the world, but especially in the supply chain sector. The effect was predicted to influence the groups indefinitely; hence, it was less likely for the supply chain to recover its pre-COVID-19 position.

Gupta et al. (2021) expressed the effect of the pandemic on the supply chain, its disturbances, related obstacles, and its trends. A study of different stages of the supply chain, its causes, strength, weakness, opportunities, and threats caused through the pandemic were also evaluated in this study. An analytical study of pandemic COVID-19 and factors to cope with this situation was also mentioned in this manuscript. The gap through identification and categorisation by framing the vital elements and their relationship was also framed in this study. The studies about the pandemic affected different sectors national as well as international in the supply chain were also studied in this manuscript. Equilibrium remained a distant dream. Most organizations pronounced SC disruptions out of their capacity due to the pandemic. It had proved tough for any planner to put together for crises, and that too for a crisis like the COVID-19 pandemic requires fast selections in a complex and tough environment, given SC expenses and quality.

Yin et al. (2021) explained that COVID-19 seriously broke the social and monetary value of the international supply chain. This unforeseeable pandemic acted as a backlash of the sustainable development goals which were to be executed via any way of means by 2030. This study of the disruption because of the COVID outbreak explored the function of feasible finance in attaining sustainability in the supply chain. It had focused on the essential elements which could be critical to evolving the supply chain sustainability. In this manuscript, a study of meals enterprise was done on the priority of the essential elements using the techniques. The main factors and sub-factors that play an important part in sustaining supply chain finance were also mentioned. This study also provided a primary role in sustainable long-run developments goals, supply chain systems and the way of financing. The findings of this study expressed that the importance or vital asset for any enterprise was social values. This study expressed the main standards to sustain the supply chain finance, finance function for attaining the social values of the enterprise harmlessly environmental values. The pandemic COVID-19 has badly affected the social values of the enterprises. The future of the supply chain of fresh agricultural products by the mathematics model applied in this study. The supply chain had improved the value of the products. During the duration of the pandemic, the supply chain of these products was improved. During this pandemic duration, the supply chain had provided a vital role to optimize the power supply of agricultural products. A three-stage supply chain model had been used in this manuscript. Goods providers (manufacturer), third-party logistics, and shops were three factors included in this model for the assurance of the supply of agricultural products. Coefficients of virus contamination and prevention efforts of this pandemic were also included in this model. The supply chain mechanism had especially focused on human health and the largest power supply of fresh agricultural products.

Tian et al. (2020) and Singhal (2020) expressed the gastrointestinal and other related symptoms in COVID-19 and its transmission. These symptoms were included anorexia, diarrhoea, vomiting, nausea, abdominal pain, and gastrointestinal bleeding. These symptoms were different as per the different age factor. Gastrointestinal symptoms were common seen in COVID-19 patients.

SARS-CoV-2 enters gastrointestinal epithelial cells, and the feces of COVID-19 patients are potentially infectious.

6.3 MATERIAL AND METHODS

6.3.1 MATERIALS

The stakeholders in this study are 450 businesspeople, 35 small-scale industrialists, and 60 start-ups. Out of given businesspeople, SSI, and start-ups, 379 businesspeople, 29 small-scale industrialists, and 57 start-ups have participated. In the given table, every level represents a 20% change in modification.

6.3.2 METHODS

For this study, a survey has been done to know about the real scenario of the market after the pandemic. After this pandemic, the ways and types of business have changed. Many industries and businesses have modified their policies. In this study, a questionnaire has been floated to the stakeholders by Google form. The factors of the questionnaire are given in the below tables.

In this chapter, the primary data of the survey from the stakeholders are taken and some secondary data are also taken for this study. This questionnaire has been filled out by stakeholders from all over India. A random sampling method has been used in this chapter.

The research design of this chapter is based on descriptive nature and random sampling method has been used in this study. The result of this study is calculated on the percentile based.

6.3.2.1 Suggestions

Supply chain jobs extend through an immense canvas. Thus, anyone can choose job profiles either locally or regionally within the country. And if you love travel-oriented jobs, many supply chain managerial roles require traveling both nationally and internationally from time to time.

As more and more companies realize the true potential of a well-articulated supply chain management system, the demand for skilled and certified supply chain professionals will increase simultaneously (Tables 6.1–6.4).

TABLE 6.1

A Survey on the Various Businesspeople's Impact on the Pandemic

Survey for Business-People (379)		No./% of Category 1 (street seller)	No./% of Category 2 (Retailer shopkeeper)	No./% of Category 3 (Whole seller shopkeeper)	No./% of Category 4 (District level distributer)	No./% of Category 5 (State level distributer)
BSM01	Supply of goods	197/57	68/17.9	48/12.67	37/9.7	29/7.65
BSM02	Transport facilities	126/33.2	47/12.4	54/14.24	65/17.15	87/22.95
BSM03	Long term relationship	37/9.7	74/19.5	63/16.6	83/21.9	122/32.2
BSM04	Improvement in quality	158/41.7	58/15.3	49/12.9	74/19.5	45/11.8
BSM05	Planning and goal setting	46/12.1	57/15	75/19.8	6717.7	134/35.3
BSM06	New product development	64/16.9	57/15	74/19.5	37/9.7	147/38.8
BSM07	Continuously solving problem process	58/15.3	76/20	84/22.1	58/15.3	103/27.1

TABLE 6.2

A Survey for Customer Satisfaction with the Product and Services after the Pandemic

Survey from Customer		No./% of Category 1 (street seller)	No./% of Category 2 (Retailer shopkeeper)	No./% of Category 3 (Whole seller shopkeeper)	No./% of Category 4 (District level distributer)	No./% of Category 5 (State level distributer)
CM01	Customer satisfaction	58/15.3	76/20	39/10.29	98/25.85	108/28.49
CM02	Frequently follow up with the customer	67/17.67	87/23	69/18.2	96/25.3	60/15.8
CM03	Frequently interaction with the customer	39/10.3	48/12.7	56/14.8	179/47.2	57/15
CM05	Future relation with the customer	136/35.9	68/17.9	77/20.3	29/7.6	69/18.2
CM06	Customer care unit	56/14.7	74/19.5	85/22.4	74/19.5	90/23.7
CM07	Feedback from customer	47/12.4	74/19.5	54/14.2	93/24.5	201/53

TABLE 6.3

A Survey of Small-Scale Industrialists after the Pandemic

Survey for Small-Scale Industriali-sts (29)		No./% of (Category 1) Annual turnover less than 10 Cr.	No./% of (Category 2) Annual turnover from 10 to 20 Cr.	No./% of (Category 3) Annual turnover from 20 to 30 Cr.	No./% of (Category 4) Annual turnover from 30 to 40 Cr.	No./% of (Category 5) Annual turnover from 40 to 50 Cr.
SSI01	Improvement in industry as per requirement of pandemic	9/31	6/20.7	7/24	3/10.3	4/13.8

TABLE 6.3 (Continued)
A Survey of Small-Scale Industrialists after the Pandemic

Survey for Small-Scale Industriali-sts (29)		No./% of (Category 1) Annual turnover less than 10 Cr.	No./% of (Category 2) Annual turnover from 10 to 20 Cr.	No./% of (Category 3) Annual turnover from 20 to 30 Cr.	No./% of (Category 4) Annual turnover from 30 to 40 Cr.	No./% of (Category 5) Annual turnover from 40 to 50 Cr.
SSI02	Start new products as per requirement of pandemic	5/17.2	4/13.8	3/10.3	6/20.7	11/37.9
SSI03	Support to the customer in pandemic	5/17.2	3/10	6/20.7	9/31	6/20.7
SSI04	Work for society welfare after pandemic	8/27.6	5/17.2	7/24.1	5/17.2	4/13.8
SSI05	Loss due to pandemic	8/27.5	7/24	9/31	3/10.3	2/6.8
SSI06	Profit due to pandemic	4/13.7	6/20.7	6/20.7	5/17.2	8/27.5
SSI07	Manufacturing delay due to pandemic	7/24	5/17.2	6/20.7	8/27.5	3/10.3
SSI08	Shortage of raw materials	3/10.3	8/27.6	5/17.2	7/24.1	6/20.7
SSI09	Lack of skilled workers	5/17.2	7/24	4/13.7	9/31	4/13.7

TABLE 6.4
A Survey of New Start-Ups Entrepreneurs after a Pandemic Occasion

Survey for Start-Ups (57)		No./% of (Category 1) Less than 5 Cr.	No./% of (Category 2) 5 to 10 Cr.	No./% of (Category 3) 10 to 15 Cr.	No./% of (Category 4) 15 to 20 Cr.	No./% of (Category 5) 20 to 25 Cr.
EN01	Lack of Fund	11/19.3	9/15.7	11/19.3	14/24.5	12/21
EN02	Lacking of demand of products	17/29.8	21/36.8	8/14	19/33	8/14
EN03	Lack of receiving of payment	19/33	18/31.5	10/17.5	6/10.5	4/7
EN04	Lacking of guidance from other agencies	8/14	11/19.3	7/12.3	13/22.8	18/31.6
EN05	Inspiration from the government organization for the continuous supply chain of the products	11/19.3	9/15.7	9/15.7	12/21	16/28

6.4 RESULTS AND DISCUSSION

For this study, 379 business people have given their feedback. This study is divided into five categories. Category 1 relates to the businesspeople who are street sellers. Category 2 contains the businesspeople who are selling in retail and situated in each area of the city. Wholesale businesspeople are in category 3. Category 4 contains district-level businesspeople and category 5 belongs to state-level businesspeople. In the survey of effects on the supply of goods on various types of businesspeople, categories 1 to 5 are affected by 57%, 17.9%, 12.67%, 9.7%, and 7.65%, respectively. During and after the pandemic, businesspeople also faced transport facilities due to a lack of labor. 33.2%, 12.4%, 14.24%, 17.15%, 22.95% businesspeople from category 1 to 5 were facing the problem of transportation. After the pandemic, long-term relationships with the customer were affected by 9.7%, 19.5%, 16.6, 21.9%, and 32.2% by categories 1 to 5, respectively. In the question of quality improvement, 41.7% of street sellers answered that they have improved the quality of their product, 15.3% retail shopkeepers, 12.9% whole seller shopkeepers, 19.5% district level distributors, and 11.8% state-level distributer improved in quality of the product. When the survey has done on planning and goal setting, 12.1% of street sellers, 15% of retail shopkeepers, 19.8% of whole seller shopkeepers, 17.7% of district-level distributors, and 35.3% of state-level have answered that after the pandemic, they have targeted some goal of sale of their product; 16.9% street sellers, 15% retail shopkeepers, 19.5% whole seller shopkeepers, 9.7% district level distributors, and 38.8% distributers state level have answered that after the pandemic, they have launched some new products. One question of the survey is regarding the continuously solving problem process. 16.9% street sellers, 15% retail shopkeepers, 19.5% whole seller shopkeepers, 9.7% district level distributors, and 38.8% distributers state level have answered that after the pandemic, they have continuously watched and solved the problems of the customer which also help in support of continuous supply chain of the products.

A survey is also done which is related to the services of customers. In this survey, again 379 businesspeople participated and gave their views. When it is asked from the street seller that they are taken care of the customer satisfaction, 15.3% of street sellers, 20% of the retail shopkeeper, 10.29% whole seller shopkeeper, 25.85% district level distributor, and 28.49% of distributers state-level reply about the customer satisfaction. The regular follow-up from the customer is taken by 17.67% of street sellers, 23% of retail shopkeepers, 18.2% of whole seller shopkeepers, 25.3% district-level distributors, and 15.8% distributers state level. In general, businesspeople are not in contact with their customers; 10.3% street sellers, 12.7% retail shopkeepers, 14.8% whole seller shopkeepers, 47.2% district-level distributors, and 15% distributers state level are continuously in contact with the customers and 35.9% of street sellers, 17.9% of retail shopkeepers, 20.3% of whole seller shopkeepers, 7.6% of district-level distributors, and 18.2% of distributers' state level are doing something for maintaining the future relationship with the customer. Customer care units are established by 14.7% street sellers, 19.5% retail shopkeepers, 22.4% whole seller shopkeepers, 19.5% district-level distributors, and 23.7% distributers state-level businesspeople. Feedback from the customer is taken from 12.4% of street sellers, 19.5% of retail shopkeepers, 14.2% of whole seller shopkeepers, 24.5% of district-level distributors, and 53% of state-level distributors.

The industrialists who have an annual turnover of up to 50 Crore will come in the series of small-scale industrialists (SSI). The survey has been done by SSI about improvement in their industry as per the requirement of the pandemic. In this survey, 29 small-scale industrialists have participated. It is observed that 31% from category 1, 20.7% from category 2, 24% from category 3, 10.3% from category 4, and 13.8% from category 5, SSI are improving their industry as per the requirement after the pandemic. However, pandemics gave bad news and results but for some people and industrialists, it gave the chance to do something different and innovative. The survey from SSI regarding starting new products as per the requirement of the pandemic gave the result as 17.2% from category 1, 13.8% from category 2, 10.3% from category 3, 20.7% from category 4, and 37.9% from category 5; 17.2% from category 1, 10% from category 2, 20.7% category 3, 31% category 4, and 20.7% category 5 SSI have supported to the customer during the pandemic. Some SSIs have also lost economically during the pandemic. It is observed that 27.5% from category 1, 24% from category 2, 31% from category 3, 10.3% from category 4, and 6.8% from category 5 SSI have economically lost during the pandemic. But many small-scale industrialists earned profit during the pandemic. Survey has shown that 13.7% from category 1, 20.7% from category 2, 20.7% from category 3, 17.2% from category 4, and 27.5% from category 5 SSI earned extra profit during the pandemic. Due to a shortage in all types of labor, SSI faced manufacturing delays due to pandemics; 24% from category 1, 17.2% from category 2, 20.7% category 3, 27.5% category 4, and 10.3% category 5 SSI faced the problem of delay in manufacturing. During the pandemic, transportation is also affected. The movement of trucks and other transport vehicles become low in number on the roads; 10.3% from category 1, 27.6% from category 2, 17.2% from category 3, 24.1% from category 4, and 20.7% from category 5 faced a problem of lacking raw materials due to transportation. During the pandemic, the labor returned to their hometowns. It became the reason for lacking skilled labor in the factories and industries. The survey about this shows that 17.2% from category 1, 24% from category 2, 13.7% from category 3, 31% from category 4, and 13.7% from category 5 SSI were facing the problem of skilled labor.

A survey of new start-ups entrepreneurs is also done after the pandemic. In this survey, 57 entrepreneurs participated. One survey has been done for the lacking of funds. The entrepreneurs have been classified into five categories: 19.3% from category 1, 15.7% from category 2, 19.3% from category 3, 24.5% from category 4, and 21% from category 5 accepted the lacking of funds during the pandemic. At the time of the pandemic, the markets were closed. The people were not moving in the market. The people required only the raw materials which are required for food preparation: 29.8% from category 1, 36.8% from category 2, 14% from category 3, 33% from category 4, and 14% from category 5 entrepreneurs were facing the problem of lack of demand of for products. As the market was closed, there was no sale done by the shopkeepers. The entrepreneurs were also facing the problems of payments from the shopkeepers. As per the survey, 33% from category 1, 31.5% from category 2, 17.5% from category 3, 10.5% from category 4, and 7% of category 5 entrepreneurs were facing the problem of receiving payments from the shopkeepers. During this challenging period, some guidance was given by the other agencies: 14% from category 1, 19.3% from category 2, 12.3% from category 3, 22.8% from category 4, and 31.6% from category 5 entrepreneurs have faced a lack of guidance from external agencies. The government of all the countries was continuously inspiring their entrepreneurs for good production and continuous supply chain of the products. As per the entrepreneur's view, 19.3% from category 1, 15.7% from category 2, 15.7% from category 3, 21% from category 4, and 28% from category 5 were feeling government support and guide for the continuous supply chain of the products.

As per one case study of Raj et al. (2022), during and after the pandemic, the supply chain of all goods was affected in many ways. The COVID-19 epidemic has wreaked havoc on global supply systems at a scale never seen before. The supply chain issues that manufacturing businesses have faced as a result of the COVID-19 pandemic, particularly in emerging markets, was examined. We provide a conceptual framework for analyzing issues and their related mitigation measures based on dynamic capacity theory. Based on a literature review, evaluation of many news stories, and consultations with experts, ten key challenges have been identified. The Grey-Decision-making Trial and Evaluation Laboratory (Grey-DEMATEL) method is also used to investigate the connections between diverse supply chain issues. Scarcity of labor is the major difficulty, followed by scarcity of materials.

6.5   CONCLUSION AND FUTURE SCOPE

Supply chain management has a wider and more significant scope. This field has many opportunities for research and experimentation work. Now in this competitive era, every sector has followed the supply chain management for the betterment and enhances their business. The pandemic timing had been stated in the world since 2019. The coronavirus had spread very rapidly from one country to another country. The pandemic period has affected all types of businesspeople, small scale industrialists and entrepreneurs. The after-effects of the pandemic in supply chain management are studied through a survey. After analyzing the results of the survey, it is concluded that businesspeople, small-scale industrialists, and entrepreneurs were affected due to the pandemic. The results of the survey concluded that the street businesspeople were affected by the supply of goods, supply transportation and quality improvement. The state-level distributors were suffering the problems of long-term relationships, planning and goal setting, new product development, and continuous solving problem process. During and after the pandemic, supply chains of the goods were generally affected. Either the street businesspeople or state-level whole-seller businesspeople were taking care of customer satisfaction: frequent follow-up with customers, frequent interacton, customer care, and feedback from the customers. As per the survey of small-scale industrialists, the improvements in industry as per the requirement of the pandemic were affected for those who had an annual turnover of fewer than 10 cr. SSI, which has a turnover of 40 to 50 cr., were facing the problem to start new product development. The SSI, who have less annual turnover, been unable to do work for social welfare. The entrepreneurs, who were middle income, were facing the problems of lack of funds, lack of demand for products, and lack of inspiration from the government organization for the continuous supply chain of the products.

Now after pandemic, India and all the other countries are ready to cope with any typos of hurdles. Now India has continuously improved in all the aspects. But if the supply chain may be more focused by the government, then the above such type of problems may be ignored or recovered.

Future Scope: In the future, this study will be very fruitful to know the market supply chain status in different aspects and these may be improved. With the help of this study, everyone will be aware of the problems faced by various forms of businesses and, by improving supply chain management, these problems may be removed.

REFERENCES

Bai, H. M. (2020). The socio-economic implications of the coronavirus pandemic (COVID 19): A review. *ComFin Research*, 8(4), 8–17. doi: 10.34293/commerce.v8i4.3293

Bartoszko, J. J., Farooqi, M. A. M., Alhazzani, W., & Loeb, M. (2020). Medical masks vs N95 respirators for preventing COVID-19 in health care workers: A systematic review and meta-analysis of randomized trials. *Influenza and Other Respiratory Viruses*, 14, 365–373.

Brodin, P. (2020). Why is COVID-19 so mild in children? *Acta Paediatrica*, 109(6), 1082–1083. doi: 10.1111/apa.15271

Chan, J. F. W., Lau, S. K. P., To, K. K. W., Cheng, V. C. C., Woo, P. C. Y., & Yuen, K. Y. (2015). Middle East respiratory syndrome coronavirus: Another zoonotic beta coronavirus causing SARS-like disease. *Clinical Microbiology Reviews*, 28(2), 465–522.

Cheng, Z. J., & Shan, J. (2020). 2019 Novel coronavirus: Where we are and what we know. *Infection*, 48(2), 155–163. 10.1007/s15010-020-01401-y

Ding, Q., & Zhao, H. (2021). Study on e-commerce logistics cost control methods in the context of COVID-19 prevention and control. *Soft Computing*, 25(2), 11955–11963. 10.1007/s00500-021-05624-5

Dona, D., Minotti, C., Costenaro, P., Da Dalt, L., & Giaquinto, C. (2020). Fecal–oral transmission of Sars-Cov-2 in children: Is it time to change our approach. *Pediatric Infectious Disease Journal*, 39, 567–571.

Fu, Y., Cheng, Y., & Wu, Y. (2020). Understanding SARS-CoV-2-mediated inflammatory responses: From mechanisms to potential therapeutic tools. *Virol Sendhanidham*, 35(3), 266–271. doi: 10.1007/s12250-020-00207-4

Gidengil, C., Parker, A., & Zikmund-Fisher, B. (2012). Trends in risk perceptions and vaccination intentions: A longitudinal study of the first year of the H1N1 pandemic. *American Journal of Public Health*, 102, 672–679.

Graham, R. L., Donaldson, E. F., & Baric, R. S. (2015). A decade after SARS: Strategies for controlling emerging coronaviruses. *Natur Review Microbiology*, 11, 836–848.

Gupta, N., & Soni, G. (2021). A decision-making framework for sustainable supply ChainFinance in Post-COVID era. *International Journal of Global Business and Competitiveness*, 16(1), 29–38. 10.1007/s42943-021-00028-6

Habibzadeh, P., & Stoneman, E. K. (2020). The novel coronavirus: A bird's eye view. *International Journal of Occupational & Environmental Medicine*, 11(2), 65–71.

Hayakawa, K., Mukunoki, H., & Urata, S. (2021). Can e-commerce mitigate the negative impact of COVID-19 on international trade? *The Japanese Economic Review*, 21(1), 1–18. 10.1007/s42973-021-00099-3

Hussain, A., Shahzad, A., Hassan, R., & Doski, S. A. M. (2021). COVID-19 impact on B2B E-commerce: A multigroup analysis of sports and surgic. *Pakistan Journal of Commerce and Social Sciences*, 15(1), 166–195.

Ivanov, D. (2020). Viable supply chain model: Integrating agility, resilience and Sustainability perspectives—lessons from and thinking beyond the COVID-19 pandemic. *Annals of Operations Research*, 1–21. 10.1007/s10479-020-03640-6

Ivanov, D., & Dolgui, A. (2020). Viability of intertwined supply networks: Extending the supply chain resilience angles towards survivability. A position paper motivated by COVID-19 outbreak. *International Journal of Production Research*, 58(10), 2904–2915, doi: 10.1080/00207543.2020. 1750727

Ivanov, D., & Dolgui, A. (2021). A digital supply chain twin for managing the disruption risks and resilience in the era of Industry 4.0. *Production Planning & Control*, 32(9), 775–788. doi: 10.1080/09537287. 2020.176845

Jabbour, A.B.L. de Sousa, Filho, A. G. A., Viana, A. B. N., & Jabbour, C. J. C. (2011). Measuring supply chain management practices. *Journal Measuring Business Excellence*, 15(2), 18–31. doi: 10.1108/13 683041111131592

Jeng, M. J. (2020). COVID-19 in children: Current status. *Journal of the Chinese Medical Association*, 83(6), 527–533.

Kim, Y., Cheon, S., & Min, C. (2016). Spread of mutant middle east respiratory syndrome coronavirus with reduced affinity to human CD26 during the South Korean outbreak. *mBio*, 7, 236–241.

Kumar, R., & Mishra, R. S. (2020). COVID-19 global pandemic: Impact on management of supply chain. *International Journal of Emerging Technology and Advanced Engineering*, 10(4), 132–139.

Lai, C. C., Shih, T. P., Ko, W. C., Tang, H. J., & Hsueh, P. R. (2020). Severe acute respiratory syndrome coronavirus 2 (SARS-CoV-2) and coronavirus disease-2019 (COVID-19): The epidemic and the challenges. *International Journal of Antimicrobial Agents*, 55, 105924–105930.

Lee, P. I., Hu, Y. L., Chen, P. Y., Huang, Y. C., & Hsueh, P. R. (2020). Are children less susceptible to COVID-19? *Journal of Microbiology, Immunology and Infection*, 53(3), 371–372. 10.1016/j.jmii. 2020.02.011

Li, X., Xu, W., Dozier, M., He, Y., Kirolos, A., & Theodoratou, E. (2020). The role of children in transmission of SARS-CoV-2: A rapid review. *Journal of Global Health*, 10(1), 234–241. doi: 10.7189/ jogh.10.011101

Liu, W., Zhang, Q., Chen, J., Xiang, R., Song, H., Shu, S., Chen, L., Liang, L., Zhou, J., You, L., Wu, P., Zhang, B., Lu, Y., Xia, L., Huang, L., Yang, Y., Liu, F., Semple, M. G., Cowling, B. J., Lan, K., Sun, Z., Yu, H., & Liu, Y. (2020). Detection of Covid-19 in children in early January 2020 in Wuhan, China. *New England Journal of Medicine*, 382(14), 1370–1371. doi: 10.1056/NEJMc2003717

Lu, R., Zhao, X., Li, J., Niu, P., Yang, B., Wu, H., Wang, W., Song, H., Huang, B., Zhu, N., Bi,Y., Ma, X., Zhan, F., Wang, L., Hu, T., Zhou, H., Hu, Z., Zhou, W., Zhou, L., Chen, J., Meng, Y., Wang, J., Lin, Y., Yuan, J., Xie, Z., Ma, J., Liu, W. J., Wang, D., Xu, W., Holmes, E. C., Gao, G. F., Wu, G., Chen, W., Shi, W., & Tan, W. (2020). Genomic characterization and epidemiology of 2019 novel coronavirus: Implications for virus origins and receptor binding. *Lancet*, 395(10224), 565–574. doi: 10.1016/S0140-6736(20)30251-8

Lu, X., Zhang, L., Du, H., Zhang, J., Li, Y. Y., Qu, J., Zhang, W., Wang, Y., Bao, S., Li, Y., Wu, C., Liu, H., Liu, D., Shao, J., Peng, X., Yang, Y., Liu, Z., Xiang, Y., Zhang, F., Silva, R. M., Pinkerton, K. E., Shen, K., Xiao, H., Xu, S., & Wong, G. W. K. (2020). SARS-CoV-2 infection in children. *New England Journal of Medicine*, 382(17), 1663–1665. doi: 10.1056/NEJMc2005073

Ma, X., Su, L., Zhang, Y., Zhang, X., Gai, Z., & Zhang; Z. (2020). Do children need a longtime to shed SARS-CoV-2 in stool than adults. *Journal of Microbiology, Immunology and Infection*, 53(3), 373–376. doi: 10.1016/j.jmii.2020.03.010

Magableh, G. M. (2021). Supply chains and the COVID-19 pandemic: A comprehensive framework. *European Management Review*, 18, 363–382. doi: 10.1111/emre.12449

Oboho, I. K., Tomczyk, S. M., & Al-Asmari, A. M. (2015). MERS-CoV outbreak in Jeddah—A link to health care facilities. *New England Journal of Medicine*, 372, 846–854.

Pathak, E. B., Salemi, J. L., Sobers, N., Menard, J., & Hambleton, I. R. (2020). COVID-19 in children in the United States: Intensive care admissions, estimated total infected, and projected numbers of severe pediatric cases in 2020. *Journal of Public Health Management and Practice*, 26(4), 325–333. doi: 10.1097/PHH.0000000000001190

Paul, S. K., Chowdhury, P., Moktadir, M. A., & Lau, K. H. (2021). Supply chain recovery challenges in the wake of COVID-19 pandemic. *Journal of Business Research*, 136(56), 316–329. 10.1016/j.jbusres.2021.07.056

Raj, A., Mukherjee, A. A., Jabbour, A. B. L., & Srivastava, S. K. (2022). Supply chain management during and post-COVID-19 pandemic: Mitigation strategies and practical lessons learned. *Journal of Business Research*, 142, 1125–1139. doi: 10.1016/j.jbusres.2022.01.037

Reardon, T., Heiman, A., Lu, L., Nuthalapati, C. S. R., Vos, R., & Zilberman, D. (2021). Pivoting by food industry firms to cope with COVID-19 in developing regions: E-commerce and co pivoting delivery intermediaries. *Agricultural Economics*, 52, 459–475. doi: 0.1111/agec.12631

Russell, C., Millar, J., & Baillie, J. (2020). Clinical evidence does not support corticosteroid treatment for 2019-nCoV lung injury. *Lancet*, 395(10223), 473–475.

Russell, D., Ruamsook, K., & Roso, V. (2020). Managing supply chain uncertainty by building flexibility in container port capacity: A logistics triad perspective and the COVID-19 case. *Maritime Economics & Logistics*, 24, 1–22. 10.1057/s41278-020-00168-1

Salami, E., Aydinli, S., & Oral, E. L. (2016). Barriers to the implementation of supply chain management-case of small to medium sized contractors in Turkey. *International Journal of Science and Research*, 5(9), 516–520.

Shen, K., Yang, Y., Wang, T., Zhao, D., Jiang, Y., Jin, R., Zheng, Y., Xu, B., Xie, Z., Lin, L., Shang, Y., Lu, X., Shu, S., Bai, Y., Deng, J., Lu, M., Ye, L., Wang, X., Wang, Y., & Gao, L. (2020). Diagnosis, treatment, and prevention of 2019 novel coronavirus infection in children: Experts consensus statement. *World Journal of Pediatrics*, 16(3), 223–231. doi: 10.1007/s12519-020-00343-7

Singhal, T. (2020). A review of coronavirus disease-2019 (COVID-19). *Indian Journal of Pediatrics*, 87(4), 281–286. doi: 10.1007/s12098-020-03263-6

Stower, H. (2020). Clinical and epidemiological characteristics of children with COVID19. *Nature Medicine*, 26(4), 465–470. 10.1038/s41591-020-0849-9

Su, L., Ma, X., Yu, H., Zhang, Z., Bian, P., Han, Y., Sun, J., Liu, Y., Yang, C., Geng, J., Zhang, Z., & Gai, Z. (2020). The different clinical characteristics of coronavirus disease cases between children and their families in China—the character of children with COVID-19. *Emerging Microbes and Infections*, 9(1), 707–713. doi: 10.1080/22221751.2020.1744483

Sun, D., Li, H., Lu, X.-X., Xiao, H., Ren, J., Zhang, F. R., & Liu, Z. S. (2020). Clinical features of severe pediatric patients with coronavirus disease 2019 in Wuhan: A single center's observational study. *World Journal of Pediatrics*, 16(3), 251–259. doi: 10.1007/s12519-020-00354-4

Tian, Y., Rong, L., Nian, W., & He, Y. (2020). Review article: Gastrointestinal features in COVID-19 and the possibility of faecal transmission. *Alimentary Pharmacology & Therapeutics*, 51, 843–851.

Van Doremalen, N., Bushmaker, T., Morris, D. H., Holbrook, M. G., Gamble, A., Williamson, B. N., Tamin, A., Harcourt, J. L., Thornburg, N. J., Gerber, S., Lloyed-Smith, J. O., De Wit, E., & Munster, V. J. (2020) Aerosol and surface stability of SARS-CoV-2 as compared with SARS-CoV-1. *New England Journal of Medicine*, 382(16), 1564–1567. doi: 10.1056/NEJMc2004973

Vander, H. L. (2007). Human coronaviruses: What do they cause. *Antiviral Therapy*, 12, 651–658.

Velavan, T. P., & Meyer, C. G. (2020). The COVID-19 epidemic. *Tropical Medicine & International Health*, 7, 278–280.

Wang, Y., Wang, Y., Chen, Y., & Qin, Q. (2020). Unique epidemiological and clinical features of the emerging 2019 novel coronavirus pneumonia (COVID-19) implicate special control measures. *Journal of Medical Virology*, 92(6), 568–576. doi: 10.1002/jmv.25748

Xia, W., Shao, J., Guo, Yu, Peng, X., Li, Z., & Hu, D. (2020). Clinical and CT features in pediatric patients with COVID-19 infection: Different points from adults. *Pediatric Pulmonology*, 55(5), 1169–1174. doi: 10.1002/ppul.24718

Yeo, C., Kaushal, S., & Yeo, D. (2020). Enteric involvement of coronaviruses: Is faecal-oral transmission of SARS-CoV-2 possible. *Lancet Gastroenterol Hepatol*, 5, 335–337.

Yin, S., Bai, L., & Zhang, R. (2021). Prevention schemes for future fresh agricultural products (FAPs) supply chain: Mathematical model and experience of guaranteeing the supply of FAPs during the COVID-19 pandemic. *Journal Science Food Agriculture*, 101, 6368–6383, doi: 10.1002/jsfa.11308

Yvonne, X. L., Yan, L. N., James, P. T., & Ding, X. L. (2016). Human corona viruses: A review of virus–host interactions. *Diseases*, 4, 1–28.

Zhang, H., Penninger, J. M., Li, Y., Zhong, N., & Slutsky, A. S. (2020). Angiotensin- Converting enzyme 2 (ACE2) as a SARS-CoV-2 receptor: Molecular mechanisms and potential therapeutic target. *Intensive Care Medicine*, 46(4), 586–590. doi: 10.1007/s00134-020-05985-9

Zhu, N., Zhang, D., Wang, W., Li, X., Song, J., Zhao, X., Huang, B., Shi, W., Lu, R., Niu, P., & Zhan, F. (2020). A novel coronavirus from patients with pneumonia in China, 2019. *New England Journal of Medicine*, 382, 727–733. 10.1056/NEJMoa2001017

Zimmerman, P., & Curtis, N. (2020). Coronavirus infections in children including COVID19: An overview of the epidemiology, clinical features, diagnosis, treatment and prevention options in children. *Pediatric Infectious Disease Journal*, 39(5), 355–368. doi: 10.1097/INF.0000000000002660

7 Artificial Intelligence from Vaccine Development to Pharmaceutical Supply Chain Management in Post-COVID-19 Period

Abhishek Dadhich
School of Allied Health Sciences and Management, New Delhi, India

Priyanka Dadhich
Department of Computer Science and Engineering, Delhi Technical Campus, Guru Gobind Singh Indraprastha University New Delhi, India

CONTENTS

7.1 INTRODUCTION TO INDIAN PHARMACEUTICAL INDUSTRY

Indian Pharmaceuticals stands as the world's third largest medicine producer by volume and rank fourteen in term of value. Globally, India is the largest generic medicine manufacturer with efficient and advanced technology. The Indian pharmaceutical industry supplies more than half of global demand for vaccines, 40% of generic demand in the United States, and 25% of all pharmaceuticals in the United Kingdom. The domestic pharmaceutical market includes more than 10,500 medicine manufacturing units, among most of them manufacturing units are U.S. Food and Drug Administration (USFDA), European Medicine Agency (EMA), and many other pharmaceutical regulatory bodies approved to produce affordable and best quality of drugs to meet the global demand. The large number of scientists and engineers have the ability to propel the sector forward to new heights. Indian pharmaceuticals companies are the major hub of antiretroviral drugs, globally 80% of antiretroviral drugs to combat AIDS (Acquired Immune Deficiency Syndrome) are manufactured and supplied by Indian pharmaceutical companies. In comparison to other industries, pharmaceutical industries operations and supply chain is complex in nature due to heavy scientific technology investment and regulatory guidelines for governing the overall pharmaceutical business. After the COVID-19 pandemic, there is a need to prepare new strategies for diagnosis, surveillance, identification, and development of therapeutics solutions against COVID-19. The present chapter is focused on the various applications of advance technologies like AI, machine learning, and blockchain to meet the demand of consumer healthcare services during COVID-19 and preventing pharmaceutical supply chain disruptions during emergencies.

7.1.1 Indian Pharmaceutical Market Size

As Indian economy survey 2021 reported that the Indian pharma market is expected to grow three times in the next decade. Presently the domestic market is at US$42 billion in 2021, which is expected to reach at US$65 billion by 2024 and with all the advance resources it may further expand up to US$130 billion by 2030. The biotechnology industry of India comprises of various sectors like bio-services, bio-agriculture, biopharmaceuticals and bioinformatics. In 2020, the Indian biotechnology sector was worth $70.2 billion, and by 2025, it is predicted to be worth $150 billion. In FY20, India's medical device market was worth US$10.36 billion. From 2020 to 2025, the market is estimated to grow at a 37 percent CAGR to reach US$50 billion.

According to CARE Ratings, Indian pharmaceutical market would grow at an annual pace of 11% over the next two years, reaching a value of more than US$60 billion. India's pharmaceutical sector is the largest generic drug manufacturer, supplying over 60% of global immunization demand and accounting for 20% of global supply by volume.

7.1.2 Pharmaceutical Export

Presently, Indian pharmaceutical exports valued US$24.44 billion in 2021 globally India stands at 12th position as a largest medical goods exporter. India exported 586.4 lakh COVID-19 vaccines to 71 countries, including grants (81.3 lakh), commercial exports (339.7 lakh), and COVAX platform exports (165.5 lakh). Indian pharmaceuticals are exported to over 200 nations around the world, United States is the largest pharmaceutical market from an export point of view. In terms of volume, generic medications account for 20% of global exports, making the country the world's largest provider of generic medicines. With a 34% stake, North America was India's largest pharma export market, with shipments to the United States, Canada, and Mexico increasing by 12.6%, 30%, and 21.4%, respectively (Figure 7.1).

7.1.3 Indian Pharma Road Ahead

Over the upcoming next five years, India's medical spending is expected to increase by 9–12%, making Indian Pharmaceuticals one of the top ten countries in terms of medical spending. For better future growth in domestic market the Indian pharma companies align their therapeutic product portfolio towards chronic diseases like antidiabetics, cardiovascular, anti-depressants, and anti-cancer which are at a high prevalence rate.

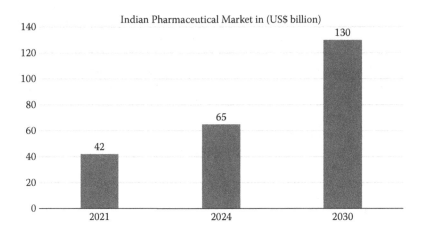

FIGURE 7.1   Indian Pharmaceutical Market in (US$ Billion).

The government of India has taken a number of initiatives to make affordable and lower the healthcare costs. The entry of generic pharmaceuticals into the market as quickly as possible has remained a priority, and it is likely to help Indian pharmaceutical companies. In addition, Indian pharmaceutical businesses will benefit from the increased focus on rural health programs, life-saving medications, and preventive immunizations.

7.1.4 PHARMACEUTICAL SUPPLY CHAIN DISRUPTION DUE TO COVID-19

The source of (Active Pharmaceutical Ingredients) APIs is a critical aspect for Indian pharma industries as a strategic plan to combat the COVID-19 epidemic, according to a report on the Indian pharmaceutical sector. India is major supplier of APIs (30%), which are used in generic medicine manufacturing by United States–based pharmaceuticals. Though, for the development of their medication formulations, Indian pharmaceuticals majorly rely on the China-based API market, with over 70% of their APIs coming from China, as globally by volume China is the world's largest manufacturer and exporter of active pharmaceutical ingredients.

With the explosion of the coronavirus pandemic, COVID-19 has exposed the dependency of Indian pharmaceutical industries reliance for APIs on China. Manpower shortages in China's pharma companies caused supply chain problems and product export restriction in India. This was caused by impose of quarantine regulations in different provinces of China by the governments. The interruption of logistic and transportation networks, which limited access and movement of goods to and from ports, had a significant impact on supplies. Moving towards globalization, the Indian pharma companies, with large-scale formulation production, propelled the API procurement demand from China due to low cost of manufacturing.

Further reliance of Indian pharmaceutical organizations on China's market poses a severe threat to national health security, forcing the Indian administration to establish a committee to examine the indigenous active pharmaceutical ingredients sector. The NITI Aayog (an Indian government policy think tank) and major pharmaceutical stakeholders recommended that more and more pharmaceutical infrastructure should be developed and approved; regulatory approval from the environment ministry, and government subsidies; and tax exemptions for the estab-lishment and development of pharmaceutical hubs can give new dimension to Indian pharma-ceutical market [1–4].

Objectives of the Chapter: The chapter assists readers to become familiar with the Indian pharmaceutical market scenario in context to implementation of artificial intelligence and blockchain from drug and vaccine development to systematic supply chain management. After reading this chapter, the readers should be able to:

- Understand the Indian pharmaceutical market and its growth prospects.
- Understand issues faced by pharmaceutical industry during supply chain disruptions due to COVID-19.
- Understand the importance of artificial intelligence in vaccine development, COVID-19 infection diagnosis, drug discovery.
- Understand the role of artificial intelligence and blockchain technology in pharmaceutical supply chain management.
- Forecast the future upcoming challenges with the implementation of AI in healthcare services.

Organization of the Chapter: Section 7.2 discusses research literatures associated with AI in vaccine development and supply chain system. Section 7.3 highlights the application of artificial intelligence (AI) in antiviral drug discovery and Section 7.4 elaborates about vaccine development with AI application. Section 7.5 discuss predicting viral mutation through AI applications. Section 7.6 emphasizes pharmaceutical supply chain transformation with AI. Section 7.7 discusses AI and blockchain technology role in implementing an intelligent supply chain. Section 7.8 elaborates on AI application challenges and limitations in health care. Section 7.9 focuses on a case study (How Blockchain Technology Preventing Counter-Fit Drugs in Pharmaceutical Supply Chain System). Section 7.10 concludes the chapter with future scope.

7.2   LITERATURE REVIEW

Chen and Decary [5] mentioned the guide to understanding the fundamentals of artificial intelligence. Authors discussed the basic concepts of AI technologies in health care, which includes natural language, machine learning, and AI voice assistance in improving and transforming into digital healthcare support. Authors also emphasized the performance evaluations indicators to measure the success of AI in clinical services for enhancing quality, safety, and efficacy.

Shah et al. [6] discussed the new pathways for the adaptation of AI and ML in clinical development; the authors proposed a real-world data (RWD) based simulation to allow to design clinical studies, preparing models that show different types of study impacts within a timeline study.

Zeng et al. [7] emphasized the network-based prediction of drug target interaction with the use of embedded deep forest; with the help of a heterogenous network, authors integrated 15 networks like chemical, phenotypic, genomic, and network profile among drugs protein/targets and disease. The study results in identifying high accurate molecular targets among known drugs arbitrary-order proximity embedded deep forest (AOPEDF).

Deo [8] discussed both supervised and unsupervised machine learning can be applied in clinical data sets for redefining patient classes and preparing robust risk model. The study emphasizes that, with the help of curated data or training, data machine learning solves the problems by predictions and decision-making process.

Baldominos et al. [9] made a comparison of deep learning and machine learning for activity recognition with help of mobiles. The activity-based chain recognition sequence was performed and finally they observe that deep learning techniques are not always best option to deal with ML and sensory data models as they are based on topology and chosen parameters.

Abbasi et al. [10] tried to predict the severity of COVID-19 with raw digital X-ray images; they used external validation and 10-fold cross validation in real settings to evaluate the severity of COVID-19 infection in the state-of-the-art methods for diagnostic purposes.

Khanday et al. [11] approached COVID-19 detection through ML. In their research study, textual clinical reports were employed to predict the prevalence of COVID-19 in patients. To find the best feature in the textual dataset, term frequency/inverse document frequency, report length, and bag of words (BOW) were applied as advanced engineering algos. Traditional and ensemble ML classifiers were trained using a set of features. The findings revealed that multinomial Naïve Bayes and logistic regression shows 96.2% accuracy rate in recognizing COVID-19 positive cases.

Banerji et al. [12] in their study discussed the development of vaccine during COVID-19 is the biggest milestone, but safety like side effects and allergic reactions are the major concern in vaccine administration and to avoid these reactions, a risk stratification schema care guide should be implemented.

Moore and Nishimura [13] discussed the presence of antigen peptide by MHC-II is a major strategy in vaccine development that can be performed by the MARIA (Major Histocompatibility Complex Analysis with Recurrent Integrated Architecture) and MoDec programs and are extremely based on AI algorithms.

Asgary et al. [14] proposed a machine learning model based on a big data set that can be run-through large vaccination simulation mean and further can predict output of the simulation tool and assist end-users and policy managers to assess the effects of several policy options especially for immunization programs.

Hie et al. [15] developed the NLP-based algorithm model to forecast physical escape trends of various viruses that also include COVID-19 infectious virus. The study also emphasizes on how viruses escape from the immune system with preserving pathogenic capabilities of a virus strain.

Toyoda et al. [16] proposed to develop an anti-counterfeit drug system in a supply chain with the help of RFID tag information to authorize access to the system with the help of blockchain.

Leng et al. [17] discussed public blockchain with a double-chain design system to improve the efficiency of supply chain systems; they demonstrated that proposed technology allows public service platforms to modify their rent-seeking and matching procedures. It ensures transaction information transparency and security, as well as the privacy of company data. The vastness of the underlying blockchain network, as well as performance difficulties, are the biggest disadvantages. The transparency, privacy, and fairness under the AI algorithm are required to be monitored and governed by the agencies.

Lalmuanawma et al. [18] elaborated about the application of AI and ML in the COVID-19 pandemic; they emphasized various developments of AI and ML in screening, forecasting, and drug/vaccine development during the COVID-19 pandemic, but authors also focused on the ground reality that many of the AI models during COVID-19 are theoretical based, which are never tested for their real utility.

7.3 APPLICATION OF ARTIFICIAL INTELLIGENCE (AI) IN ANTIVIRAL DRUG DISCOVERY

In recent years, AI has revolutionized many fields of science and engineering. It has a significant impact on our daily lives, from speech and facial recognition to large-scale customization of targeted advertisements. In medical sciences, AI also plays a crucial role in screening and diagnosis of viral diseases like COVID-19, as shown in Figure 7.2.

The design of sophisticated and advance drug development can be enabled with the application of AI algorithms that can speed up the process and lower the price of traditional drug research method. The various research studies shows that there are two main forms of compounds examined against COVID-19 are antiviral agents and immunomodulators. The preliminary clinical studies done for repurposing of medicines like ivermectin, remdesivir, ritonavir, lopinavir, and many other antiviral medicines to treat COVID-19 infection. Presently, only limited drug molecules have exposed capability as positive treatment results towards

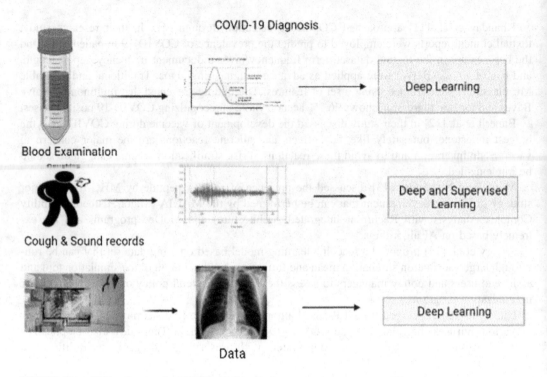

FIGURE 7.2   AI Application in COVID-19 Diagnosis.

COVID-19. The application of machine learning (ML) and deep learning (DL) on experimental data AI-based techniques were found very valuable in identifying repurposed drug molecule candidate as an effective antiviral drug. The "Arbitrary-Order Proximity Embedded Deep Forest approach" (AOPEDF) also can assist in identifying novel drug target interactions, Zeng et al. [7]. According to authors [19–22], they predicted that drugs can target SARS-CoV-2-related proteins and are commercially available. The neural network-based model (DeepCE), a deep learning algorithm, can be used to identify the repurposed drug molecule. The (DeepCE) enables to predicts potential leads for COVID-19 infection. On the basis of 1330 positive drug-disease associations, and further prediction of new indications for available drugs molecule and herbal compounds, machine learning model was constructed, which was not directly concerned with COVID-19. The potential drug candidate requires sophisticated bioassay and infrastructure to assess their bioavailability, metabolism, efficacy, toxicity, and drug interaction with another biomolecule. The majority of drug candidates failed during pharmacokinetic assessment in various phases of clinical trials. The combination of ML algorithms and cheminformatics information can be very useful in identification of potential drug candidate against SARS-CoV-2 treatment. In the screening of millions of compounds, ML algorithms are very useful. Authors in [23–25] proposed that screening of small molecules and peptides against SARS-CoV-2 pathological proteins deep learning–based pipeline can be implemented. In some studies, a "Densely Fully Connected Neural Network" (DFCNN) allows faster drug screening to extract more features from data.

The studies also showed the PDBBIND database used to train the DFCNN renders structural information of macromolecular complexes and protein, 66 inactive and 66 active compounds data set are included in their training. To predict the novel inhibitors a ML-based model was constructed by using inhibitors of COVID-19 SARS 3CLpro and 3CLpro proteins. In their study, they employed six different classifiers (DT, RF, LR, SVM, KNN, NB) to screen the library; the researcher also used logistic regression.

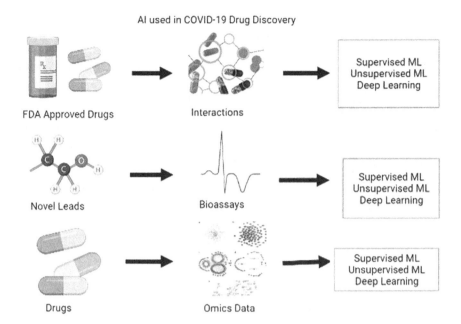

FIGURE 7.3 AI Used in COVID-19 Drug Discovery.

Kabra and Singh [26] used ML algorithms to predict antiviral peptide, that's bound with SARS-CoV-2 protease. The AI strategies used in various studies will not only open the road for COVID-19 infection treatment development, but also enables research and development of novel medications to treat other infectious diseases. To improve the clinical success rates identification of biomarkers and drugs, sensitivity predictions can be implemented with the help of ML-based methods. Clinical trials can be performed with the help of artificial intelligence–based solutions to discover novel treatments that are safe and effective (Figure 7.3).

7.4 VACCINE DEVELOPMENT WITH AI APPLICATION

Developing a vaccine is one of the most effective ways to tackle COVID-19. For the development of effective vaccines there are several virus components are used like nucleocapsid (N) protein, spike (S) protein, membrane (M) protein, and the whole virus. The vaccine candidate (Johnson and Johnson) JNJ-78436735/Ad26.COV2.S, Covidshield (Oxford-AstraZeneca) and mRNA-1273 (Moderna) are developed with all virus components which later get approval from Emergency Use Authorization (EUA). The intervention with vaccination still facing some safety issues and are not able to give 100% percent protection. Many of the vaccines reported side effects like allergic reactions after the administration. The challenges related to vaccine production, concern related to safety and efficacy, supply chain, and storage of different vaccines candidate can be resolved with the implementation of AI technology. AI algorithms are used to accelerate the development of vaccines during clinical trial phases (Figure 7.4).

The identification of proteins sequence and antigens determinant's paly important role in vaccine development which can be processed by AI basic tools like recursive feature selection (RFE), random forest (RF), and support vector machine (SVM). Valid binding prediction of major histocompatibility complex (MHC) proteins and peptides can be determined by Deep Convolutional Neural Networks (DCNN). MHC class II molecules display exogenous peptides that attach with a T-cell receptor on CD4plus T cells, whereas to trigger cytotoxic lympho-cytes, MHC class I molecule attach with CD8plus T cells. The antigen-specific responses are

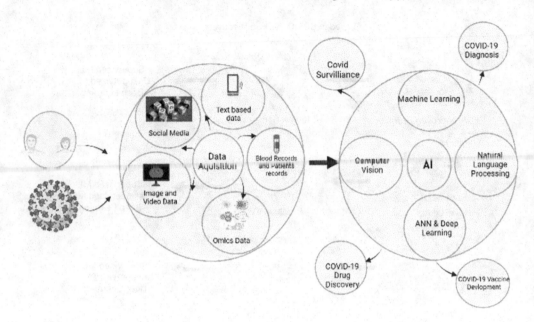

FIGURE 7.4   Application of AI in COVID-19 Pandemic.

induced by MHC class I and II molecule induced immunity with vaccine administration. Here, AI-based strategies and machine learning are used to identify the antigen peptide with MHC II molecule for vaccine development. MoDec and MARIA are the programs develop by ML to predicts the antigen presence. To understand the patient's natural immunity and presence of viral peptide SARS-CoV-2 on molecule MHC, various ML algos and AI tools are used. These basic understandings enable to determine COVID-19 definite immune response which further assisting in discovery of effective vaccine against COVID-19 infection. The reverse vacci-nology tool Vaxign-ML is used by some researchers to predicts the target for developing safe and effective vaccine. Authors [27–37] identified 174 SARS-CoV-2 epitopes that can firmly bind with highly efficient binding scores to 11 HLA allotypes, feed forward neural networks are used. Prachar et al. [38] studied for further development of COVID-19 vaccines and treatments in future along with validation study of binding and non-binding peptides and their identification play a role in the process (Figure 7.5).

For example, the public awareness and perception towards COVID-19 vaccines and to increase the outreach of vaccine awareness program the various government and healthcare regulatory bodies randomly used AI tool. On the other hand, better prediction and analysis of COVID-19 prevalence rate and various previous data are analyzed with the help of ML and AI tools. Based on the number of patients studied, like COVID positive or recovered, and the number of patients who expired from the infection, AI can also suggest paradigms for ana-lyzing and understanding future development of vaccines. In order to develop COVID-19 vaccine and treatments, the data set of compounds, peptides, and COVID-19 epitopes are used to train the models run by AI algorithms [39–41]. Many other studies were conducted to predict the efficacy of vaccine on different racial and minority populations. The use of ML-based programs like EvalMax and OptimAX study the various genetic structures of popula-tions and identification of peptide sequence for developing vaccines. According to Optivax, the SARS-CoV-2 spike protein may not be sufficient to provide full immunity to all racial ethnicities. AI-based models allow policy makers and users to assess the impact of various policy options and also assist in designing and developing the most effective COVID-19 vaccines.

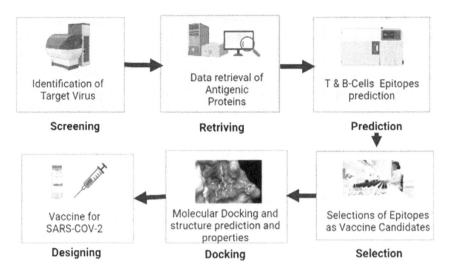

FIGURE 7.5 AI-Based Antibodies Discovery and COVID-19 Vaccine Development.

7.5 PREDICTING VIRAL MUTATION THROUGH AI APPLICATIONS

The effectiveness of antiviral treatment and vaccines are always hindered with the viral mutation. The early prediction of viral mutation with drug resistance may facilitate in the development of effective vaccines and antiviral treatments against COVID-19 [44–45]. The recent artificial intelligence–based research has provided a deep insight into how to predict various mutational landscape. The algorithm-based NLP tool can predict the viral mutation that have capacity to protect themselves from the immune system and can also reserve the infective capability of virus strain. By using these models, the researcher can predict various mutations held in different pattern of virus strains. To provide the large coverage from various COVID-19 virus strains, broad-spectrum vaccines are used that are developed on an AI algorithm model. For predicting epitope hotspot models, Malone et al. [42] assessed around 3400 SARS-CoV-2 sequence category. Malone et al. [42]. In another study, Salma et al. [43] applied neural network techniques that effectively predict the presence of nucleotides in upcoming generations after virus mutations. By the implementation of recurrent neural network (RNN) based on a long short-term memory (LSTM) model, the researcher predicted the mutation rate in the patient after exposure to COVID-19 infection. Another model based on a phylogenetic tree by Haimed et al. [46] the evolutionary behavior of virus can be captured and with the help of a viral reverse engineering process similarity pattern in proteins and genomic sequences may be identified. The worldwide availability of vaccines and treatments increases the faith and confidence in community to handle the COVID-19 infection challenges. But somehow emergence of new COVID-19 variants like Omicron and Delta variants always put healthcare structure under pressure with the uncertainty of vaccine efficacy in the future. COVID-19 treatments and preventions are also affected with the emergence of new variants. To predict and forecast these mutational changes AI methods are found effective and key strategic step in controlling the prevalence rate of COVID-19 infections. (Bansal, A., et al. [47], Philomina, et al. [48]).

7.6 PHARMACEUTICAL SUPPLY CHAIN TRANSFORMATION WITH AI

Digitalization trends are continuing to transform the business world, with offering new opportunities in a variety of industries. Artificial intelligence is seen as a significant driver of the digital revolution with the potential to create new revenue streams. Apart from AI, recent developments in

FIGURE 7.6 Application of AI in Pharmaceutical Supply Chain Management.

automation and machine learning have spawned an entirely new business environment. The pharmaceutical supply chain generates a large amount of internal and external data on a regular basis, yet this data has been underutilized in the past. But with the implication of AI technologies, these data are utilized in increasing the operational efficiency and creating a cost-effective supply chain process. The AI technologies allows pharmaceutical supply chain fully data driven process by providing and processing real-time data for decision making by reducing human-biased decisions. In the pharmaceutical supply chain, AI plays a critical role in demand forecasting, logistic optimization, inventory management, and workforce planning and resulted in a more efficient and visible process during supply chain (Figure 7.6).

The pharmaceutical supply chain system is complex in nature where the organizations need to respond quickly which may also impacts the outcome. To perform the right decision and mitigate the diversification end to end visibility is most important foundation in supply chain system. It enables it to access all the data related to all transactions and demand generated. AI and blockchain technologies are the applications that improve the visibility across the supply chain system. End-to-end visibility across the pharmaceutical interconnected supply chain will enable data to be safely extracted in real time using AI techniques, resulting in actionable insights and an improved decision-making process.

AI tools and techniques enables organizations to generate accurate demand forecast to manage the inventory by pharmaceutical companies. It analyzes multiple sources of information and patterns. To improve the process efficiency and productivity without human interference with operational data AI tools such as computer vision, NLP and IoT platform are used. To find out the different pattern and interdependent variables missed in traditional methods, AI technologies provide real-time data for optimizing maintenance, minimum downtime, and maximum productivity. For protecting the integrity of pharmaceutical supply chain, some advance technologies with AI are used, such as blockchain technology that creates a system that is immutable, transparent, secure, and protected to prevent the entrance of counterfeit and substandard drugs in supply chain system.

7.7 AI AND BLOCKCHAIN TECHNOLOGY ROLE IN IMPLEMENTING AN INTELLIGENT SUPPLY CHAIN

The supply chain process integrates many checkpoints involved from product assembly to product distribution. A supply chain nowadays can consist of hundreds of stages and geological zones. Because there are data errors and roadblocks in each step of the supply chain process and it's difficult to follow a fool proof framework. The security in supply chain process is extremely important in pharmaceutical supply chain system. Due to the complex nature of pharmaceutical supply chain management, there will be chance of uploading of counterfeit drugs or substandard drugs in the supply chain process that is harmful for every nation. Some of the major illnesses or deaths of patients may be occur if counterfeit drugs are not prevented in whole supply chain systems. All of these issues come as a result of several flaws in the pharmaceutical industry's supply chain. Pharmaceutical supply chain system includes a business to customer and business to business network that is complex in nature. With the flow of pharmaceutical products, the product-related information is also flow among all stakeholders. This complex nature and long supply chain process makes the process vulnerable towards unethical practices.

AI and blockchain technology with its innovative qualities provides productive responses for the current vulnerabilities discovered in pharmaceutical supply chain management system. Blockchain technology has been providing fascinating research areas to improve the supply chain's traceability and security. Blockchain technology is a decentralized database technology that keeps track of an ever-growing list of data entries created by the nodes involved. The data is kept in a public ledger that contains information from every completed transaction. With the absence of a middleman to control agreement and inalterability, blockchain technology also exhibits qualities such as decentralization of database, security, anonymity, and data integrity. AI and blockchain architecture are more robust, safe, and scalable than existing solutions to provide improved transaction privacy, and possible options for securing the pharmaceutical supply chain.

7.7.1 Tracing Medicine in Pharmaceutical Supply Chain System

In the present scenario of pharmaceuticals, the verification system and supply chain must meet the following criteria:

- Allow numerous parties to update the data,
- Allow numerous parties to communicate the data,
- Verification is performed to ensure the validity of information,
- Collaboration with national and European medicines verification systems to ensure that all stakeholders are aware of the drug's legitimacy.

With the present supply chain system, there are several significant issues like:

- More the intermediaries will lead to more complexity due to which may lead to higher the risk of damage of interruptions
- These interactions are time sensitive and the supply chain medicines should be performed within a time frame as the pharmacy should able to provide prompt service to patients
- The pharmaceutical manufacturers, wholesalers, retailers, and logistics companies are not able to perform complete transparency during the supply chain for authenticity and validity of drugs quality (Figure 7.7).

The pharmaceutical supply chain could benefit significantly from a blockchain system. Packaging containing barcodes would be scanned and recorded at each stage of the procedure onto a blockchain ledger system, which would then record and generate an audit trail of the medicine journey. For temperature-sensitive medicines, sensors can also be used in the supply chain, with temperature and humidity recorded in the ledger system. This is especially critical for medicines that need to be kept in the refrigerator, such as insulin or some precision medicine. The medicine can be traced from any point of supply chain system to dispensing of drugs which can be recorded onto the blockchain ledger.

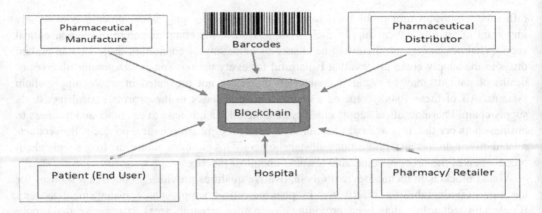

FIGURE 7.7   Blockchain Technology in Pharmaceutical Supply Chain.

7.7.2 ADVANTAGES OF BLOCKCHAIN SYSTEM IN THE PHARMACEUTICAL SUPPLY CHAIN PROCESS

There are many advantages of using blockchain in pharmaceutical supply chain management

1. Simplifying supply chain process in economic value
   - Medicines are easier to track as the supply chain system becomes traceable
   - Allowing better stock control and rotations of medicines based on expiry information
2. Decreases number of errors
   - Drugs can be traced with their quality and authenticity
   - As the drugs validated with authenticity patient harm with counterfeit drugs are also reduced
3. Improve security
   - Minimizing counterfeit drugs in pharmaceutical supply chain system
   - Increase data protection
   - Reduction in patients' health hazards with substandard medicines
4. Transparency in sharing and transactions
   - When required, all parties have easy access to drug information in order to ensure the quality and authenticity of the drug being dispensed
   - All the stakeholders in supply chain can trust the quality of medicine with available data.
5. Creates the trail of evidence
   - Medicines can be traced readily and easy to recall
   - The different stakeholders can see if the process of pharmaceutical supply chain is conceded at any point.
6. Transparency between authorised parties is improved:
   - All stakeholders can observe every step of the medicine journey to confirm its authenticity.
   - Pharmaceutical regulatory bodies, such as drug inspectors, can more accurately track the rate of counterfeit drugs entering into the supply chain.

The medicine journey can be made more secure and efficient by utilising blockchain technology. Every delivery can be tracked, and the delivery driver can be identified using biometric data. Using biometric measures, 2d barcode scanning, or sensor technology, every checkpoint involving the medication is logged and traced. Biometric measures, 2d barcode scanning, and sensor technology are used to track and record every checkpoint involving in the medicine manufacturing point to distribution point. The entire medication path becomes seamless, accurate, audited, and secure as the drug is traced from creation to patient. With blockchain technology the end user has greater responsibility and control over his own health and treatment now patient can know where his drug came from, who created it, and what it's made of. This could help patient to adhere with his medication and treatment plan, which allowing him to live a longer and quality life.

7.7.3 DRUG INFORMATION SYSTEM DESIGN TO ENABLE A BLOCKCHAIN SYSTEM

To operationalize the blockchain technology it requires all drug information's from different stages of drug life cycle which are as follows:

1. The regulatory bodies should register the medicine on the basis of the specification of the medicinal product which also categorise the type of medicinal product. To perform the manufacturing of the medicinal product the manufacturer has to take a production license that is registered as a participant manufacturer that indicates the link to his account and the same license number is printed on final product packaging for tracing and validating purpose.
2. Medicinal product manufactured by pharmaceutical companies can be identified with unique code which reflect the information about type of medicine with its expiry date.
3. The pharmacy store produces order with required quantity should be link with manufacturer where unique unit number is generated from the warehouse. On delivery of the unit, the owner should file the drug unit number and recipient ID that is the pharmacy.
4. Finally, when doctors prescribe the medicine to patients. The patients buy medicine with digital medical policy which will be later tied to the asset of his account. This also contains the detailed treatment profile of patients. In retail pharmacies, transactions takes place with real-time inventory and the patient becomes owner of in both real and virtual units of medicine.

As a result, each licenced unit of medicine leaves a distinct code in the register that can be easily followed when any disputes occur. The implementation of blockchain system in pharmaceutical supply chain make the task easier and enables monitoring and controlling throughout the process. The transparency in supply chain process motivates all stakeholders to abide with rules established under law.

7.8 AI APPLICATION CHALLENGES AND LIMITATIONS IN HEALTH CARE

The recent development of AI technology has demonstrated various success in the field of healthcare and pharmaceutical services such as application and development of big data, neural networks, brain sciences, and other technologies. But there are some major challenges existing in making AI-based decisions in health care and medical sciences such as fairness, interpretability, generalization, validation, risk mitigation, and inclusiveness. The most common ethical concern is unfairness driven by prejudice in data sources. Any data set will be skewed to some degree due to characteristics such as gender, sexual orientation, race, sociologic, environmental, and economic factors. The data protection, privacy, ownership, objectivity and transparency are all important aspects of data ethics, Maher et al. [49] and Jiang et al. [50]. The ability of AI-based algorithms to function well in a variety of situations is referred to as generalization (Beil et al. [51]). Wang and Preininger [52] observed narrow context AI models with different data set always fail at the broad level aspects. The patient data privacy, confidentiality, and safety are the major challenges in terms of regulatory and ethical aspects especially in context to health care and clinical setups. To enhance AI algorithms, fairness, accountability, and transparency the two major factors should be covered under governance that is interpretability and explainability. AI technology and governance should be able to deals with bias and lack of transparency among different stakeholders. Handling the clinical ethical issues such as fairness, privacy, and transparency should be primary focus of ethical governance. In the real condition, the utility of the AI model is largely unverified as they are based on theoretical aspects. But the role of AI in connecting public health with a supply chain system is now adopted with many of the organizations, which requires continuous monitoring and evaluations. In the upcoming days, the advance AI tools and technologies may be implemented to perform real-time decision-making process in medical sciences and supply chain, population risk assessment, and also execute up to a certain level in treatment and clinical care. The prevention of anti-counterfeit drugs in a pharmaceutical supply chain can be done with the implementation of AI and blockchain technology that includes product registration and transferring with smart contracts, the transferring records of the products recorded in unchangeable ledger. The system allows to processed decentralized which only allows authorize access of stake holders but the system has limitation that it can work only on real time database to create more transparency in the system Humanyun et al. [77], [53–76].

7.9 CASE STUDY: HOW BLOCKCHAIN TECHNOLOGY PREVENTING COUNTERFEIT DRUGS IN PHARMACEUTICAL SUPPLY CHAIN SYSTEM

Effective supply chain system is biggest challenge in pharmaceutical industry. As many more risks are associated with the healthcare supply chain and due to complex healthcare structure, the compromised supply chain system can also affect the patient safety. With globalization and advance technology adaptation, there is a rise in interference of multiple stakeholders which make pharmaceutical supply chain system complex in nature. By the end of 2023 it is expected that global pharmaceutical industry will reach USD$1.5 trillion. The Organization for Economic Co-operation and Development (OECD) has reported that worldwide 3.3% of counterfeit drugs are available in the pharmaceutical market, which is a severe issue for all developed and developing nations.

The blockchain technology in the pharmaceutical supply chain process provides various potential benefits:

- Security
- Data provenance
- Integrity
- Functionality

The global healthcare organizations like Red Cross, Global Fund, and UNICEF are tracing the distributed donated medicine globally with prevention of counterfeit drugs with the help of blockchain technology.

To increase the visibility and adherence with regulatory compliance in pharmaceutical, blockchain technology plays a vital role which enforce the guidelines to make pharmaceutical products safe, effective and up to quality parameter. The blockchain technology increases the visibility of movement and players involved in the supply chain of pharmaceuticals and medicines with efficient inventory management and traceability. The digitalized transactions between manufacturer to patients enable clear visualization of the healthcare product journey. The technology detects the possible vulnerable points in supply chain system and also reduces the chances of fraud and loss takes place during supply chain process. At the time of safety issues in products, batch recall management can also be performed with the help of blockchain.

The pharmaceutical company like Merck plan and develop SAP Pharma Blockchain POC app in partnership with SAP that provide advance track and trace system with essential information like serial number, item number, batch number, and expiry date that allows organization to enhance accountability and transparency among all stakeholders. The world pharmaceutical leader Novartis also launched IoT and blockchain-based technology to identify counterfeit medicines and a temperature tracking process for sensitive medicines and vaccines during logistic distributions.

From the above case, identify the sensitive issues pertaining to pharmaceutical supply chain that can be resolved by the implementation of blockchain technology.

7.10 CONCLUSION AND FUTURE SCOPE

After the worldwide COVID-19 outbreak, the urge for new vaccines and drug development to prevent and treat COVID-19 infection rises faster and increases in demand. The use of advanced technology like machine learning and artificial intelligence significantly accelerates the drug development and approval process. The computational methods with predictive model are implemented with the help of AI for early detection and diagnosis of cases and combat the rise in cases of COVID-19 infection. Human biological data like recordings of cough, blood samples, and clinical images with AI technology enable faster and reliable assessment and diagnosis of critical and severe diseases. The patient with asymptomatic COVID-19 infections can also be diagnosed and enables clinicians to take real time decisions for treatment and preventions of severe infections. AI-based technology provides better solutions to fight COVID-19 infections by tracing cases, evaluating safety, and efficacy profile of various therapeutics and assisting in biomedical reports. AI algorithm models enable researchers to detect the mutational changes with alteration in the sequence of peptides and protein structures and also assist to develop future vaccines with large broad-spectrum coverage.

Pharmaceutical supply chain systems with incorporation of AI and blockchain technology become transparent and also optimize the process with maximum utilization of resources. Blockchain improves the traceability and security of pharmaceutical supply chain process with advanced barcoding and data security system that also prevents the entrance of counterfeit or substandard drugs in supply chain system. However, AI technologies overcome the barriers existing between vaccine development to regulatory authorizations but somehow the full utilization of AI technology is missing due to some challenges like data handling, data privacy, and governmental oversight.

In the future, we expect advanced AI and ML technology will provide better healthcare services and vaccine development process and also improve the technical issues pertaining to data privacy, risk of algorithm bias, and better output generation and data integration. We believe that AI, ML, and blockchain technology will play a significant role in future healthcare products. It is the key capability underpinning the development of vaccines and medicine, which is universally acknowledged as a much-needed advancement in health care. In disease diagnosis and treatment, AI has proven its value in healthcare services. The X-ray diagnosis and speech recognition are the valuable properties of AI to provide better treatment by the prescribers.

Although many more advanced technologies are adopted in clinical practices but real challenges are regarding transparency, confidentiality and privacy of data while using AI is a major challenge faced by the organizations and in future research, we expect these issues may also mitigate with advancement in AI technology. In pharmaceutical sectors, organizations are launching AI and blockchain technology for supply chain and inventory management, but future studies may be focused on how AI and blockchain technology can be implemented to safeguard the quality, safety, and efficacy during medicine development and manufacturing. Perhaps after a pandemic like COVID-19, organizations and prescribers realize that with the augmentation of artificial intelligence in healthcare services, they can provide better treatment with accurate diagnosis.

REFERENCES

[1] Yadav, L. (2022). Industry captains urge govt to urgently introduce robust OTC drug policy to reduce out-of-the-pocket healthcare spend, 2022, accessed from http://pharmabiz.com/NewsDetails.aspx?aid=147743&sid=1 (accessed March 12, 2022).

[2] Pharmaceutical annual reports 2022 by department of pharmaceuticals, https://pharmaceuticals.gov.in/sites/default/files/english%20Annual%20Report%202020-21.pdf (accessed March 24, 2022). (2020–21).

[3] Pharmaceutical industry report 2022 by IBEF, https://www.ibef.org/industry/pharmaceutical-india (accessed April 02, 2022). (2022).

[4] Dadhich, A., & Gurbani, N. (2021). COVID-19: Unprecedented challenges for Indian pharmaceutical industry and pharma emerging markets. *Int. J. Sci. Rep.*, 7(6), 319–324.

[5] Chen M., & Decary M. (2020). Artificial intelligence in healthcare: An essential guide for health leaders. *Healthcare Manage. Forum*, 33(1), 10–18. doi: 10.1177/0840470419873123

[6] Shah, P., Kendall, F., Khozin, S., Goosen, R., Hu, J., Laramie, J., Ringel, M., & Schork, N. (2019). Artificial intelligence and machine learning in clinical development: A translational perspective, *NPJ Digit. Med.*, 2, 69.

[7] Zeng, X., Zhu, S., Hou, Y., Zhang, P., Li, L., Li, J., Huang, L. F., Lewis, S. J., Nussinov, R., & Cheng, F. (2020). Network-based prediction of drug-target interactions using an arbitrary-order proximity embedded deep forest. *Bioinformatics*, 36, 2805–2812.

[8] Deo, R. C. (2015). Machine learning in medicine. *Circulation*, 132, 1920–1930.

[9] Baldominos, A., Cervantes, A., Saez, Y., & Isasi, P. (2019). A comparison of machine learning and deep learning techniques for activity recognition using mobile devices. *Sensors*, 19, 521.

[10] Abbasi, W. A., Abbas, S. A., & Andleeb, S. (2022). Covidx: Computer-aided diagnosis of COVID-19 and its severity prediction with raw digital chest X-ray images. arXiv:2012.13605.

[11] Khanday, A. M. U. D., Rabani, S. T., Khan, Q. R., Rouf, N., & Mohi Ud Din, M. (2020). Machine learning based approaches for detecting COVID-19 using clinical text data. *Int. J. Inf. Technol.*, 12, 731–739.

[12] Banerji, A., Wickner, P. G., Saff, R., Stone, C. A., Jr., Robinson, L. B., Long, A. A., Wolfson, A. R., Williams, P., Khan, D. A., & Phillips, E., et al. (2021). mRNA vaccines to prevent COVID-19 disease and reported allergic reactions: Current evidence and suggested approach. *J. Allergy Clin. Immunol. Pract.*, 9, 1423–1437.

[13] Moore, T. V., & Nishimura, M. I. (2020). Improved MHC II epitope prediction—A step towards personalized medicine. *Nat. Rev. Clin. Oncol.*, 17, 71–72.

[14] Asgary, A., Valtchev, S. Z., Chen, M., Najafabadi, M. M., & Wu, J. (2021). Artificial intelligence model of drive-through vaccination simulation. *Int. J. Environ. Res. Public Health*, 18, 268.

[15] Hie, B., Zhong, E. D., Berger, B., & Bryson, B. (2021). Learning the language of viral evolution and escape. *Science*, 371, 284–288.

[16] Toyoda, K., Mathiopoulos, P. T., Sasase, I., & Ohtsuki, T. (2017). A novel blockchain-based product ownership management system (POMS) for anti-counterfeits in the post supply chain. *IEEE Access*, 5, 17465–17477.

[17] Leng, K., Bi, Y., Jing, L., Fu, H. C., & Nieuwenhuyse, I. (2018). Research on agricultural supply chain system with double chain architecture based on blockchain technology. *Future Gener. Comput. Syst.*, 86, 641–649.

[18] Lalmuanawma, S., Hussain, J., & Chhakchhuak, L. (2020). Applications of machine learning and artificial intelligence for COVID-19 (SARS-CoV-2) pandemic: A review. *Chaos Solitons Fractals*, 139, 110059.

[19] Beck, B. R., Shin, B., Choi, Y., Park, S., & Kang, K. (2020). Predicting commercially available antiviral drugs that may act on the novel coronavirus (2019-nCov), Wuhan, China through a drug-target interaction deep learning model. *Comput. Struct. Biotechnol. J.*, 18, 784–790.

[20] Kim, E., Choi, A. S., & Nam, H. (2019). Drug repositioning of herbal compounds via a machine-learning approach. *BMC Bioinform.*, 20, 247.

[21] Pham, T.-H., Qiu, Y., Zeng, J., Xie, L., & Zhang, P. (2021). A deep learning framework for high-throughput mechanism-driven phenotype compound screening and its application to COVID-19 drug repurposing. *Nat. Mach. Intell.*, 3, 247–257.

[22] Bowick, G. C., & Barrett, A. D. T. (2010). Comparative pathogenesis and systems biology for biodefense virus vaccine development. *J. Biomed. Biotechnol.*, 2010, 236528.

[23] Zhang, H., Saravanan, K. M., Yang, Y., Hossain, M. T., Li, J., Ren, X., Pan, Y., & Wei, Y. (2020). Deep learning-based drug screening for novel coronavirus 2019-nCov. *Interdiscip. Sci.*, 12, 368–376.

[24] He, Y., Xiang, Z., & Mobley, H. L. T. (2010). Vaxign: The first web-based vaccine design program for reverse vaccinology and applications for vaccine development. *J. Biomed. Biotechnol*, 2010, 297505.

[25] Shah, P., Kendall, F., Khozin, S., Goosen, R., Hu, J., Laramie, J., Ringel, M., & Schork, N. (2019). Artificial intelligence and machine learning in clinical development: A translational perspective. *NPJ Digit. Med.*, 2, 69.

[26] Kabra, R., & Singh, S. (2021). Evolutionary artificial intelligence-based peptide discoveries for effective COVID-19 therapeutics. *Biochim. Biophys. Acta Mol. Basis Dis.*, 1867, 165978.

[27] Ahmed, Z., Mohamed, K., Zeeshan, S., & Dong, X. (2020). Artificial intelligence with multi-functional machine learning platform development for better healthcare and precision medicine. Database.

[28] Callaway, E., & Ledford, H. (2021). How to redesign Covid vaccines so they protect against variants. *Nature*, 590, 15–16.

[29] Livingston, E. H., Malani, P. N., & Creech, C. B. (2021). The Johnson & Johnson vaccine for COVID-19. *JAMA*, 325, 1575.

[30] Madkaikar, M., Gupta, N., Yadav, R. M., & Bargir, U. A. (2021). India's crusade against COVID-19. *Nat. Immunol.*, 22, 258–259.

[31] Wise, J. (2021). COVID-19: Pfizer biontech vaccine reduced cases by 94% in israel, shows peer reviewed study. *BMJ*, 372, n567.

[32] McMurry, R., Lenehan, P., Awasthi, S., Silvert, E., Puranik, A., Pawlowski, C., Venkatakrishnan, A. J., Anand, P., Agarwal, V., O'Horo, J. C., et al. (2021). Real-time analysis of a mass vaccination effort confirms the safety of FDA-authorized mRNA vaccines for COVID-19 from Moderna and Pfizer/Biontech. medRxiv.

[33] Moore, T. V., & Nishimura, M. I. (2020). Improved MHC II epitope prediction—A step towards personalized medicine. *Nat. Rev. Clin. Oncol.*, 17(2), 71–72.

[34] Racle, J., Michaux, J., Rockinger, G. A., Arnaud, M., Bobisse, S., Chong, C., Guillaume, P., Coukos, G., Harari, A., Jandus, C., et al. (2019). Robust prediction of HLA class II epitopes by deep motif deconvolution of immunopeptidomes. *Nat. Biotechnol.*, 37, 1283–1286.

[35] Ong, E., Wong, M. U., Huffman, A., & He, Y. (2020). COVID-19 coronavirus vaccine design using reverse vaccinology and machine learning. *Front. Immunol.*, 11, 1581.

[36] Abd El-Aziz, T. M., & Stockand, J. D. (2020). Recent progress and challenges in drug development against COVID-19 coronavirus (SARSCoV-2)—An update on the status. *Infect. Genet. Evol.*, 83, 104327.

[37] He, Y., Xiang, Z., & Mobley, H. L. T. (2010). Vaxign: The first web-based vaccine design program for reverse vaccinology and applications for vaccine development. *J. Biomed. Biotechnol*, 2010, 297505.

[38] Prachar M., Justesen S., Steen-Jensen D. B., et al. (2020). Identification and validation of 174 COVID-19 vaccine candidate epitopes reveals low performance of common epitope prediction tools. *Sci. Rep.*, 10(1), 20465.

[39] Keshavarzi Arshadi, A., Webb, J., Salem, M., Cruz, E., Calad-Thomson, S., Ghadirian, N., Collins, J., Diez-Cecilia, E., Kelly, B., Goodarzi, H., et al. (2020). Artificial intelligence for COVID-19 drug discovery and vaccine development. *Front. Artif. Intell.*, 3, 65.

[40] Liu, G., Carter, B., Bricken, T., Jain, S., Viard, M., Carrington, M., & Gifford, D. K. (2020). Computationally optimized SARS-CoV-2 MHC class I and II vaccine formulations predicted to target human haplotype distributions. *Cell Syst.*, 11, 131–144.e6.

[41] Liu, G., Carter, B., & Gifford, D. K. (2021). Predicted cellular immunity population coverage gaps for SARS-CoV-2 subunit vaccines and their augmentation by compact peptide sets. *Cell Syst.*, 12, 102–107.e4.

[42] Malone, B., Simovski, B., Moliné, C., Cheng, J., Gheorghe, M., Fontenelle, H., Vardaxis, I., Tennøe, S., Malmberg, J.-A., Stratford, R., et al. (2020). Artificial intelligence predicts the immunogenic landscape of SARS-CoV-2 leading to universal blueprints for vaccine designs. *Sci. Rep.*, 10, 22375.

[43] Salama, M. A., Hassanien, A. E., & Mostafa, A. (2016). The prediction of virus mutation using neural networks and rough set techniques. *EURASIP J. Bioinform. Syst. Biol.*, 2016, 10.

[44] Wilson, B., Garud, N. R., Feder, A. F., Assaf, Z. J., & Pennings, P. S. (2016). The population genetics of drug resistance evolution in natural 2 populations of viral, bacterial, and eukaryotic pathogens. *Mol. Ecol.*, 25, 42–66.

[45] Baranovich, T., Wong, S., Armstrong, J., Marjuki, H., Webby, R., Webster, R., & Govorkova, E. (2013). T-705 (Favipiravir) induces lethal mutagenesis in influenza a H1N1 viruses in vitro. *J. Virol.*, 87(7), 3741–3751.

[46] Haimed, A. M. A., Saba, T., Albasha, A., Rehman, A., & Kolivand, M. (2021). Viral reverse engineering using artificial intelligence and big data COVID-19 infection with long short-term memory (LSTM). *Environ. Technol. Innov.*, 22, 101531.

[47] Bansal, A., Padappayil, R. P., Garg, C., Singal, A., Gupta, M., & Klein, A. (2020). Utility of artificial intelligence amidst the Covid 19 pandemic: A review. *J. Med. Syst.*, 44, 156.

[48] Philomina, J. B., Jolly, B., John, N., Bhoyar, R. C., Majeed, N., Senthivel, V., Cp, F., Rophina, M., Vasudevan, B., Imran, M., et al. (2021). Genomic survey of SARS-CoV-2 vaccine breakthrough infections in healthcare workers from Kerala, India. *J. Infect.*, 83, 237–279.

[49] Maher, M. C., Bartha, I., Weaver, S., di Iulio, J., Ferri, E., Soriaga, L., Lempp, F. A., Hie, B. L., Bryson, B., Berger, B., Robertson, D. L., Snell, G., Corti, D., Virgin, H. W., Kosakovsky Pond, S. L., & Telenti, A. (2022). Predicting the mutational drivers of future SARS-CoV-2 variants of concern. *Sci. Translational Med.*, 14(633), eabk3445.

[50] Jiang L., Wu Z., Xu X., et al. (March 2021). Opportunities and challenges of artificial intelligence in the medical field: Current application, emerging problems, and problem-solving strategies. *J. Int. Med. Res.*, 14(633).

[51] Bell, M., Proft, I., van Heerden, D., Svırı, S., & van Heerden, P. V. (2019). Ethical considerations about artificial intelligence for prognostication in intensive care. *Intensive Care Med. Exp.*, 7, 70.

[52] Wang, F., & Preininger, A. (2019). AI in health: State of the art, challenges, and future directions. *Yearb. Med. Inform.*, 28, 16–26.

[53] Goodman, K., Zandi, D., Reis, A., & Vayena, E. (2020). Balancing risks and benefits of artificial intelligence in the health sector. *Bull. World Health Organ.*, 98, 230–230A.

[54] Gerke, S., Minssen, T., & Cohen, G. (2020). Ethical and legal challenges of artificial intelligence-driven healthcare. In *Artificial Intelligence in Healthcare.* Elsevier: Amsterdam, The Netherlands, pp. 295–336.

[55] Majumdar S., Nandi S. K., Ghosal S., et al. (2021). Deep learning-based potential ligand prediction framework for COVID-19 with drug-target interaction model. *Cognit Comput.* doi: 10.1007/s12559-021-09840-x

[56] Ingraham, N. E., Lotfi-Emran, S., Thielen, B. K., Techar, K., Morris, R. S., Holtan, S. G., Dudley, R. A., & Tignanelli, C. J. (2020). Immunomodulation in COVID-19. *Lancet Respir. Med.*, 8, 544–546.

[57] Lv, H., Shi, L., Berkenpas, J. W., Dao, F. Y., Zulfiqar, H., Ding, H., Zhang, Y., Yang, L., & Cao, R. (November 2021). Application of artificial intelligence and machine learning for COVID-19 drug discovery and vaccine design. *Briefings in Bioinformatics*, 22(6).

[58] Bocek T., Rodrigues B. B., Strasser T., & Stiller B. (2017). Blockchains everywhere - A use-case of blockchains in the pharma supply-chain. Proceedings of the IM IFIP/IEEE International Symposium on Integrated Network and Service Management, pp. 772–777.

[59] Accurate, audited and secure How blockchain could strengthen the pharmaceutical supply chain, PWC report, December (2017).

[60] Sunny, J., Undralla, N., & Pillai, V. M. (2020). *Supply Chain Transparency through Blockchain-Based Traceability: An Overview with Demonstration.* Computers & Industrial Engineering, Computers & Industrial Engineering Elsevier.

[61] Shi, J., Yi, D., & Kuang, J. (2019). *Pharmaceutical Supply Chain Management System with Integration of IoT and Blockchain Technology.* Springer Nature Switzerland AG 2019 M. Qiu (Ed.). SmartBlock, LNCS.

[62] Akhtar, M. M., & Rizvi, D. R. (2020). *Traceability and Detection of Counterfeit Medicines in Pharmaceutical Supply Chain Using Blockchain-Based Architectures.* Sustainable and Energy Efficient Computing Paradigms for Society, Springer Innovations in Communication and Computing.

[63] Haq, I., & Esuka, O. M. (2018). Blockchain technology in pharmaceutical industry to prevent counterfeit drugs. *Int. J. Comput. Appl.*, 180(25), 8–12.

[64] Szabo, N. (1997). Formalizing and securing relationships on public networks. *First Monday*, 2. 10.5210/fm.v2i9.548

[65] Botcha, K. M., Chakravarthy, V. V., & Anurag. (2019). Enhancing traceability in pharmaceutical supply chain using Internet of Things (IoT) and blockchain. In IEEE International Conference on Intelligent Systems and Green Technology.

[66] Soundarya, K., & Pandey, P. (2018). Counterfeit solution for pharma supply chain. *EAI Endorsed Trans. Cloud Syst.*, 3, 154550.

[67] Mackey, T. K., & Liang, B. A. (2011). The global counterfeit drug trade: Patient safety and public health risks. *J. Pharm. Sci.*, 100(11), 4571–4579.

[68] da Silva, R. B., & de Mattos, C. A. (2019). Critical success factors of a drug traceability system for creating value in a pharmaceutical supply chain (PSC). *Int. J. Environ. Res. Public Health*, 16(11), 1972.

[69] Raza, K. (2020). Artificial intelligence against COVID-19: A meta-analysis of current research. In *Big Data Analytics and Artificial Intelligence against COVID-19: Innovation Vision and Approach* (Volume 78, pp. 165–176). Hassanien, A.-E., Dey, N., & Elghamrawy, S. (Eds.). Cham, Switzerland: Springer International Publishing.

[70] Shi, F., Wang, J., Shi, J., Wu, Z., Wang, Q., Tang, Z., He, K., Shi, Y., & Shen, D. (2021). Review of artificial intelligence techniques in imaging data acquisition, segmentation, and diagnosis for COVID-19. *IEEE Rev. Biomed. Eng.*, 14, 4–15.

[71] Bachtiger, P., Peters, N. S., & Walsh, S. L. (2020). Machine learning for COVID-19-asking the right questions. *Lancet Digit. Health*, 2, e391–e392.

[72] Acharya, A., Agarwal, R., Baker, M. B., Baudry, J., Bhowmik, D., Boehm, S., Byler, K. G., Chen, S. Y., Coates, L., Cooper, C. J., et al. (2020). Supercomputer-based ensemble docking drug discovery pipeline with application to COVID-19. *J. Chem. Inf. Model.*, 60, 5832–5852.

[73] Swayamsiddha S., Prashant K., Shaw D., & Mohanty C. (2021). The prospective of artificial intelligence in COVID-19 pandemic. *Health Technol (Berl)*, 11(6), 1311–1320.

[74] Naseem, M., Akhund, R., Arshad, H., & Ibrahim, M. T. (2020). Exploring the potential of artificial intelligence and machine learning to combat COVID-19 and existing opportunities for LMIC: A scoping review. *J. Prim. Care Community Health*, 11, 2150132720963634.

[75] Alsharif, M. H., Alsharif, Y. H., Albreem, M. A., Jahid, A., Solyman, A. A. A., Yahya, K., Alomari, O. A., & Hossain, M. S. (2020). Application of machine intelligence technology in the detection of vaccines and medicines for SARS-CoV-2. *Eur. Rev. Med. Pharmacol. Sci.*, 24, 11977–11981.

[76] D'souza, S., Nazareth, D., Vaz, C., & Shetty, M. Blockchain and AI in pharmaceutical supply chain (May 24, 2021). In Proceedings of the International Conference on Smart Data Intelligence (ICSMDI 2021), Available at SSRN: https://ssrn.com/abstract=3852034 or 10.2139/ssrn.3852034

[77] Humayun M., Jhanjhi N. Z., Niazi M., Amsaad F., & Masood I. (2022). Securing drug distribution systems from tampering using blockchain. *Electronics*, 11(8), 1195. 10.3390/electronics11081195

8 Blockchain for SCM: A Prospective Study Based on a Panel of Literature Reviews

Mostafa Qandoussi

Faculty of Economics and Management, Laboratory of Economics and Management of Organizations, University Ibn Tofail, Kenitra, Morocco

Abdellah Houssaini

Faculty of Economics and Management, University Ibn Tofail, Kenitra, Morocco

CONTENTS

8.1 INTRODUCTION

Recently, the industrial world has undergone a great change, the use of digital tools to manage upstream and downstream activities of the supply chain is booming. A few of the difficulties that have prompted the need to modernize and upgrade the current SCM include the globalization of production and distribution flows, the diversity of stakeholders in SCM processes and activities, the absence of mechanisms for collaboration and coordination between internal and external actors of the same SCM, and extremely demanding consumers.

As a result, SCM becomes a more intricate network with a variety of issues, including rising product demand, stock outs, and customer satisfaction. In order to supply consumers with a high-quality product that meets their needs, efficient management of all SCM processes, products flows, and information flows becomes important. so it requires the cooperation of all SC parties involved in an activity.

The aforementioned SCM issues as well as additional issues could all be solved by blockchain.

A progressive application of blockchain technology in the SCM domain will positively impact the quality of current systems, particularly with regard to the nature of the information introduced into the system, the reliability of the information circulating between the actors, and the access to the information by the various participants. Thus, the blockchain presents unavoidable advantages, such as resistance to data modification and increased efficiency, allowing real-time data analysis (*OECD, 2019*).

In 2008, the blockchain was initially used to secure the transfer of the cryptographic currency named Bitcoin. The Bitcoin blockchain is the first application of blockchain distributed ledger technology. This application has fully disrupted the economic sphere. The world has begun to move from a traditional economy to a digital economy through the progressive digitization of physical, informational, and logistical processes (Dujak & Sajter, 2019). The Blockchain distributed ledger was developed by a person or a group of people under the pseudonym of Satoshi Nakamoto to solve the problems of loss of trust in central systems and financial intermediaries, the high cost of transaction operations, and the complexity of financial procedures.

The adaptation of this new decentralized, independent, global, and autonomous system by non-financial companies, such as insurance, health care, and supply chain management (SCM), is becoming a strategic necessity and an important competitive practice in order to survive in a rapidly changing ecosystem. Until today, several attempts to create blockchain applications for SCM have been registered, as a result of collaborative acts between SCM companies and information and communication companies.

Several SCM actors, academic authors, and researchers in the field of SCM believe that the blockchain platforms created do not allow companies to benefit from all the mechanisms and features of blockchain technology. They represent only simple technical illustrations of the technology, that needs to be put to development projects and reconfigurations.

Multiple scientific research has been published in recent years dealing with the problem of the use of blockchain in the SCM, most of them describing the uses of the existing platforms or dealing with the problems of the current SCM and presenting the blockchain as a practical and adequate solution to overcome the anomalies that the SCM suffers. Some of them have been enough to list the existing applications, the companies that create them, and the industrial companies that use them. Only a few articles have critically examined the possibility of using blockchain technology in SCM, presenting opinions and bringing ideas.

Using a panel of literature reviews, pre-selected according to well-defined evaluation indexes, classified into groups according to the criteria of quality of information processed, the membership of experts interviewed or used in the study as control agents, and the future projections presented in the studies.

Scope of the Study: The work aims at incorporating all the literature that has been published as articles, review papers, and short surveys pertaining to the applications, integration, and implementation of blockchain technology in SCs and logistics. We formulate projections from a panel of literature available using a systematic literature searching methodology to capture significant data from available literature. The Chapter highlights the current applications of blockchain technology, future prospects and trends, and the benefits that this technology could bring combined with emerging technological tools (IOT, AI ...) and Industry 4.0 to the current SCM system. Through this chapter, guidelines are provided for further research on the technology and assist managers and executives who are considering implementing blockchain for their respective companies.

Organization of the Chapter: The chapter is organized as: Section 8.2 entails conceptual framework of the study and specifies problem definition and ecosystem and blockchain innovations in SCM. Section 8.3 highlights research methodology. Section 8.4 discusses basic features and factors influencing the adoption of blockchain technology in SCM. Section 8.5 lays the foundation of basic projections. Section 8.6 discusses analysis and results. Section 8.7 elaborates on the limitations of the study. Section 8.8 concludes the chapter with future scope.

8.2 CONCEPTUAL FRAMEWORK OF THE STUDY

8.2.1 PROBLEM DEFINITION

As a result of the globalization of production and distribution chains, as well as the expansion of the retail sector, SCMs are becoming increasingly complicated and time consuming. An electronic product, for example, can be manufactured, packaged, and sold in several countries in different geographical areas. This creates many difficulties and challenges in terms of quality and regulatory requirements. Because the SCM has a number of internal and external stakeholders, consumers today are increasingly demanding and want to have confidence in the authenticity of the products as well as the fact that the products are produced in accordance with ethical and environmental standards.

However, traditional SCMs are increasingly facing huge problems and drawbacks, such as:

- Traditional SCMs are unable to guarantee product safety at any stage of the chain.
- Detailed information about the origin of the product may no longer be available.
- Failure of transparency and traceability.
- Product control and life cycle management.

Advanced technologies are being developed rapidly by the day to address the complex strategic challenges facing SCM. One of the most emerging technologies is the decentralized ledger technology, blockchain. It is based on cryptographic hashing. The SCM and logistics sectors are among the most obvious applications of blockchain technology. In logistics (Madumidha et al., 2019), blockchain offers a wealth of information related to shipping data and product status, which can help prepare for any expected delays and eliminate the need for trusted third-party representatives.

Consumers today are increasingly demanding and want to know that the products they buy are genuine and that they are produced in accordance with ethical and environmental standards (Feng Tian). Thus, in the absence of practical studies on the topic, it is very difficult for the SCM industry players to clearly understand how blockchain could be used in the SCM situation, to secure the operations of transferring physical and information flows, improve the traceability of the product throughout the SCM and track it in real time. The other emerging IT technologies must combine to achieve the visibility envisaged behind the adoption of the blockchain project in SCM.

Retail, the size of the retail network, and their geographical locations, are one of the main strategic sectors of the SCM where blockchain technology must be implemented in order to improve traceability. The potential of blockchain remains so far untapped in the service and industry sectors, as the applications used do not meet the expectations of the industry players and do not allow for anomaly correction and contingency planning. If a participant refuses to use the blockchain application that the SCM companies have embraced, it has an impact on the entire SCM's visibility.

Academic papers, company reports, expert opinions, and notes from industry participants and SC participants are examined and a prospective study is conducted on blockchain and its potential use in SCM to determine the evaluation tools needed to understand the effectiveness of blockchain technology in managing and improving SCM traceability.

8.2.2 Ecosystem and Determining Blockchain Innovations in SCM

Thematic interests in studying the diffusion of technology are quite diverse among researchers and practitioners. The organizational and environmental technological contexts of blockchain are among the main criteria that determine the place of blockchain in SCM systems. Increasingly, we are also interested in using what we know about SCM information fusion processes to formulate adoption scenarios, strategic potentials, and quality of innovation implementation, support the use of valuable blockchain innovations, and, as ultimate outcomes, demonstrate the effectiveness of blockchain adoption at the SCM network level.

- **Comparative advantage:** Comparative advantage is a determinant that largely affects the organizational adoption of blockchain-based SCM (Brandon-Jones & Kauppi, 2018). In terms of blockchain adoption in SCM, the comparative advantage is equivalent to performance expectations. The potential usage of blockchain technologies in SCM might boost its efficiency and effectiveness. Transparency in information quality, efficiency, product quality, and supply chain management procedures are all encouraged by blockchain technology, which improves participant performance (Tönnissen & Teuteberg, 2020; Babich & Hilary, 2020; Fosso Wamba et al., 2020).
- **The complexity level:** Complexity is the degree to which an invention is seen as being relatively challenging to comprehend and apply (Everett & Rogers, 2003). The adoption of the blockchain makes it possible to reduce the degree of complexity of the existing SCM systems, it makes it possible to eliminate the central trusted third parties (Kim & Laskowski, 2017). Nevertheless, the lack of standardization (Makhdoom et al., 2019) and regulation is a challenge for the adoption of blockchain in the SCM industry, to this is added the absence of auditing, management, and control standards.
- **Compatibility and scalability:** Compatibility is an important concept for innovation adoption (Imeri et al., 2019). Blockchain compatibility with the organization's current SCM system makes it a key factor influencing the use of blockchain in SCM, while scalability can affect the adoption of blockchain-based SCMs because system scalability can lead to extremely complicated costs (Wang et al., 2016), related to adding new data and validation and storage processes that can cause inefficiency and slow transactions (Makhdoom et al., 2019).
- **Trust and confidentiality:** Trust is an important factor in the adoption of blockchain technology in SCM, in terms of system performance, availability, and possible risks associated with the adoption of the system (Şener et al., 2016). The confidentiality of the stakeholders and the effective operation of the SCM must not be impacted by the blockchain's encryption feature.
- **Organisations sizes and infrastructure systems:** The size of the organization is a key determinant of blockchain adoption (Mendling et al., 2018) in SCM. The larger the company, the more inclined it is to adapt to blockchain in order to set itself apart from competitors. the adoption of blockchain in SCM systems is a strategic decision by the executive members of the organization and requires new regulatory requirements in the organization (Wang et al., 2016). Thus, the adoption of the blockchain requires a solid IT infrastructure, consist of physical infrastructure, information system, competent human resources able to support the use of the system and financial resources to finance all the necessary implementation procedures. Generally, the blockchain is an interconnected computing infrastructure capable of consolidating existing systems into a single system, and allowing each entity to keep a copy of immutable transactions (Francisco & Swanson, 2018).

- Regulations, and external pressure: The organization gains a competitive edge and can improve efficiency and transparency by implementing blockchain technology (Swan, 2015.). But, the adoption of new technologies is significantly impacted by regulations between SCM partners. Since blockchain technology is still in its infancy and rules are still being defined, this has a substantial influence on adoption policy and, in some situations, hampers the adoption of blockchain in SCM (Gökalp et al., 2018; Wang et al., 2016).

8.3   RESEARCH METHODOLOGY

8.3.1   PROSPECTIVE STUDY

The field of the research is still young, uncertain, and complex. The main research question aims to investigate the possible prospects of using blockchain technology in the SCM industry. The prospective study technique is based on an exploratory prospective analysis (*Chapitre III. La Prospective Exploratoire et Normative: Entre Possibles et Souhaitables*, n.d.). The literature reviews and articles published on the subject by researchers and experts in the fields of blockchain and SCM.

In a first step, we studied and analyzed existing academic research (journals, conference proceedings, and scientific publications) and practitioners' opinions (discussions and online blogs) on blockchain and SCM (Dujak & Sajter, 2019) to fully understand the recent developments and build a knowledge base on the topic. Then we selected the articles to be analysed; taking into account the relevance of the article to the research topic, the date the article appeared, the bibliography of the article authors, their experience in the fields of blockchain and SCM, and the nature of the content. This step allowed us to build a panel of articles that we studied in-depth to build the projections for the study. After that, we selected the studies that have a future or prospective vision of the application of blockchain technology for SCM. Finally, we analyzed and discussed the results to draw information about the potential role of blockchain in the SCM industry (Figure 8.1).

8.3.2   METHODOLOGY CHOOSING ARTICLES

Research on the topic was initially conducted on key core concepts such as "blockchain, SCM, SCM challenges, traceability, transparency, SCM security …" in the Scopus database, because of its reliability, its broad coverage compared to other academic databases, and the greater quantity of publications it indexes. And in other academic databases in a second phase to capture all potentially relevant publications that are not indexed in Scopus.

Smart contracts, IoT, and DLT are the most popular keywords used in research to highlight the potential of blockchain-based smart contracts in reforming SCM and logistics operations.

Because most case studies in blockchain and SCM research published to date are based on the deductive technique, it's impossible to guarantee their generalizability.

Textual documents, research articles, reports, and patents have become a rich source of information, where several researchers use these new tools to express ideas about various technological

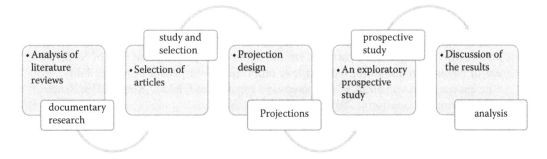

FIGURE 8.1   Steps of the Research.

advancements and applications in the form of theories, hypotheses, methods, approaches, and results of research projects experienced by researchers and among interested parties. Indeed, scientific articles are becoming a popular strategy to keep up with the rapid growth of information because they allow different stakeholders to express and present their ideas in critical and academic ways about the uses of technologies as well as future trends.

An extensive review was conducted to select articles that were relevant to our work. The selection of articles was based on the four recommendations of (Rowley & Slack, 2004): The article must be relevant to the research topic, the article must be new and up to date, the article must have extensive source references, and the article must be evaluated as relevant and of scientific quality.

However, thematic modeling is a promising research method, and is becoming more and more adopted by blockchain researchers and supply chain management specialists. As a result, article analysis provides a useful supplementary alternative approach for investigating the use of blockchain technology in supply chain management.

The biggest challenge found in picking the articles utilized in this study is figuring out how to extract meaningful information from entire texts and a large database.

8.4 BASIC FEATURES AND FACTORS INFLUENCING THE ADOPTION OF BLOCKCHAIN TECHNOLOGY IN SCM

8.4.1 BLOCKCHAIN: AN UNDERLYING STRUCTURE OF TRUST

A blockchain is a form of shared, distributed, decentralized, immutable, and secure database structure. The blockchain uses a computer consensus protocol to validate transactions on the blockchain network (Bahri & Girdzijauskas, 2019). Miner nodes in the blockchain network work competitively to solve the computational problem corresponding to each transaction. A hash function in the form of code can then be easily sent over an Iinternet network without fear that someone can modify it. The first node to solve the problem would receive a "cryptocurrency" token as a reward for their work. The blockchain can be used as an alternative network to the traditional banking system, a new monetary system without hierarchy and proper authorization for transactions, which is the case for the Bitcoin network. It can also be used as an authentic database to store valid transaction blocks or as a digital asset management platform against forgery attempts. The unprecedented exploration of this new currency makes blockchain technology a deterministic structure of the global economy, and it requires propagation in areas of the economy other than the financial field, especially in the SCM sector, in order to overcome the current problems and needs in terms of trust (Qian & Papadonikolaki, 2021), security, transparency, traceability, and value storage (Figures 8.2 and 8.3).

8.4.2 BLOCKCHAIN'S FUNCTIONALITY AND ADOPTION ECOSYSTEM IN THE SCM INDUSTRY

The study thus demonstrated that blockchain through these strategic features can improve the relevance of SCM practices and SCM performance. There are six main features, namely:

- **Real-time information sharing:** Real-time information sharing allows companies to know the location and status of assets at any stage of the SCM process. This enhances SCM integration, strategic planning, and assists SCMs in creating strong and efficient partnerships amongst SCM partners. On the operational performance side of SCM, the use of blockchain allows for reduced lead times, improved planning and flexibility, and efficient use of an organization's resources and capabilities (Hald & Kinra, 2019; Kshetri, 2018; Mylrea & Gourisetti, 2018; Queiroz et al., 2020).
- **Cyber security:** The actual problems with translational SCM networks revolve around security and confidentiality of data shared among SCM members. They are the main challenges that hinder the development of a strong and integrated global SCM. The blockchain raises the

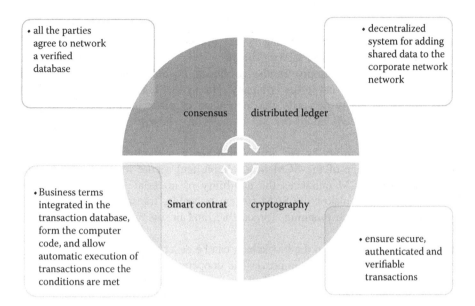

FIGURE 8.2 Pillars of Blockchain Technology.

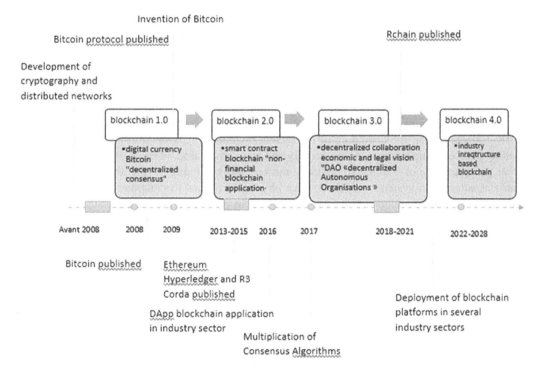

FIGURE 8.3 Evolution of Blockchain Technologies.

level of trust between SCM participants while enhancing security. It provides a secure, immutable, and shared transaction platform. Because data stored on the shared ledger must be consistent and immutable, the blockchain improves inter- and intra-organizational secrecy and trust (Cole et al., 2019; Hald & Kinra, 2019; Kim & Shin, 2019, Mylrea & Gourisetti, 2018).

- **Transparency:** In SCM-based blockchain, all SCM participants may access the product information at any moment thanks to the blockchain's ability to foster open communication among the chain's parties. It allows, thus, a real-time management and audit of the product in an automated way via a time-stamping system, immutable, and cryptographic (Cole et al., 2019; Kim & Shin, 2019; Yoo & Won, 2018).

- **Reliability:** Blockchain improves the resiliency of SCM functions and practices, boosting security and trust in the company (Hasan, 2017; Kshetri, 2018). Blockchain allows any SCM entity to verify the legitimacy of other members, and to have all the information it needs about the SC actors with whom to try to create a business relationship.
- **Traceability:** Traceability is the ability to trace a product from origin to the end customer at any stage of the SCM in a reliable and secure manner. The application of blockchain in SCM enhances the audibility of material, product, and SCM information flows and makes the SCM system traceable, transparent, and accountable. This improves resource planning, forecasting, and business flexibility (Hasan, 2017; Song et al., 2019).
- **Visibility:** SCM visibility in the blockchain can be described by identifying the location of products in real time. This improves the cooperation between SCM actors and the visibility of organizational processes and documents (Kim & Shin, 2019; Kshetri, 2018; Rogerson & Parry, 2020). As a result, SCM businesses will be able to clearly authenticate faulty or fake products as well as promptly and precisely identify faults that prohibit the SC from functioning as it should.

8.4.3 HISTORY AND FACTORS INFLUENCING INDUSTRY 4.0 AND SCM

The industrial revolution began in the 18th century, with the introduction of new production methods that reshaped the workplace. Great Britain was the first industrial nation, and its transition lasted almost a century from the 1750s to the 1850s. Over a period of nearly 200 years, three industrial revolutions took place in three waves. In the late 1700s, the innovation of steam engines and the centralization of production in factories led to an extremely high increase in production. Nearly 100 years later, the beginning of the second revolution marked the use of assembly lines, resulting in an extreme increase in productivity and mass production. In the 1950s, the paradigm of digital programming of automated systems marked the beginning of the third industrial era. Industrial automation systems, which connect physical activities to computer infrastructure and communications via interconnected systems, ushered in the fourth and last industrial revolution (Demir, 2021) (Figures 8.4 and 8.5 and Table 8.1).

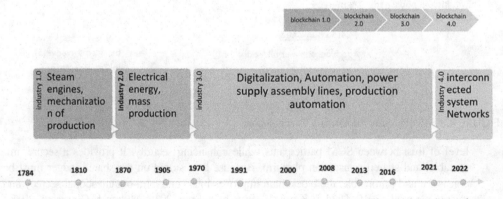

FIGURE 8.4 History of Blockchain and Industry.

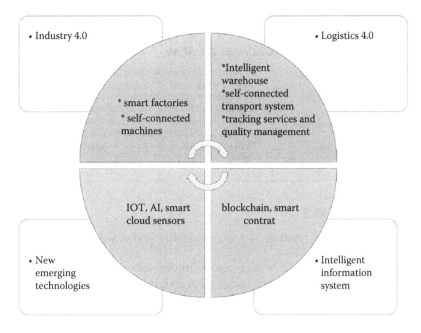

FIGURE 8.5   Ecosystèm and Adoption Technologies of Blockchain in SCM.

TABLE 8.1
Benefits Blockchain Technology Provides to the SCM

Blockchain & SCM	The contribution of blockchain to SCM	Description
	• Transparency • Responsibility • Flexibility	Blockchain should increase transparency and accountability in SCM networks, making the SCM more flexible (Ahram et al., 2017; Kshetri, 2017a; Kshetri, 2017b; Kshetri, 2018; O'Leary, 2017).
	• Visibility • Optimization • Request	Applications based on blockchain technology have the potential to generate breakthroughs in three areas: visibility, optimization, and demand.
	• Fight against counterfeiting • Reduce paperwork • Product tracking	Blockchain will improve logistics and SCM, identifying counterfeit products, decreasing paper load processing, and facilitating origin tracking (Hackius et al., 2017; Kennedy et al., 2017; Toyoda 2017; Wee et al., 2018).
	• Transactions without a trusted third party	Buyers and sellers can make direct transactions without the help of a trusted intermediary (Subramanian, 2018).
	• Security	The use of blockchain-based applications in SCM can ensure security (Dorri et al, 2017; Ahmed & Broek, 2017).
	• Fighting against information asymmetry	More efficient management of more robust contracts between logistics (3PL, 4PL) (Lesueur-Cazé et al., 2022).
	• Tracking and tracing	Blockchain-based applications have the potential to improve tracking and tracing mechanisms (Apte & Petrovsky, 2016; Dûdder & Ross, n.d.; Heber, 2017; Q. Lu & Xu, 2017; Tian, 2017).
	• Information management	Blockchain-based applications offer better SCM information management (Banerjee, 2018; O'Leary, 2017; Turk & Klinc, 2017).
	• Protection of intellectual property	Blockchain-based applications can improve intellectual property (Herbert & Litchfield, 2015; Holland et al., 2017; Tsai et al., 2017).

(*Continued*)

TABLE 8.1 (Continued)
Benefits Blockchain Technology Provides to the SCM

• Advanced data analysis	Blockchains offer a new encrypted and decentralized recommendation system (Frey et al., 2016; Frey et al., 2016).
• Inventory management	Blockchain improves inventory management and performance in complex SCMs (Madhwal & Panfilov, 2017).
• Intelligent transport system	Blockchain can improve the current transportation system and move it towards an intelligent system (Yuan & Wang, 2016; Lei et al., 2017).
• New manufacturing architectures	Blockchain proposes new smart manufacturing architectures (SyncFab, 2018).

8.5 BASIC PROJECTIONS

Projection 1: Blockchain technology is a very valuable and scalable technology:

Blockchain technology is a scalable technology, the number of transactions and network nodes is growing rapidly, resulting in increased capacity requirements. The coordination between SCM members, and the tracking of transactions along the chain in real time requires connecting a large number of machines, which greatly increases the number of transactions per second. The evolution of transactions and the increase of connected devices could have limits and drawbacks in terms of security.

Projection 2: The creation of new international standards compatible with blockchain is important to manage the SCM domain:

To encourage the use of blockchain technology in SCM, make application development simpler, or facilitate the quick interchange of information and data amongst SCM players, standards must be established. This is crucial for interoperability across autonomous machines. The SCM industry is different from other industries using blockchain, especially in the financial sector, where the first blockchain application was established. This, strongly encourages the idea of establishing standards and government structures of its own for the SCM industry, to meet the growing regulatory needs and encourage interoperability and heterogeneity of blockchain systems.

Projection 3: The Blockchain technology enables a collaborative trust ecosystem:

The widespread use of blockchain technology to manage SCMs requires the establishment of a global and unified regulatory system. The current system relies heavily on the laws and regulations of individual countries. This poses regulatory and organizational challenges, hindering the smooth operation of global SCMs and the desire to apply blockchain technology to manage SC processes and logistics operations. For example, if an organization deletes or does not enter reliable and correct data, which is contrary to the immutability of blockchain and controversial to the compliance of blockchain.

Projection 4: There is a legal framework for fraudulent acts arising from illicit transactions.

In the digital world, the computerized transaction protocols that execute and enforce the terms of a computer contract are called "smart contracts" (Nick Szabo, 1994). It is not clear how strong the codes of the smart contract are in the real world. It is always debatable whether there are errors in the contract that allow the transaction through fraudulent acts, if a hacker acted in a guilty manner (O'hara, 2017).

Projection 5: SCMs using blockchain technology are no longer controlled by humans

The developers begin to think of developing more autonomous applications for the SCM that act independently and are able to take, based on the information received from the devices and intelligent tools integrated into the system, adequate decisions to correct the probable problems. Shortly, the SCM will become fully autonomous, self-controlled, and human intervention could be limited to certain strategic tasks that they cannot assign to machines.

Projection 6: The majority of blockchain applications are related to the financial industry's economy.

More than 50% of blockchain applications and use cases come from the financial services industry. Many researchers argue that the use of blockchain in SCM requires the use of other technological tools such as IOT. The literature available to date has discussed the benefits of an interconnected and intelligent SCM. This indicates a rising desire to adopt blockchain as an underlying technology in the SCM 4.0 business on a large scale.

Projection 7: There is no blockchain for SCM; all that has been developed so far are tracking and tracing applications using some principles of blockchain technology.

All that is being produced at the moment are decentralized applications using the principles of blockchain technology to create performance in the SCM and ensure traceability and tracking of products throughout the supply chain. To date, several applications have emerged in several sectors. Most of the applications aim to trace the product from the origin to the final customer and to provide detailed information about the product in an immutable way.

Projection 8: Blockchain technology and IoT tools allow for the decentralized and automatic transfer of money between bank accounts

The use of electronic sensors and intelligent IOT tools is increasingly transforming machines into intelligent and autonomous objects. As a result of the rapidly increasing number of SCMs, towards an SCM 4.0, the new intelligent factories 4.0 are capable of detecting their surroundings, making decisions, and carrying out the necessary payments autonomously.

Projection 9: The decentralized exchange of information, data, and electronic payments without trusted intermediaries has become an everyday business.

SCM based on blockchain technology and new Industry 4.0 tools takes advantage of smart tools such as IOT, AI and electronic sensors to improve the transparency and traceability of SCM and to ensure real-time tracking of products transferred on the chain. In addition, direct payment without the use of a trusted third party is vitally important and allows for faster transfer of operations and asset management and minimizes costs related to operations. The crypto-currency bitcoin is the first electronic currency used for payment, but it still presents some challenges for micro-payments. This necessitates the search for alternative blockchain consensus mechanisms that allow for low or no transaction costs.

Projection 10: All acts governing SCM partner relationships and transaction regimes are written as algorithmic code in smart contracts

In the SCM of tomorrow, the influence of smart contracts on companies and actors in the same SCM chain is controversial. On the one hand, a smart contract, according to a number of researchers, is a magical instrument for automated management of property transfer and payment operations between SCM members. These digital instruments can induce cost reductions and increase the efficiency and transparency of processes, so organizations can act in autonomous and decentralized ways as market participants without human interference. On the other side, some experts believe that relying only on smart contracts can generate issues for businesses, particularly due to smart contracts' immutability (Jones & Vigliotti, 2020).

Projection 11: The. The third-generation blockchain and combined technologies are the main dominant protocols for the management of the SCM

The goal is to develop applications that combine blockchain and smart tools to track and trace products along the supply chain. These applications promise to enable micro-payments using cryptocurrencies and communication between machines without incurring transaction costs. Applications using solutions such as Ethereum are the most suitable for the SCM context, according to several practitioners and researchers.

Projection 12: The blockchain technology is strongly combined with IoT and AI technology to be the fundamental technology of the future SCM

Several research works in the field of SCM have critically discussed how blockchain can significantly improve the transformation of SCM from a traditional SCM using classical methods for managing and monitoring operations under different SCM processes to a digital SCM based on decentralized and interconnected systems. The use of IoT in combination with blockchain technology is supposed to improve the performance of logistics processes and consensus protocol (Francesco Corea, 2017). In addition, the use of intelligent tools makes it possible to recognize complete information about the location, environment and storage conditions of each product transferred by the chain in a permanent way, which is allowed him to take corrective actions and security measures in real time. For the time being, no blockchain application would meet the challenges or allow the SCM system to benefit from all the advantages brought by blockchain.

Projection 13: The blockchain technology provides total security for transactions throughout the SCM.

In a digital environment, SC members are interconnected and continuously share product information, based on machine-to-machine communication linking all SCM processes (Popov, 2018). In addition, emerging collaborative IoT technologies allow for detailed and up-to-date product information through the use of QR tagging technology, FRID technology, or smart sensors. Along with the SC process, applications installed on smartphones allow users to read the information written on the product label and transmit it to the interested parties at any time. In this digital ecosystem, security issues are becoming more critical and systems are increasingly vulnerable to attacks. To address these threats, several researchers and practitioners consider blockchain a promising technology (Cullen et al., 2020; Sharma et al., 2017; IBM, 2022; Dorri et al., 2017; Tayal et al. 2021, Vora et al., 2018), and several studies discuss attempts to use blockchain in various sectors other than money, and they point out that blockchain itself has security challenges that need to be addressed before adding value to the SCM industry, such as how blockchain can change AI and how decentralized intelligence affects our future (Francesco Corea, 2017), especially in terms of optimizing warehouses and monitoring the supply chain. Additionally, SCMs are expanding geographically and in scale, therefore the players must master their digital transition to prevent any human error. In addition, blockchain has become an essential solution to secure product transfers and ensure reliability and traceability throughout the SCM. The NFT "Non-Fungible Token" (a digital asset that takes the form of an image, a sound, a video, etc.) can better serve each stage of the SC, clarify the responsibility of each actor, and ensure customer satisfaction due to their unchangeable characters (Table 8.2).

TABLE 8.2

Subject Groups and Disruption of Future Influence of the Blockchain in SCM

Subject Groups	Subject Subgroups	Description	Future Influence of the Blockchain in SCM
Types of Blockchains	• Public Blockchain • Private blockchain • consortium Blockchain	• Public blockchain (the user can be a user, miners, developers) • Private blockchain (users need authorization to join the networks) • A blockchain consortium (combines the advantages of Confidentiality of an authorized and private blockchain with the advantages of security and transparency of a public blockchain).	• Privacy policy • Absence of a centralized authority. • Reliability of the data. • A quick and easy audit. • The Consortium blockchain is the most suitable for the SCM context, and it increases the security of transactions. • A standardized form of blockchain that allows SCM actors to maintain the necessary communication, flexibility, transparency, and trust.
Traceability	• Product life cycle • Traceability of the logistics system • Information on the origin and provenance of the product • Real-time product and resource tracking	The distributed nature of blockchain technology could help companies track and trace SCM products from origin to end customer. The use of blockchain technology in combination with other emerging technologies (smart mobile devices, IOT, artificial intelligence, cloud, 5G, RFID, smart sensors …) for SCM management can effectively improve traceability.	• Detailed information on the origin of the product. • Data reliability and fast authentication of fraud and threats. • Best quality/service ratio. • Improve customer satisfaction. • Limit the use of a trusted third party.
Technological developments and blockchain tools	• Blockchain application development platforms • Ethereum blockchain platform • Smart contract • Digital identity	Blockchain is the underlying technology of the Bitcoin currency. The first blockchain application was in the financial field for managing financial transactions. In 2015, the world saw the first attempts to use blockchain in industrial sectors, including SCM with the Ethereum blockchain. In addition, applications use smart contracts similar to translational contracts to facilitate transaction operations. Smart contracts are automatically executed once the predefined conditions of the contract are established.	From the origin to the completion of the product, blockchain applications can effectively facilitate the connection, transparency, improve efficiency, and illuminate the responsibility of each participant. It could also help consumers have a detailed and complete record of the products, which can help them choose their products to consume. Using the Industry 4.0 and IoT paradigms as well as smart contracts to create an automated, integrated, and virtualized SCM.

(Continued)

TABLE 8.2 (Continued)
Subject Groups and Disruption of Future Influence of the Blockchain in SCM

Subject Groups	Subject Subgroups	Description	Future Influence of the Blockchain in SCM
Transparency	• Visibility • Data reliability • Sustainable SCM • Interconnected process	The issue of transparency is crucial and poses several problems for SCM managers and users. Transparency is an important criterion today for the evaluation of the quality of SCM, as well as for the choice of the distribution chain by companies using SCM. Consumers are also becoming more discerning in their product selection, and they seek the most transparent SCM that provides a greater amount of information and visibility on the product as well as the conditions and ecosystem of production.	Blockchain could improve the transparency of product and process information throughout the supply chain (SC). Increase responsible consumption and sustainability in SCM. The blockchain creates SC visibility, real-time information sharing about the location and status of an object along the SCM. This increases customer satisfaction, the immutability of product offerings, and compliance through transparency. The blockchain provides transparency of activities, but not complete access by users to all the information included in the blockchain, allowing users to access only the information they need without exceeding the data privacy limit.
Security	• Security and blockchain: two similar concepts • Data quality, reliability and properties • Secure data sharing and exchange • Confidence in the network. • Illumination of trusted third parties	Blockchain provides a secure transaction mechanism, respecting the principle of data confidentiality and limiting access to data to authorized parties (Imeri & Khadraoui, 2019).	Blockchain improves the quality, reliability, and ownership of data shared between SCM actors. Secure exchange of value between participants without intermediaries. Immutable storage of transaction data. The immutable nature of the blockchain creates a high level of security. Privacy is becoming a major concern for SCM managers and users, and it should be at a high level. User roles should be clear to all parties on the blockchain. Verifiable information sharing could help overcome fraudulent actions.

| An innovative information systemof

Applications in the field of technology and computing | • Applications integrating IoT, IoT, and blockchain
• Artificial intelligence supported by blockchain
• Machine to machine interactions
• Autonomous materials and machines
• Intra-active and interconnected information systems | Newly emerging IoT and distributed systems technologies represent a driver for SCM innovation. And provide promising solutions to critical SCM problems.
The integration of emerging technologies generally involves high costs and slow diffusion. New distributed systems and associated blockchain application platforms have created better integration between the physical, information, and computing spheres, and enable active correlation between industry sectors, especially in the fourth Industrial Revolution. | RFID and sensor technologies that are integrated can provide secure and reliable data on the product.
The combination of new developing technologies and blockchain could make it easier to collect product data throughout the product life cycle.
New emerging technologies can provide promising solutions to critical SCM problems, improve it and make it competitive.
• Integrating IoT with a blockchain-based SCM platform has the potential to boost overall SCM efficiency.
• Collaborative technologies: IoT devices can send data to blockchain registries in real time about the location, arrival time, and shipping status of products and environmental conditions.
• The combined use of RFID and blockchain allow for tracking and tracing of products in real time and provides information that could be imported directly into the SCM blockchain (Wang et al., 2022).
The IoT allows for integrated management of SCM processes and helps to achieve traceability and transparency throughout the SCM (e.g. use of electronic sensors). |

8.6 ANALYSIS AND RESULTS

A diverse source of information has provided us with valuable insights into the potential roles of blockchain technology in the SCM industry. We discuss our derived concepts in light of existing literature reviews using the authors' perspectives. The adoption of blockchain in SCM is no longer an easy operation, as it is important to take into account certain factors, such as political, legal, and commercial preferences, in addition to consumer preferences.

There is a consensus among the authors of the selected articles in the second phase of the study that the scalability of blockchains has a strong impact on the role of blockchain in SCM and that technological advances are creating an entirely new business environment full of opportunities for companies and accelerate the pace of global competition (projection 1).

Recording and sharing data on a blockchain, has the potential to improve product visibility, security, solve counterfeiting, equity, and social responsibility issues, by making the processes of tracing and tracking the product throughout the SC more efficient and transparent (Kshetri, 2021). Many companies have partnered to develop blockchain-based solutions for SCM. Tracking can be done by blockchain platform or any other authorized ledger. All delivery activities will be recorded on the blockchain such as product origin, shipment data, transportation license, temperature; generally speaking all updates related to the status of the product in the SC.

According to opinion writers, the need to adopt a new information system is the best way to achieve an effective and real-time integration between the physical and digital to improve performance, involving recent advanced technologies, such as IOT, robotics, big data, artificial intelligence, industry 4.0, IoT, and blockchain (projections 12 and 11). The idea of creating an integrated system for SCM is discussed in much researches and it presents challenges like a lack of security. Using an integrated and open platform to manage transactions is much riskier and can cause problems in terms of privacy (projection 2). This has been the subject of several research studies on the problem of cyber security. Most of the opinions described in the articles have considered blockchain, because of its properties and the management mechanisms it offers, as the best solution to some of these challenges (Aslam et al., 2021). Thus, the blockchain can be used to support SCM practices by increasing the integration between all SCM functions and consequently improving operational performance (projection 9). SCMs experts and managers confirm that the lack of a decision-making and managerial framework is another challenge that needs to be overcome to better decide whether to invest in blockchain implementation or not, the cost that may entail the adoption of blockchain in small to medium level SCM, the changes that may bring blockchain to traditional SCM networks, and the global instability, knowing that SCM practices may vary from one SCM to another or even from one sector to another (Casado-Vara et al., 2018; Madhwal & Panfilov, 2017; Y. Wang et al., 2019; Helo & Hao, 2019).

The blockchain will give all stakeholders accurate, timely, real-time, and immediate information on the product's state, any issues that have arisen, and any effects that have resulted from any SCM process step. This will help to improve the problem resolution report. Thus, the principle of immutability will help to preserve the integrity of the blockchain network applied to SCM (projections 9 and 8). The blockchain networks with authorization remain the most suitable for the SCM context because of their high confidentiality (projection 6) (Xu et al., 2018).

Due to the deployment of both modern technology and antiquated SCM practices, SCM managers assert that this industry is very exposed to risks that develop from both internal and external sources. This has a detrimental effect on each link in the chain and makes it exceedingly challenging to handle the monitoring, control, and audit activities. Blockchain technology, through functionalities and features such as traceability, transparency, and smart contracts, can help to identify and solve SCM risks and offer great data protection (projections 1 and 3) (Fu & Zhu, 2019; Barron et al., 2016; Lu et al., 2017). The smart contract will make transactions and the transfer of ownership of an object from one entity to another less costly and eliminate the need for a trusted intermediary (Kshetri & Loukoianova, 2019).

SCMs based on blockchain technology will be more automated through the use of smart contracts in the management of processes. This gives more security to exchanges, cyber-resilience, improved automatic management and corrective acts, and confidentiality and trust through the use of POA consensus. The use of smart contracts associated with the blockchain immutability principle will guarantee the integrity of the information (projections 10 and 5) (Mylrea & Gourisetti, 2018).

Blockchain smart contracts can help overcome the shortcomings of traditional contracts in terms of efficiency, reduction of paperwork, and paper documents. In the traditional state, this information transfer is done via paper documentation exchanges, which are difficult to synchronize due to different data formats and architectures that do not provide timely information. In addition, information is difficult to verify and mitigate (OECD, 2019) security, tracking and traceability of the product. Through automatic management based on blockchain technology, contracts will be activated automatically when predefined conditions are met. A smart contract is an algorithmic language based on a code that allows the agreed actions to occur spontaneously, immediately and without intermediary upon satisfaction of the terms of the contract (projections 10 and 5) (Madumidha et al., 2019).

Several developers foresee the use of advanced smart contracts to handle complex transactions and enforce contract terms and conditions. At this level, smart contracts can also cover the negotiation phase between users, which would lead to an agreement between the parties. This means that in the near future, users could conclude contracts on the basis of variable supply and demand, or through auctions (Courbe et al., 2021). The frameworks and decision agents give the smart contract the tools it needs to carry out distributed artificial intelligence-based optimization and decision-analysis tasks. Through artificial intelligence algorithms, the smart contract will be able to modify rules and behaviors to improve the overall objectives of a community (projections 3, 4, 12, 11, 10 et 5).

Other statements consider that the immutable character and the cryptographic consensus protocol of the blockchain, will make the blockchain a trusted transaction platform, allowing it to create traceability and transparency between the activities of logistics and the processes of the SCM (projections 1 and 3). Thus, the distinctive features of blockchain such as real-time information sharing, cyber security, transparency, reliability, traceability, and visibility. They can improve the efficiency and operational performance of SCM by reducing time (manufacturing, delivery, and inventory turns) and costs (technical, quality, and manufacturing) (Behnke & Janssen, 2020; Cole et al., 2019; Helo & Hao, 2019; Phadnis, 2018).

Deploying blockchain is a good practice by which SCM retailers can maintain buyers' trust in their products and avoid the risk associated with contaminated products. Companies using emerging technologies such as QR combined with blockchain to track their products through SCM can be considered more credible and trustworthy. The use of new technological tools (QR codes, RFID …) has made counterfeiting impossible or extremely costly. Each QR code is unique and cannot be duplicated. Once the buyer receives the product, he can easily verify the legitimacy of the product using a QR code scanner, the consumer therefore in a single click can have all the history of the product (projection 13) (Wang et al., 2022).

The use of blockchain in combination with other technologies such as RFID, QR, and IoT has the potential to improve the creativity and efficiency of SCMs. Blockchain-based applications can help provide instant, real-time answers to shipping status questions such as (projections 13, 12 and 11): where is the current location of the shipment? When will it arrive at its destination? Were there any problems during the journey? and which SC participant is responsible for the problem at hand? (Kshetri et al., 2021).

According to experts, the market for smart devices is rapidly growing. Sensors located on top of machines allow them to collect and process massive raw data and make intelligent decisions autonomously according to the requirements of the new Industry 4.0 smart factories and without the need for human intervention (projection 5). The use of these new intelligent tools in future SCMs must take into consideration certain issues and challenges related to the quality and processing speed of the collected data and the security challenges (integrity and authentication). Some researchers believe that blockchain combined with IOT technologies will become imposed as a key technology to process the data collected by the decentralized architecture of the IOT network in full security. This can make the SCM more effective, efficient, intelligent, and automated.

Recently, Co-founder and CEO Eric Jennings told (Pete Rizzo, 2021): "Almost all these companies have the same concern – 'What is my IoT strategy?' Many of these companies are good at what they build but they don't have a lot of expertise in mesh networking or blockchains, but they know they need to connect these networks to gain efficiencies or risk going out of business." Thus, it also said 'Why it does it matter using the blockchain?' to create a decentralized communication platform.

Experts agree that blockchain alone will not be able to provide the secure infrastructure for SCM transactions. It will require a technology mix that includes AI and wireless sensors to connect the network machines together. This is likely to enable enterprises to efficiently manage identification and intercommunication operations between network nodes in real time. The blockchain will allow us to keep the identity of each participating node in a reliable way while avoiding threats such as identity theft. Blockchain-based applications involve sensors connected to a decentralized system and using smart contracts to exchange values and automatically execute actions. These applications thus help address challenges and predict maintenance issues, which is going to allow organizations to avoid equipment failures and save millions of dollars per year (venturebeat, 2022; nasdaq, 2017; Abe Eshkenazi, 2018).

The current health crisis, caused by the COVID-19 pandemic, reveals the need to digitize SCMs, and to face the problems of information sustainability between SCM processes. Modern SCMs are characterized by extremely complex complexity, lack of visibility, and flexibility. This hinders the ability of companies to respond appropriately to changes that affect SC and to various health, economic, and political risks, such as the disruptions resulting from the COVID-19 pandemic (Abdullah et al., 2021; Zamiela et al., 2022). Because of the undesirable logistical and SCM disruption consequences of the COVID-19 pandemic, several automotive manufacturers in Europe and North America have been forced to close their plants due to the difficulty in sourcing parts manufactured in China. In order for SCMs to continue to function even in the event of an unforeseen crisis, the various SCM processes must be connected. Blockchain technology is largely likely to address this concern, as the use of blockchain combined with new information and communication technologies such as AI, IoT, has the potential to transform the SCM and make it more flexible, visible, and traceable (projection 5) (Hod Fleishman, 2020). SCM actors can view a product's history at any time and take corrective actions to problems that affect the SC at the time of occurrence. End-to-end traceability of goods will help fight against theft and counterfeiting by tagging and tracking products throughout the SCM, and sharing the data on a blockchain accessible by all participants in full confidentiality (Alotaibi, 2019).

The authors describe how blockchain applications offer significant security advantages for the SCM industry, through encrypted peer-to-peer data networks for more secure communication more greater privacy (projection 1) (Alotaibi, 2019).

Blockchain experts say also, that the use of blockchain will allow a great saving in cost and reduce the time needed to trace a product on the SC. They announce that for some products, as in the case of Walmart, that it took only 2.2 s to find out an individual fruit's weight, variety, growing location, time it was harvested, date it passed through customs, when and where it was sliced, which cold-storage facility the sliced mango was held in, and for how long it waited before being delivered to a store. (Kshetri, 2021) (projection 3).

Hese days, most consumers are increasingly concerned about the sources of their food. Consumers are no longer convinced by certificates and brand quality attributed by companies to products, and are looking for critical information about the origin of the product, the cultures of production, and the conditions of social and environmental responsibility (Schifferstein et al., 2021). Thus, there is a communication problem related to the lack of channels and information flows between manufacturers, consumers, and SCM partners. The blockchain presents operational and efficient solutions to make SC actors accountable and improve consumer trust and satisfaction. The decentralized blockchain system allows consumers to track, verify, and trace the product they buy from the place of its origin (projection 1).

Despite the enormous benefits that blockchain technology brings to the SCM field, there are some obstacles that prevent companies from adopting blockchain or integrating a blockchain system created by another organization (Agi & Jha, 2022). Most businesses operate in a complex environment that requires different parties to comply with various laws, regulations and institutions. Implementing blockchain-based SCM solutions can be an extremely complex and limited task. Due to the increased globalization of production and distribution flows, the high level of information required by blockchain technology to be implemented effectively, the high costs of deploying blockchain, and the need for cooperation between multiple companies belonging to multiple countries, especially developing countries.

In addition, current business processes are subject to numerous regulations and standards of governance and standardization. The new era of digitization, through the deployment of blockchain technologies, may lead to the creation of new data structures that may be incompatible with existing regulatory standards (Casino et al., 2019). This necessitates the creation of new standards for blockchain data structures for at least two reasons: first, it is important to create an interaction from a technical point of view between the SC data carried by the different blockchain platforms. Second, from an information perspective, the creation of data structure standards is necessary to comply with different regulatory requirements.

Finally, in translational SCMs, especially bilateral SCMs, the underprivileged classes are strongly exploited by the dominant and powerful groups. For example, in some African countries, a large part of the population lives below the poverty line while they hold a large untapped mineral reserve and do not benefit from the revenues of the mining industry. In addition to the problem of wealth distribution, there is also the problem of equity and social responsibility, the exploitation of miners and girls in mining, and the share of jobs that the informal sector represents. The use of blockchain in the mining industry can ensure social equity and improve social responsibility towards disadvantaged groups, guaranteeing the fair value of their work and reducing health risks and exploitation of miners. The transparency guaranteed by blockchain forces downstream participants of SCMs to be more accountable, and people working in the mining industry can be entitled to the fair value of their work.

8.7 LIMITATIONS OF THE STUDY

This study provides support for understanding the current state of blockchain technology, future projections, and the potential application of blockchain for tomorrow's digital SCM. This study only considered opinions and explanations extracted from literature journal articles, conference proceedings to study, according to prospective analysis, the potential roles of blockchain technology application for SCM management. Therefore, in a future prospective study, data and opinions from a survey study with domain experts in SCM, blockchain, and logistics could potentially be included to address the shortcomings of the present study.

8.8 CONCLUSION AND FUTURE SCOPE

Blockchain can help companies achieve key SCM goals. There is a difference between the use of blockchain in the financial sector and for SC activities. In the financial sector (Nakamoto, n.d.), blockchain is used to solve the database problem. Whereas, in SCM, in addition to database problems, the expected prospects of applying blockchain technology are related to communication problems, where a number of entities need to be communicated with efficiently and in the same format, in order to obtain approval from various authorities. So, in practice, in addition to the database problem, the messaging problem is another problem solved by blockchain in SCM.

The study showed that blockchain has a much higher potential in SCM activities compared to the banking and financial sector. In the absence of other alternatives, it is likely to become an attractive and profitable option. The use of blockchain in supply chain management might automate numerous procedures and make them more effective, not to mention the cost reductions that could be realized.

Blockchain solutions can also help address quality issues in SC, including those related to fraud and counterfeit product issues. In particular, industries facing counterfeit risk are more likely to adopt blockchain in SC to ensure the quality of their products. Blockchain improves product transparency, traceability, and reliability. It allows products to be tracked throughout SC and puts pressure on SC partners to be more accountable for their actions.

Today, following the SCM blockchains created between 2015 and 2022. SC domain companies having a wide range of solutions. There are several comprehensive and easy-to-use platforms for companies to create a blockchain project for SCM. For example, Hyperledger Fabric is one of the most well-known platforms. Many large technology companies, such as IBM, AWS, SAP, Oracle, and Microsoft, offer enterprise blockchain solutions based on Hyperledger Fabric.

A recent study by Hackius and Petersen, 2017, examined the use of blockchain in the SCM and logistics industry, based on a sample of 152 expert logistics departments. The results show that companies are still reluctant to devote resources to potential blockchain applications.

Thus, the blockchain is a unique source of truth for all SCM stakeholders, but not yet fully exploitable. And requires a lot of research to be effectively applicable to SCM. Its character as a distributed ledger means that each information output from one step is an input for the next. A blockchain is a high value-added technology and an unprecedented revolutionary driving factor.

Future Scope: The adoption of blockchain for SCM management needs to be further studied in future research to better identify the most important factors for its implementation in multinational supply chain management. It should be explored in several industries belonging to various countries of the world. The development of a solid preferential contains all the processes and identifies the way of adoption of blockchain in SCM has become a demanding reality, in order to take advantage of the unique opportunities offered by the blockchain distributed ledger technology. Future research must investigate how to address some of the major challenges that strongly hinder the adoption of blockchain technology in SCM, namely: the lack of understanding of the technical benefits, the absence of implementation references, and the existing confusion between traditional databases and ledger technology. The problems of data integrity, system interoperability, and how to convince all SCM entities to adopt blockchain are issues that need to be explored in future research.

On the other hand, despite the fact that the idea of using blockchain in SCM has been around for more than four years, and despite the works that explain the potential benefits of this technology for the SCM field, there is still a significant legal gap, despite some works that address security guarantees and privacy policies. Regulatory uncertainty and the immature and vulnerable nature of blockchain technology make companies uncertain of the benefits of this technology. A widespread adoption of blockchain in the coming years could have an influence on employment sectors by removing trusted intermediates, resulting in social difficulties and job losses.

Among other things, it is necessary to study in depth the development of a common standard (combination of blockchain, IoT, AI) to link the different platforms or the creation of a single global platform. Re-engineering of SCM is needed and deserves further study to examine the implementation of blockchain in SCM, including security and visibility of shared data across all SCM partners, resilience of blockchain-based SCM to events such as virus outbreaks, and a practical exploration of scalability issues and how blockchain enables inter-organizational digital trust through the effective use of smart contracts to ensure trust among SC partners and eliminate corruption issues.

REFERENCES

Abe Eshkenazi. (2018). Real Benefits from Digital Twins | APICS Blog. http://www.apics.org/sites/apics-blog/thinking-supply-chain-topic-search-result/thinking-supply-chain/2018/09/21/real-benefits-from-digital-twins

Agi, M. A. N., & Jha, A. K. (2022). Blockchain technology in the supply chain: An integrated theoretical perspective of organizational adoption. *International Journal of Production Economics*, 247, 108458. (pp.1–15). 10.1016/j.ijpe.2022.108458

Ahmed, S., & Broek, N. ten. (2017). Blockchain could boost food security. *Nature*, 550(7674), 43–43. 10.1038/550043e

Ahram, T., Sargolzaei, A., Sargolzaei, S., Daniels, J., & Amaba, B. (2017). Blockchain Technology Innovations.IEEE Technology & Engineering Management Conference (TEMSCON). (pp.1–5). 10.1109/TEMSCON.2017.7998367

Alotaibi, B. (2019). Utilizing blockchain to overcome cyber security concerns in the Internet of Things: A review. *IEEE Sensors Journal*, 19(23), 10953–10971. 10.1109/JSEN.2019.2935035

Apte, S., & Petrovsky, N. (2016). Will blockchain technology revolutionize excipient supply chain management? *Journal of Excipients and Food Chemicals*, 7(3), 76–78.

Aslam, J., Saleem, A., Khan, N. T., & Kim, Y. B. (2021). Factors influencing blockchain adoption in supply chain management practices: A study based on the oil industry. *Journal of Innovation and Knowledge*, 6(2), 124–134. 10.1016/j.jik.2021.01.002

Babich, V., & Hilary, G. (2020). Distributed ledgers and operations: What operations management researchers should know about blockchain technology. *Manufacturing and Service Operations Management*, 22(2), 223–240. 10.1287/MSOM.2018.0752

Bahri, L., & Girdzijauskas, S. (2019). Trust Mends Blockchains: Living up to Expectations.

Banerjee, A. (2018). Integrating blockchain with ERP for a transparent supply chaiNn. https://www.infosys.com/oracle/white-papers/Documents/integrating-blockchain-erp.pdf

Barron, S., Cho, Y. M., Hua, A., Norcross, W., Voigt, J., & Haimes, Y. (2016). Systems-based cyber security in the supply chain. In *IEEE Systems and Information Engineering Design Symposium, SIEDS 2016*, 20–25. 10.1109/SIEDS.2016.7489299

Behnke, K., & Janssen, M. F. W. H. A. (2020). Boundary conditions for traceability in food supply chains using blockchain technology. *International Journal of Information Management*, 52, (pp.1–10). 10.1016/j.ijinfomgt.2019.05.025

Brandon-Jones, A., & Kauppi, K. (2018). Examining the antecedents of the Technology Acceptance Model within e-procurement.

Casado-Vara, R., Prieto, J., La Prieta, F. De, & Corchado, J. M. (2018). How blockchain improves the supply chain: Case study alimentary supply chain. *Procedia Computer Science*, 134, 393–398. 10.1016/j.procs.2018.07.193

Casino, F., Dasaklis, T. K., & Patsakis, C. (2019). A systematic literature review of blockchain-based applications: Current status, classification and open issues. *Telematics and Informatics*, 36(November), 55–81. 10.1016/j.tele.2018.11.006

Chapitre III. La prospective exploratoire et normative: Rntre possibles et souhaitables. http://www.pays-de-la-loire.developpement-durable.gouv.fr/IMG/pdf/ChapitreIII_epbl_cle1f89c5-1.pdf

Cole, R., Stevenson, M., & Aitken, J. (2019). Blockchain technology: Implications for operations and supply chain management. *Supply Chain Management*, 24(4), 469–483. 10.1108/SCM-09-2018-0309

Courbe, Thomas Tucci- Piergiovanni, Sara. Memmi, Gérard. Lanusse, Agnès. Jacovetti, Gilles. Gonthier, Georges. Duvaut, Patrick. Dalmas, Stéphane. Rapport Avril. (2021). *Les Verrous Technologiques Des Blockchains*.BCom de la DGE. (pp.1–106).

Cullen, A., Ferraro, P., Sanders, W., Vigneri, L., & Shorten, R. (2020). *Access Control for Distributed Ledgers in the Internet of Things: A Networking Approach*. https://arxiv.org/pdf/2005.07778.pdf

Dahlan Abdullah, Untung Rahardja, Fitra Putri Oganda. Covid-19: Decentralized Food Supply Chain Management. 12(3)(2021):142–152.

Demir Sercan Paksoy Turan Kochan Cigdem Gonul. A Conceptual Framework for Industry 4.0 (How is it Started, How is it Evolving Over Time?). Logistics 4.0: digital transformation of supply chain management. (2021)1–14.

Dorri, A., Kanhere, S. S., & Jurdak, R. (2017). *Towards an Optimized BlockChain for IoT*. IoTDI 2017, 227–232.

Dorri, A., Kanhere, S. S., Jurdak, R., & Gauravaram, P. (2017). Blockchain for IoT security and privacy: The case study of a smart home. In *IEEE International Conference on Pervasive Computing and Communications Workshops, PerCom Workshops*, 618–623. 10.1109/PERCOMW.2017.7917634

Dûdde̶r, B., & Ross, O. (2017). *Timber Tracking: Reducing Complexity of Due Diligence by Using Blockchain Technology*. (pp.1–6). 10.2139/ssrn.3015219

Dujak, D., & Sajter, D. (2019). *Blockchain Applications in Supply Chain* (pp.21–46). 10.1007/978-3-319-91668-2_2

Everett, M., & Rogers. (2003). Diffusion of Innovations. 5th Edition.

Feng Tian, An agri-food supply chain traceability system for china based on RFID & blockchain technology, *13th International Conference on Service Systems and Service Management (ICSSSM)*.

Fosso Wamba, S., Queiroz, M. M., & Trinchera, L. (2020). Dynamics between blockchain adoption determinants and supply chain performance: An empirical investigation. *International Journal of Production Economics*, 229(April), 107791. 10.1016/j.ijpe.2020.107791

Francesco Corea. (2017). The convergence of AI and Blockchain: what's the deal?.https://francesco-ai. medium.com/the-convergence-of-ai-and-blockchain-whats-the-deal-60c618e3accc

Francisco, K., & Swanson, D. (2018). The supply chain has no clothes: Technology adoption of blockchain for supply chain transparency. *Logistics*, 2(1), 2. 10.3390/logistics2010002

Frey, R. M., Vučkovac, D., & Ilic, A. (2016). A secure shopping experience based on blockchain and beacon technology. *CEUR Workshop Proceedings*, 1688(October).

Frey, R. M., Wörner, D., & Ilic, A. (2016). Collaborative filtering on the blockchain: A secure recommender system for E-commerce. In *Surfing the IT Innovation Wave – 22nd Americas Conference on Information Systems*, (pp.1–5).

Fu, Y., & Zhu, J. (2019). Big production enterprise supply chain endogenous risk management based on blockchain. *IEEE Access*, 7, 15310–15319. 10.1109/ACCESS.2019.2895327

Gökalp, E., Gökalp, O., Çoban, S., & Erhan Eren, P. (2018). *Analysing Opportunities and Challenges of Integrated Blockchain Technologies in Healthcare*. 10.1007/978-3-030-00060-8_13

Hackius, N. Petersen, Moritz (2017). *Blockchain in Logistics and Supply Chain: Trick or Treat?* 23, (pp. 3–18). 10.15480/882.1444

Hald, K. S., & Kinra, A. (2019). How the blockchain enables and constrains supply chain performance. *International Journal of Physical Distribution and Logistics Management*, 49(4), (pp.376–397). 10.1108/IJPDLM-02-2019-0063

Hasan, M. (2017). *Sustainable Supply Chain Management Practices and Operational Performance. March*. 10.4236/ajibm.2013.31006

Heber, D. (2017). *Towards a digital twin: How the blockchain can foster E / E-traceability in consideration of model-based systems engineering*. 3(August), (pp.321–330).

Helo, P., & Hao, Y. (2019). Blockchains in operations and supply chains: A model and reference implementation. *Computers and Industrial Engineering*, 136, (pp.242–251). 10.1016/j.cie.2019.07.023

Herbert, J., & Litchfield, A. (2015). A novel method for decentralised peer-to-peer software license validation using cryptocurrency blockchain technology. In *Conferences in Research and Practice in Information Technology Series*, 159(November), 27–35.

Hod Fleishman. (2020). How IoT, AI And Blockchain Can Trasform Supply Chains In 3 Steps. https://www. forbes.com/sites/hodfleishman/2020/05/28/how-iot-ai-and-blockchain-can-trasform-supply-chains-in-3-steps/?sh=29fd97164c83

Holland, M., Nigischer, C., & Stjepandic, J. (2017). Copyright protection in additive manufacturing with blockchain approach. *Advances in Transdisciplinary Engineering*, 5(July), (pp.914–921). 10.3233/978-1-61499-779-5-914

IBM (2022). Blockchain use cases | IBM Blockchain.https://www.ibm.com/blockchain/use-cases/

Imeri, A., Khadraoui, D., & Agoulmine, N. (2019). *Blockchain Technology for the Improvement of SCM and Logistics Services: A Survey* (pp. 349–361). 10.1007/978-3-030-03317-0_29

Jones, H. , & Grazia Vigliotti, M. (2020). The Executive Guide to Blockchain Using Smart Contracts and Digital Currencies in your Business.

Kennedy, Z. C., Stephenson, D. E., Christ, J. F., Pope, T. R., Arey, B. W., Barrett, C. A., & Warner, M. G. (2017). Enhanced anti-counterfeiting measures for additive manufacturing: Coupling lanthanide nanomaterial chemical signatures with blockchain. *Journal of Materials Chemistry C*, (37). 10.1039/C7TC03348F

Kim, H., & Laskowski, M. (2017). A perspective on blockchain smart contracts: Reducing uncertainty and complexity in value exchange. 2017. 26th International Conference on Computer Communications and Networks, ICCCN 2017

Kim, J. S., & Shin, N. (2019). The impact of blockchain technology application on supply chain partnership and performance. *Sustainability (Switzerland)*, 11, 6181(21). (pp.1–17). 10.3390/su11216181

Kshetri, N. (2017a). Can blockchain strengthen the Internet of Things? *IT Professional*, 19(4), (pp.68–72). 10.1109/MITP.2017.3051335

Kshetri, N. (2017b). Blockchain's roles in strengthening cybersecurity and protecting privacy. *Telecommunications Policy*, 41(10), (pp.1027–1038). 10.1016/j.telpol.2017.09.003

Kshetri, N. (2018). 1 Blockchain's roles in meeting key supply chain management objectives. *International Journal of Information Management*, 39, (pp.80–89). 10.1016/j.ijinfomgt.2017.12.005

Kshetri, N. (2021). Food and beverage industry supply chains. *Blockchain and Supply Chain Management* (pp.1–37). Elsevier. 10.1016/B978-0-323-89934-5.00001-5

Kshetri, N. et al. (2021). Blockchain and Supply Chain Management, CHAPTER 7 Opportunities, barriers, and enablers of blockchain in supply chains, pp. 163–191, 10.1016/B978-0-323-89934-5.00005-2

Kshetri, N., & Loukoianova, E. (2019). Blockchain adoption in supply chain networks in Asia. *IT Professional*, 21(1), 11–15. 10.1109/MITP.2018.2881307

Lei, A., Cruickshank, H., Cao, Y., Asuquo, P., Ogah, C. P. A., & Sun, Z. (2017). Blockchain-based dynamic key management for heterogeneous intelligent transportation systems. *IEEE Internet of Things Journal*, 4(6), 1832–1843. 10.1109/JIOT.2017.2740569

Lu, G., Koufteros, X., & Lucianetti, L. (2017). Supply chain security: A classification of practices and an empirical study of differential effects and complementarity. *IEEE Transactions on Engineering Management*, 64(2), 234–248. 10.1109/TEM.2017.2652382

Lu, Q., & Xu, X. (2017). Adaptable blockchain-based systems: A case study for product traceability. *IEEE Software*, 34(6), 21–27. 10.1109/MS.2017.4121227

Madhwal, Y., & Panfilov, P. B. (2017). Blockchain and supply chain management: Aircrafts' parts' business case. In *Annals of DAAAM and Proceedings of the International DAAAM Symposium*, 1051–1056. 10.2507/28th.daaam.proceedings.146

Madumidha, S., Siva Ranjani, P., Vandhana, U., & Venmuhilan, B. (2019). A theoretical implementation: Agriculture-food supply chain management using blockchain technology. In *Proceedings of the 2019 TEQIP - III Sponsored International Conference on Microwave Integrated Circuits, Photonics and Wireless Networks, IMICPW 2019*, 174–178. 10.1109/IMICPW.2019.8933270

Makhdoom, I., Abolhasan, M., Abbas, H., & Ni, W. (2019). Blockchain's adoption in IoT: The challenges, and a way forward. *Journal of Network and Computer Applications*, 125, 251–279. 10.1016/j.jnca.2018.10.019

Mathieu Lesueur-Cazé, Laurent Bironneau, Gulliver Lux, Thierry Morvan. (2022). Réflexions sur les usages de la blockchain pour la logistique et le Supply Chain Management: une approche prospective. Reflections on the uses of blockchain for logistics and Supply Chain Management: a prospective approach. Réflexions sur les usages de la blockchain pour la logistique et le Supply Chain Management. *Revue Française de Gestion Industrielle*, 36(1), 60–82.

Mendling, J., Weber, I., Van Der Aalst, W., Brocke, J. V., Cabanillas, C., Daniel, F., Debois, S., Di Ciccio, C., Dumas, M., Dustdar, S., Gal, A., García-Bañuelos, L., Governatori, G., Hull, R., La Rosa, M., Leopold, H., Leymann, F., Recker, J., Reichert, M., …Zhu, L. (2018). Blockchains for business process management - Challenges and opportunities. *ACM Transactions on Management Information Systems*, 9(1), 1–16. 10.1145/3183367

Mylrea, M., & Gourisetti, S. N. G. (2018). Blockchain for supply chain cybersecurity, optimization and compliance. In *Proceedings – Resilience Week 2018, RWS 2018*, January, 70–76. 10.1109/RWEEK.2018.8473517

Nakamoto, S. (n.d.). Bitcoin: A peer-to-peer electronic cash system. www.bitcoin.org

Nasdaq. (2017). Energizing the Blockchain—A Canadian Perspective. https://www.nasdaq.com/articles/energizing-blockchain-canadian-perspective-2017-01-26

Nick Szabo. (1994). Smart Contracts. https://www.fon.hum.uva.nl/rob/Courses/InformationInSpeech/CDROM/
    Literature/LOTwinterschool2006/szabo.best.vwh.net/smart.contracts.html

O'hara, K. (2017). *Smart Contracts-Dumb Idea.*

O'Leary, D. E. (2017). Configuring blockchain architectures for transaction information in blockchain con-
    sortiums: The case of accounting and supply chain systems. *Intelligent Systems in Accounting, Finance
    and Management*, 24(4), 138–147. 10.1002/isaf.1417

OECD. (2019). Is there a role for blockchain in responsible supply chains? www.oecd.org/finance/oecd-
    blockchain-policy-forum.htm

Pete Rizzo. (2021). Filament Nets $5 Million for Blockchain-Based Hardware. https://www.coindesk.com/
    markets/2015/08/18/filament-nets-5-million-for-blockchain-based-internet-of-things-hardware/

Phadnis, S. (2018). Internet of Things and supply chains: Framework for identifying oppertunities for
    improvement and its application. *Emerging Technologies for Supply Chain Management*, (2), 7–24.
    http://hdl.handle.net/1721.1/109375

Popov, S. (2018) *The Tangle* https://assets.ctfassets.net/r1dr6vzfxhev/2t4uxvsIqk0EUau6g2sw0g/45eae33637ca
    92f85dd9f4a3a218e1ec/iota1_4_3.pdf

Qian, Xiaoning (Alice), & Papadonikolaki, Eleni (2020). Shifting trust in construction supply chains through
    blockchain technology. Engineering, Construction and Architectural Management, 28, 584–602. 10.
    1108/ecam-12-2019-0676.

Queiroz, M. M., Telles, R., & Bonilla, S. H. (2020). Blockchain and supply chain management integration: A
    systematic review of the literature. In *Supply Chain Management* (Vol. 25, Issue 2, pp. 241–254).
    Emerald Group Holdings Ltd. 10.1108/SCM-03-2018-0143

Rogers, E. M. (2003). *Diffusion of innovations, 5th Edition.* Free Press.

Rogerson, M., & Parry, G. C. (2020). Blockchain: Case studies in food supply chain visibility. *Supply Chain
    Management*, 25(5), 601–614. 10.1108/SCM-08-2019-0300

Rowley, Jennifer, & Slack, Frances (2004). Conducting a literature review. Management Research News, 27,
    31–39. 10.1108/01409170410784185.

Schifferstein, H. N. J., de Boer, A., & Lemke, M. (2021). Conveying information through food packaging:
    A literature review comparing legislation with consumer perception. In *Journal of Functional Foods*
    (Vol. 86). Elsevier Ltd. 10.1016/j.jff.2021.104734

Şener, U., Gökalp, E., & Erhan Eren, P. (2016). *Cloud-Based Enterprise Information Systems: Determinants
    of Adoption in the Context of Organizations.* 10.1007/978-3-319-46254-7_5

Sharma, P. K., Singh, S., Jeong, Y. S., & Park, J. H. (2017). DistBlockNet: A distributed blockchains-based
    secure SDN architecture for IoT networks. *IEEE Communications Magazine*, 55(9), 78–85. 10.1109/
    MCOM.2017.1700041

Song, J. M., Sung, J., & Park, T. (2019). Applications of blockchain to improve supply chain traceability.
    Procedia Computer Science, 162, 119–122. 10.1016/j.procs.2019.11.266

Subramanian, B. Y. H. (2018). Decentralized Blockchain Based Electronic Marketplaces. Communications of
    the ACM, Vol.1, 1.

Swan, M. (2015). *Blockchain: blueprint for a new economy.* 1–152. 1st Edition.

SyncFab. (2018). *Decentralized manufacturing: Creating the world's first peer-to-peer manufacturing supply
    chain and incentivized token system adapted for public and private blockchains.* https://blockchain.
    syncfab.com/SyncFab_MFG_WP.pdf

Tayal, A., Solanki, A., Kondal, R., Nayyar, A., Tanwar, S., & Kumar, N. (2021). Blockchain-based efficient
    communication for food supply chain industry: Transparency and traceability analysis for sustainable
    business. *International Journal of Communication Systems*, 34(4), e4696.

Tian, F. (2017). A Supply Chain Traceability System for Food Safety Based on HACCP, Blockchain &
    Internet of Things.2017 International Conference on Service Systems and Service Management (IEEE).
    10.1109/ICSSSM.2017.7996119

Tönnissen, S., & Teuteberg, F. (2020). Analysing the impact of blockchain-technology for operations and
    supply chain management: An explanatory model drawn from multiple case studies. *International
    Journal of Information Management*, 52. (pp.1–10). 10.1016/j.ijinfomgt.2019.05.009

Toyoda, K., Mathiopoulos, P. T., Member, S., Sasase, I., Member, S., Ohtsuki, T., & Member, S. (2017).
    *A Novel Blockchain-Based Product Ownership Management System (POMS) for Anti-Counterfeits in
    The Post Supply Chain.*IEEE. (5), 1–13. 10.1109/ACCESS.2017.2720760

Tsai, W. T., Feng, L., Zhang, H., You, Y., Wang, L., & Zhong, Y. (2017). Intellectual-property blockchain-
    based protection model for microfilms. In *Proceedings - 11th IEEE International Symposium on
    Service-Oriented System Engineering, SOSE 2017*, April, 174–178. 10.1109/SOSE.2017.35

Turk, Ž., & Klinc, R. (2017). Potentials of blockchain technology for construction management. *Procedia Engineering*, 196(June), 638–645. 10.1016/j.proeng.2017.08.052

venturebeat. (2022). How blockchain can change the future of IoT | VentureBeat. https://venturebeat.com/2016/11/20/how-blockchain-can-change-the-future-of-iot/

Vora, J., Nayyar, A., Tanwar, S., Tyagi, S., Kumar, N., Obaidat, M. S., & Rodrigues, J. J. (2018, December). BHEEM: A blockchain-based framework for securing electronic health records. In *2018 IEEE Globecom Workshops (GC Wkshps)* (pp.1–6). IEEE.

Wang, H., Chen, K., & Xu, D. (2016). A maturity model for blockchain adoption. *Financial Innovation*, 2(12). (pp.1–5). 10.1186/s40854-016-0031-z

Wang, L., He, Y., & Wu, Z. (2022). Design of a blockchain-enabled traceability system framework for food supply chains. *Foods*, 11(5). (pp.1–19). 10.3390/foods11050744

Wang, Y., Singgih, M., Wang, J., & Rit, M. (2019). Making sense of blockchain technology: How will it transform supply chains? *International Journal of Production Economics*, 211, 221–236. 10.1016/j.ijpe.2019.02.002

Wee Kwan Tan, A., Zhao, Y., & Halliday, T. (2018). *A Blockchain Model for Less Container Load Operations in China*, 11(2). (pp.39–53) 10.4018/IJISSCM.2018040103

Xu, Chen, L., Gao, Z., Chang, Y., Iakovou, E., & Shi, W. (n.d.). Binding the physical and cyber worlds: A blockchain approach for cargo supply chain security enhancement. In *IEEE International Symposium on Technologies for Homeland Security (HST)*, *2018*, pp. 1–5.

Yoo, M., & Won, Y. (2018). A study on the transparent price tracing system in supply chain management based on blockchain. *Sustainability (Switzerland)*, 10(11). (pp.1–15). 10.3390/su10114037

Yuan, Y., & Wang, F. Y. (2016). Towards blockchain-based intelligent transportation systems. In *IEEE Conference on Intelligent Transportation Systems, Proceedings, ITSC*, 2663–2668. 10.1109/ITSC.2016.7795984

Zamiela, Christian, Hossain, Niamat Ullah Ibne, & Jaradat, Raed (2022). Enablers of resilience in the healthcare supply chain: A case study of U.S healthcare industry during COVID-19 pandemic. Research in Transportation Economics, 93, 101174 10.1016/j.retrec.2021.101174

9 A Blockchain-Based Framework for Circular Plastic Waste Supply Chain Management in India: A Case Study of Kolkata, India

Biswajit Debnath

Department of Chemical Engineering, Jadavpur University, Kolkata, India

Consortium of Researchers for Sustainable Development (C.R.S.D.), Agra, India

Ankita Das

Consortium of Researchers for Sustainable Development (C.R.S.D.), Agra, India

Department of Data Science and Cyber Security, Institute of Leadership, Entrepreneurship and Development (iLead), Kolkata, India

Adrija Das

A.K. Choudhury School of Information Technology, University of Calcutta, Kolkata, India

Abhijit Das

Department of Information Technology, Institute of Leadership, Entrepreneurship and Development (iLead), Kolkata, India

Department of Data Science and Cyber Security, Institute of Leadership, Entrepreneurship and Development (iLead), Kolkata, India

CONTENTS

DOI: 10.1201/9781003264521-9

9.1 INTRODUCTION

Plastics are ubiquitous in our daily lives, appearing in everything from grocery bags and cutlery to water bottles and sandwich wrap. However, our search for ease has gone too far, and we are wasting valuable resources and harming the environment by failing to use plastics efficiently (Silva 2021). Excessive consumption of plastic and poor waste management is becoming a growing problem, causing landfills to overflow, rivers to choke, and ecosystems to be threatened. This has a negative impact on industries such as tourism, shipping, and fisheries, which are important to many economies (Kataki et al. 2022).

According to WHO report, the response to the COVID-19 pandemic has put tens of thousands of tons of extra medical waste on healthcare waste management systems around the world, endangering human and environmental health and exposing a dire need to improve waste management practices (WHO 2022). COVID-19 crisis led to a 2.2% decrease in plastics use in 2020 as economic activity slowed (OECD 2022). But a rise in demand of face shields, gloves, food takeaway packaging, bubble wrap for online shopping, and plastic medical equipment such as masks has driven up littering (Debnath et al. 2021). As economic activity resumed in 2021, plastics consumption has also rebounded. Because the majority of that cannot be recycled, the waste has increased (Vanapalli 2021). However, there is another ramification. The pandemic has exacerbated the oil industry's price war between recycled and new plastic. Price data and interviews with more than two dozen businesses from five continents show that recyclers worldwide are losing the war (Brock 2020).

Plastic production has quadrupled in the last 30 years, owing to rising demand in emerging markets. Between 2000 and 2019, global plastics production more than doubled to 460 million tons. Plastics contribute 3.4% of total global greenhouse gas emissions (Zheng and Suh 2019). Globally, plastic waste is recycled in only 9% of cases (15% is collected for recycling but 40% of that is disposed of as residues). Another 19% is incinerated, 50% is disposed of in landfills, and 22% escapes waste management systems and ends up in uncontrolled dumpsites, open pits, or in terrestrial or macrophytes environments, particularly in developing countries (OECD 2022). Plastic waste leaking into aquatic environments totalled 6.1 million tons (Mt) in 2019, with 1.7 Mt flowing into oceans. Plastic waste has accumulated in seas and oceans to the tune of 30 million tons, with another 109 million tons in rivers (OECD 2022) (Figure 9.1).

The Ministry of Environment informed the Rajya Sabha over 34 lakh tons of plastic waste were generated in the financial year 2019–2020, an increase of more than 10 lakh tons over 2017–18 in India (The Economic Times 2021). According to OECD, 13% of the total plastic waste generated is recycled, 4% is incinerated, 36% is landfilled, and 46% is mismanaged and littered (OECD 2022). It is estimated that Kolkata generates anywhere between 500 to 1200 tons of plastic waste per annum (Earth5r.org 2021).

Plastic waste management is a programme aimed at reducing the amount of plastic waste in the environment through the use of circular economy and other environmentally friendly disposal methods (Payne et al. 2019). It aims to prohibit the use of plastic products for which affordable alternatives are available, encourage the use of circular materials in plastic production, and increase the use of recycled plastics (Dangis 2018). Several initiatives to control plastic waste and reduce the impact of plastic pollution have already been announced around the world (Shin et al. 2020). In developing countries like India, the problem of mismanaged and unmanaged municipal solid waste, including plastic waste, is particularly acute (Bhattacharya et al. 2018). This is due to a lack of infrastructure, an unequal distribution of economic resources, systemic poverty,

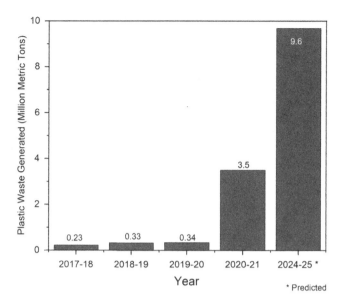

FIGURE 9.1   Plastic Waste Generation in India.

environmental injustice caused by the developed world's disposal of plastic waste, and a variety of other instances where there is a lack of accountable, equitable, and effective political governance and capacity (Rafey & Siddiqui 2021). Door-to-door collection of plastic waste by garbage collectors is totally unplanned without any systematic procedure results overflowing, with mountains of plastic piling up in the trash and ecosystems (Hossain et al. 2022).

Many industries have been transformed by digital technologies over the last two decades (Andreoni et al. 2021). It started with the spread of the Internet, then moved on to mobile technologies, and finally smartphones. AI, IoT, 3D printing, and blockchain technologies are now delivering potentially disruptive solutions for sustainability and the environment (Li & Found 2017; Rosário & Dias 2022). Digitalization is closing the loop on the plastic circular economy by removing turbulence from supply chains, providing real-time information, and more efficiently collecting or sorting waste (Bekrar et al. 2021). (Saberi et al. (2019)) examined how blockchain technology and smart contracts can be helpful in addressing supply chain sustainability. Blossey et al. (2019) developed a framework to analyse 53 applications of blockchain in supply chain management. Recently, a detailed review on application of blockchain technology in supply chain management revealed that most productive research is from the USA, China, and India. The study also provides future direction of blockchain research in supply chain management (Rejeb et al. 2021). As can be seen, there exist a plethora of articles on blockchain application in supply chain management. However, studies on application of blockchain in waste management supply chain is scant. Hence, study of blockchain application in plastic waste management supply chain is of relevance.

The objectives of this chapter:

- To develop a blockchain-based solution to the plastic waste management supply chain.
- To ratify the developed solution with the case of Kolkata city.
- To propose a solution framework for ensuring circular economy in plastic waste management system.

Organization of the Chapter: The chapter is organized as: Section 9.2 describes the methodology adopted. Section 9.3 describes the existing plastic waste management chain in India along with

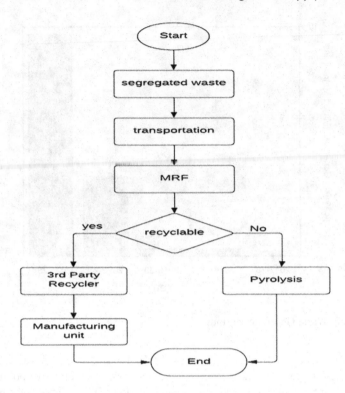

FIGURE 9.2   Flowchart Showing the Path Covered by Blockchain Considered in This Study.

issues and challenges pertaining to it. In Section 9.4, the case study of the Kolkata region is provided. Section 9.5 discusses the results. Section 9.6 concludes the chapter with future scope.

9.2   METHODOLOGY

The study focuses on finding a solution to the plastic waste supply chain problem. It considers primarily the waste generated from household. It also assumes that the waste collected is segregated at source. The details are presented in the flowchart below (Figure 9.2). At first the segregated waste is sent to the material recycling facility (MRF) through transportation. The blockchain network holds separate blocks for each segment. At MRF the wastes are checked on the basis of recyclability. Recyclable plastic wastes are sent to third party recycler and non-recyclable wastes are sent for pyrolysis. Pyrolysis is an eco-friendly alternative to incineration, which allows resource recovery as well (Debnath et al. 2022). After processing the recyclables, wastes are delivered to third-party recycling units.

9.3   EXISTING PLASTIC WASTE SCN IN INDIA

The supply chain network (SCN) of Plastic Waste Management System (PWMS) is complex and interesting. It is essential to understand the overall SCN in order to solve the issues pertaining to it. In India, the Plastic Waste Supply Chain Network (PW-SCN) starts from the consumer-generators who are the generators of the plastic waste. There are two categories of generators: bulk generators and low scale generators. Bulk generators include PW generated in malls, housing complexes, temples, restaurants, and other such public areas. The low-scale generators are the individual generators, take-away shops, street vendors, etc. The PW generated is either disposed mixed with wet waste or separately in segregated bins. These are collected by either designated personnel from the ULBs or contracted third-party recyclers depending on the source of generation. The main issue arises with individual consumer-generators who either throw them away as mixed waste or sell them to the Kabaddiwalas. Now these Kabaddiwalas are informal collectors of dry waste, often referred to as the informal sector. They usually roam door to door to collect dry waste including plastic waste, paper waste, metals, glass, e-waste, etc. in exchange for money. People tend to sell their PW to these informal collectors for the money, which is quite similar to e-waste. The Kabaddiwalas sell the collected waste to some small and medium scrap dealers. They partially segregate the waste streams and sell them to big scrap dealers. These big scrap dealers are in contact with several third-party PW recyclers. They recycle the PW. In this process often the single use plastics (SUPs), toothbrushes, styrofoam objects, earbuds, e-plastic waste, etc. are neglected and they end up in landfills. On the other hand, PW collected mixed with MSW are taken to MRFs where they are segregated and channelised to respective third-party recyclers. Collection agencies often collect from bulk consumer-generators. They usually employ their own workforce to segregate and channelise the waste to third-party recyclers.

9.3.1 Issues and Challenges in the Plastic Waste SCN in India

There are multiple issues and challenges in the plastic waste SCN in India. The very basic issue is the mismanagement and un-management of plastic waste. This can be primarily attributed to lack of infrastructure, economic resources and poverty in the country (Browning et al. 2021). Concepts such as cleaner production and circular economy which are quite appropriate for the OECD countries, are not of much help here. The big reason behind that is the lack of awareness among the common people (Baidya et al. 2016). Though with amended government policies, some action is in place, the overall scenario has not changed very much (Debnath et al. 2022).

In India, one of the primary challenges is to tackle the informal sector. The informal sector dominates the collection of plastic waste from the households (Bhattacharya and Kesar 2018). Usually, their existence is not symbiotic with the formal recycling chain, which is a big concern. On the other hand, the due to lack of awareness often plastics are disposed mixed with municipal solid waste. Hence, due to lack of source segregated waste, the ULBs spend a lot of money on segregating them in the MRF (Malav et al. 2020). Lack of collection infrastructure, lack of economic resources for governments to build, manage, and enforce plastic waste management policies play a major role (Browning et al. 2021). In addition to that, waste management setups are frequently incumbered by political issues (Ferronato and Toretta, 2019). Also, the market of recycled plastic is quite restricted, which hinders the overall cash flow and often seen as a barrier to proper plastic recycling (Figure 9.3).

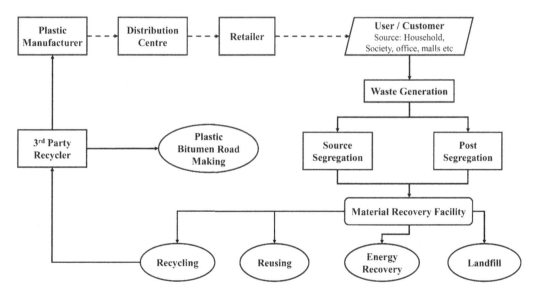

FIGURE 9.3 Generic Plastic Waste Management Supply Chain Network in India (Adopted from Shetty et al. 2022).

9.4   CASE STUDY

Name: Kolkata
Location: 22.5726° N, 88.3639° E (shown in Figure 9.4)
Status of Plastic Waste Management: Developing

Kolkata, also known as the "City of Joy," is the capital of West Bengal and one of the very important cities in India with a population of about 5 million. It is the major business, commercial, and financial hub of eastern India and also of the northeastern states. In general, the waste management system of the city is developing. There are some compactor stations they have employed which compacts mixed MSW before dispatching to landfills (Baidya et al. 2016). However, the Bidhannagar Municipal Corporation and New Town Kolkata Development Authority (NKDA) are working well with collection and transportation of segregated waste from households. They have dedicated MRF for further segregation as well. Our study is based on this developing supply chain of plastic waste in Kolkata, where complete tracking of is possible through blockchain implementation.

FIGURE 9.4   Map Showing Location of Kolkata.

9.5 DISCUSSION AND ANALYSIS

9.5.1 Chain Description

In the study, we have considered a certain extent of the whole SCN. Due to absence of reliable information, we have not considered the chain thrived by the informal sector. Additionally, the section controlled by the collection agencies are already working well. Hence, we have focused on the ULB run section of the PWSCN. Plastic chain is a blockchain that enables manufacturers and producers to disclose their means of production in order to provide value chain traceability and visibility. Attributes of plastic wastes are then wrapped as transactions written on blockchains, where consensus processes in its associated sub-chain will check and validate to establish a new block. The information in the transaction is stored in a Merkle tree, and the header contains the hash of previous blocks, index, timestamp, nonce, and so on.

9.5.2 Transaction and Block Structure

All activities are recorded as transactions on the blockchain. It generates a candidate block for the consensus layer to choose by combining transactions.

9.5.2.1 Transaction Definition

Blockchain is an add-on ledger. It is distributed among numerous nodes, which means that data is copied and saved in real time on each node in the system. When a transaction is registered in the blockchain, details like as the transaction's price, asset, and ownership are recorded, validated, and settled across all nodes in seconds. A validated change registered on any one ledger is also registered on all other copies of the ledger at the same time. There is no need for third-party verification because each transaction is transparently and permanently recorded across all ledgers and is visible to everyone. In this proposed framework the transaction can be defined using the following tuples:

$$T_{X_{mn}} = \{Id_{mn}, H_{mn}, CT_{mn}, Ts_{mn}, I/O_{mn}\} \qquad (9.1)$$

Here a typical transaction: Tx_{mn} consists of hash value H_{mn} that enables encryption to the transaction. The transaction index assigned by the blockchain is Id_{mn}, and the timestamp for the time lock of this specific transaction being received is Ts_{mn}. The current transaction is defined as CT_{mn}. The I/O_{mn} are the transaction content's input and output messages. In equation (9.1), m and n are identifiers which denotes the transaction nodes and plastic waste respectively.

The format of any I/O_{mn} message is described in following:

$$I/O_{mn} = \{\text{"Prev\_out": } [\ldots]$$
$$\text{"out": } [\{\text{"hash": "} 6da9e1c2468d5\ldots..\text{"}$$
$$\text{"checkval": "a89d97e8a9a44}\ldots.\text{"}$$
$$\text{"get\_tran": } [\ldots]\}]$$
$$\} \qquad (9.2)$$

A typical input contains information of previous block in the network which is denoted by *Prev_out* and output consists of *hash, checkval,* and *get_tran* attributes. *Hash* and *checkval* are encoded in hexadecimal and *get_tran* is a function that returns all the transactions in the network.

9.5.2.2 Block Structure

Network peers will check the signature and encrypt the transactions received in one timeframe into a block for verification, and nodes will broadcast their transactions in the network. The general structure of the blocks in the network is represented below in equation (9.3) except the genesis or initial block. Each block contains a flag attribute that ensures transaction transparency.

$$Blk_m = \{Id_m, H_{prev}, Ts_m, Array\,[CT_m], \; nonce\} \qquad (9.3)$$

Where,

$$H_{prev} = Hash\,(Blk_{m-1})$$

The initial or genesis block does not contain any previous hash value. A random hexadecimal string is explicitly provided as hash value for the first block of the network. An arbitrary array has assigned to the genesis block as the current transaction.

9.5.3 Systematic Analysis

In this section, the three most important methods are discussed. These are the building methods of blockchain. *Checkval, add_new_block,* and *get_Tran* are used to generate hash to secure each block in the chain, adding new blocks to the chain and getting all the transactions of the network respectively. Figure 9.5 highlights the snapshot of pseudocode for the above methods.

The above procedure yields the following results. According to some uncertainty situations, different results can be obtained. Such uncertainty use cases are described below. Uncertainty issues are addressed using a flag attribute, which contains binary value. If it shows 'Y,' then the chain has been interrupted somewhere in the chain; otherwise it will show its value as 'N.'

```
function checkval(self reference, blk:Blk):
        difficulty=1
        blk.nonce=0
        hash_result=blk.hash_computation()
        while not(hash_result. endswith ('0'*difficulty) and ('44'*difficulty)in hash_result) :
                blk.nonce+=1
                hash_result=blk.hash_computation()
        return hash_result
end function
    function add_new_blk(self,dt):
        blk=Blk(len(self.chain),self.chain[-1],dt,'datetime.now().timestamp()',0)
        blk.hash=self.checkval(blk)
        self.chain.append(blk.hash)
        self.transactions.append(blk.__dict__)
        return json.loads(str(blk.__dict__).replace('\'','\"'))
    end function

    function get_Tran(self,id):
        labels=['Household','Transportation','MRF','3rd_Party_Recycler','Pyrolysis']
        while True:
            try:
                if id=='all':
                    for i in range(len(self.transactions)-1):
                        print('{}:\n{}\n'.format(labels[i],self.transactions[i+1]))
                    break
                elif type(id)==int:
                        print(self.transactions[id])
                        break
            except Exception as e:
                print(e)
    end function
```

FIGURE 9.5 Snapshot of the Code.

CASE 1   BASE CASE

Base case is the utopian situation when everything goes well; hence no uncertainties arising along the plastic waste supply chain. The code shows that all the attributes are preserved through the entire path of the supply chain network considered in this study (Figure 9.6).

Household:

{'index': 1, 'previous_hash': '82fc0b46f83fb54aef9484919f9edf74d16fdc90b165c16d7a667b79c1b00285', 'current_transaction': {'transactions': [{'timestamp': 1650212039.566184, 'Name': 'Plastic', 'cost per Kg(INR)': 8, 'Characteristics': 'Dry', 'Recyclability': 0.95, 'from': 'Household x', 'to': 'Transportation x', 'Digital signature': 'approved', 'msg': 'This waste is in good order', 'FLAG': 'N'}, {'timestamp': 1650212039.566184, 'Name': 'Glass', 'cost per Kg(INR)': 10, 'Characteristics': 'Dry', 'Recyclability': 1.0, 'from': 'Household x', 'to': 'Transportation x', 'Digital signature': 'approved', 'msg': 'This waste is in good order', 'FLAG': 'N'}, {'timestamp': 1650212039.566184, 'Name': 'Organic', 'cost per Kg(INR)': 3, 'Characteristics': 'wet', 'Recyclability': 0.9, 'from': 'Household x', 'to': 'Transportation x', 'Digital signature': 'approved', 'msg': 'This waste is in good order', 'FLAG': 'N'}, {'timestamp': 1650212039.566184, 'Name': 'E-waste', 'cost per Kg(INR)': 20, 'Characteristics': 'Dry', 'Recyclability': 0.9, 'from': 'Household x', 'to': 'Transportation x', 'msg': 'This waste is in good order', 'FLAG': 'N'}, {'timestamp': 1650212039.566184, 'Name': 'Metal', 'cost per Kg(INR)': 12, 'Characteristics': 'Dry', 'Recyclability': 1.0, 'from': 'Household x', 'to': 'Transportation x', 'Digital signature': 'approved', 'msg': 'This waste is in good order', 'FLAG': 'N'}]}, 'timestamp': 'datetime.now().timestamp()', 'nonce': 69, 'hash': 'cd111d07a16c25c2df95dece4163f2042cbc2e4425f150c8346db871f5a30310'}

Transportation:

{'index': 2, 'previous_hash': 'cd111d07a16c25c2df95dece4163f2042cbc2e4425f150c8346db871f5a30310', 'current_transaction': {'transactions': [{'timestamp': 1650212039.566184, 'Name': 'Plastic', 'cost per Kg(INR)': 8, 'Characteristics': 'Dry', 'Recyclability': 0.95, 'from': 'Transportation x', 'to': 'MRF', 'Digital signature': 'approved', 'msg': 'This waste is in good order', 'FLAG': 'N'}, {'timestamp': 1650212039.566184, 'Name': 'Glass', 'cost per Kg(INR)': 10, 'Characteristics': 'Dry', 'Recyclability': 1.0, 'from': 'Transportation x', 'to': 'MRF', 'Digital signature': 'approved', 'msg': 'This waste is in good order', 'FLAG': 'N'}, {'timestamp': 1650212039.566184, 'Name': 'Organic', 'cost per Kg(INR)': 3, 'Characteristics': 'wet', 'Recyclability': 0.9, 'from': 'Transportation x', 'to': ' MRF', 'Digital signature': 'approved', 'msg': 'This waste is in good order', 'FLAG': 'N'}, {'timestamp': 1650212039.566184, 'Name': 'E-waste', 'cost per Kg(INR)': 20, 'Characteristics': 'Dry', 'Recyclability': 0.9, 'from': 'Transportation x', 'to': 'MRF', 'Digital signature': 'approved', 'msg': 'This waste is in good order', 'FLAG': 'N'}, {'timestamp': 1650212039.566184, 'Name': 'Metal', 'cost per Kg(INR)': 12, 'Characteristics': 'Dry', 'Recyclability': 1.0, 'from': 'Transportation x', 'to': 'MRF', 'Digital signature': 'approved', 'msg': 'This waste is in good order', 'FLAG': 'N'}]}, 'timestamp': 'datetime.now().timestamp()', 'nonce': 118, 'hash': 'f405a339f81e87529852a86a047c0173a18d64f16d5e0e6145341eb44efaa730'}

MRF:

{'index': 3, 'previous_hash': 'f405a339f81e87529852a86a047c0173a18d64f16d5e0e6145341eb44efaa730', 'current_transaction': {'transactions': [{'timestamp': 1650212039.566184, 'Name': 'Plastic', 'cost per Kg(INR)': 8, 'Characteristics': 'Dry', 'Recyclability': 0.95, 'from': 'MRF central facility', 'to': 'third party recycler', 'Digital signature': 'approved', 'msg': 'This waste is in good order', 'FLAG': 'N'}, {'timestamp': 1650212039.566184, 'Name': 'Glass', 'cost per Kg(INR)': 10, 'Characteristics': 'Dry', 'Recyclability': 1.0, 'from': 'MRF central facility', 'to': 'glass factory', 'Digital signature': 'approved', 'msg': 'This waste is in good order', 'FLAG': 'N'}, {'timestamp': 1650212039.566184, 'Name': 'Organic', 'cost per Kg(INR)': 3, 'Characteristics': 'wet', 'Recyclability': 0.9, 'from': 'MRF central facility', 'to': 'Organic', 'Digital signature': 'approved', 'msg': 'This waste is in good order', 'FLAG': 'N'}, {'timestamp': 1650212039.566184, 'Name': 'E-waste', 'cost per Kg(INR)': 20, 'Characteristics': 'Dry', 'Recyclability': 0.9, 'from': 'MRF central facility', 'to': 'E-waste', 'Digital signature': 'approved', 'msg': 'This waste is in good order', 'FLAG': 'N'}, {'timestamp': 1650212039.566184, 'Name': 'Metal', 'cost per Kg(INR)': 12, 'Characteristics': 'Dry', 'Recyclability': 1.0, 'from': 'MRF central facility', 'to': 'Metal factory', 'Digital signature': 'approved', 'msg': 'This waste is in good order', 'FLAG': 'N'}]}, 'timestamp': 'datetime.now().timestamp()', 'nonce': 5, 'hash': '800a035154c2d440dfda1613b473cbc19bcbea006bc001fc35386c70f54182b0'}

3rd_Party_Recycler:

{'index': 4, 'previous_hash': '800a035154c2d440dfda1613b473cbc19bcbea006bc001fc35386c70f54182b0', 'current_transaction': {'transactions': [{'timestamp': 1650212039.566184, 'Name': 'PET', 'cost per Kg(INR)': 3, 'Characteristics': 'Dry', 'Q.I.': 9, 'from': '3rd party recycler', 'to': 'manufacturing unit', 'Digital signature': 'approved', 'msg': 'This waste is in good order', 'FLAG': 'Y'}, {'timestamp': 1650212039.566184, 'Name': 'HDPE', 'cost per Kg(INR)': 4, 'Characteristics': 'Dry', 'Q.I.': 6, 'from': '3rd party recycler', 'to': 'manufacturing unit', 'Digital signature': 'approved', 'msg': 'This waste is in good order', 'FLAG': 'N'}, {'timestamp': 1650212039.566184, 'Name': 'PVC', 'cost per Kg(INR)': 7, 'Characteristics': 'Dry', 'Q.I.': 5, 'from': '3rd party recycler', 'to': 'manufacturing unit', 'Digital signature': 'approved', 'msg': 'This waste is in good order', 'FLAG': 'N'}, {'timestamp': 1650212039.566184, 'Name': 'LDPE', 'cost per Kg(INR)': 2, 'Characteristics': 'Dry', 'Q.I.': 8, 'from': '3rd party recycler', 'to': 'manufacturing unit', 'Digital signature': 'approved', 'msg': 'This waste is in good order', 'FLAG': 'N'}, {'timestamp': 1650212039.566184, 'Name': 'PP', 'cost per Kg(INR)': 3, 'Characteristics': 'Dry', 'Q.I.': 5, 'from': '3rd party recycler', 'to': 'manufacturing unit', 'Digital signature': 'approved', 'msg': 'This waste is in good order', 'FLAG': 'N'}, {'timestamp': 1650212039.566184, 'Name': 'PS', 'cost per Kg(INR)': 2, 'Characteristics': 'Dry', 'Q.I.': 9, 'from': '3rd party recycler', 'to': 'manufacturing unit', 'Digital signature': 'approved', 'msg': 'This waste is in good order', 'FLAG': 'N'}]}, 'timestamp': 'datetime.now().timestamp()', 'nonce': 0, 'hash': 'cf0859fa6e10207fb019565315debfbaab96ee0437272e15305e8d1fbad7b440'}

Pyrolysis:

{'index': 5, 'previous_hash': 'cf0859fa6e10207fb019565315debfbaab96ee0437272e15305e8d1fbad7b440', 'current_transaction': {'transactions': [{'timestamp': 1650212039.566184, 'Name': 'E-PLASTIC WASTE', 'cost per Kg(INR)': 1, 'Characteristics': 'Dry', 'Q.I.': 4, 'Digital signature': 'approved', 'msg': 'This waste is destructed', 'FLAG': 'N'}, {'timestamp': 1650212039.566184, 'Name': 'NON-RECYCLABLE', 'cost per Kg(INR)': 1, 'Characteristics': 'Dry', 'Digital signature': 'approved', 'msg': 'This waste is destructed', 'FLAG': 'N'}]}, 'timestamp':

FIGURE 9.6   Code Outputs from Case 1.

CASE 2   UNCERTAINTY ARISING DUE TO TRANSPORTATION ISSUE

This is a situation where the chain is interrupted during its transportation. Uncertainty may happen in various ways like vehicle breakage, road closed, traffic issues, etc. A digital signature validates the transaction in a blockchain network. In this case, as the chain is interrupted due to transportation hazards, the block of transport can't validate the transaction anymore (Figure 9.7).

Transportation:

{'index': 2, 'previous_hash': 'a60608234457b095baffe1ca498642d8e5faf85c7c6c4722103b7616dbcfe300', 'current_transaction': {'transactions': [{'timestamp': 1650212355.972847, 'Name': 'Plastic', 'cost per Kg(INR)': 8, 'Characteristics': 'Dry', 'Recyclability': 0.95, 'from': 'Transportation x', 'to': 'MRF', 'Digital signature': 'approved', 'msg': 'This waste is in good order', 'FLAG': 'N'}, {'timestamp': 1650212355.972847, 'Name': 'Glass', 'cost per Kg(INR)': 10, 'Characteristics': 'Dry', 'Recyclability': 1.0, 'from': 'Transportation x', 'to': 'MRF', 'Digital signature': 'not approved', 'msg': 'This waste is not in good order', 'FLAG': 'N'}, {'timestamp': 1650212355.972847, 'Name': 'Organic', 'cost per Kg(INR)': 3, 'Characteristics': 'wet', 'Recyclability': 0.9, 'from': 'Transportation x', 'to': ' MRF', 'Digital signature': 'approved', 'msg': 'This waste is in good order', 'FLAG': 'N'}, {'timestamp': 1650212355.972847, 'Name': 'E-waste', 'cost per Kg(INR)': 20, 'Characteristics': 'Dry', 'Recyclability': 0.9, 'from': 'Transportation x', 'to': 'MRF', 'Digital signature': 'approved', 'msg': 'This waste is in good order', 'FLAG': 'N'}, {'timestamp': 1650212355.972847, 'Name': 'Metal', 'cost per Kg(INR)': 12, 'Characteristics': 'Dry', 'Recyclability': 1.0, 'from': 'Transportation x', 'to': 'MRF', 'Digital signature': 'approved', 'msg': 'This waste is in good order', 'FLAG': 'N'}]}, 'timestamp': 'datetime.now().timestamp()', 'nonce': 98, 'hash': 'c133c54523d44253058935d4b9c1304186a11d12069c8d93fb42e1e47f0fc840'}

MRF:

{'index': 3, 'previous_hash': 'c133c54523d44253058935d4b9c1304186a11d12069c8d93fb42e1e47f0fc840', 'current_transaction': {'transactions': [{'timestamp': 1650212355.972847, 'Name': 'Plastic', 'cost per Kg(INR)': 8, 'Characteristics': 'Dry', 'Recyclability': 0.95, 'from': 'MRF central facility', 'to': 'third party recycler ', 'Digital signature': 'approved', 'msg': 'This waste is in good order', 'FLAG': 'N'}, {'timestamp': 1650212355.972847, 'Name': 'Glass', 'cost per Kg(INR)': 10, 'Characteristics': 'Dry', 'Recyclability': 1.0, 'from': 'MRF central facility', 'to': 'glass factory', 'Digital signature': 'approved', 'msg': 'This waste is in good order', 'FLAG': 'Y'}, {'timestamp': 1650212355.972847, 'Name': 'Organic', 'cost per Kg(INR)': 3, 'Characteristics': 'wet', 'Recyclability': 0.9, 'from': 'MRF central facility', 'to': 'Organic', 'Digital signature': 'approved', 'msg': 'This waste is in good order', 'FLAG': 'N'}, {'timestamp': 1650212355.972847, 'Name': 'E-waste', 'cost per Kg(INR)': 20, 'Characteristics': 'Dry', 'Recyclability': 0.9, 'from': 'MRF central facility', 'to': 'E-waste', 'Digital signature': 'approved', 'msg': 'This waste is in good order', 'FLAG': 'N'}, {'timestamp': 1650212355.972847, 'Name': 'Metal', 'cost per Kg(INR)': 12, 'Characteristics': 'Dry', 'Recyclability': 1.0, 'from': 'MRF central facility', 'to': 'Metal factory', 'Digital signature': 'approved', 'msg': 'This waste is in good order', 'FLAG': 'N'}]}, 'timestamp': 'datetime.now().timestamp()', 'nonce': 143, 'hash': '8fb9c8e61acd4aebbc8091a37c656631225384058b2b02f25ec6f321ad1144c0'}

FIGURE 9.7   Outputs of Case 2 Showing Issues Due to Uncertainty.

CASE 3   UNCERTAINTIES ARISING DUE QUALITY OF PLASTIC WASTE RECEIVED

Here, disruption occurred in the main recyclable facility such as a quality issue in the plastic waste, incompetent product, etc. In this case, as the chain is interrupted due to quality of plastic waste, the block of third-party recycler can't validate the transaction anymore.

9.5.4 Proposed Framework for Circular Economy

In this section, we propose a novel circular economy framework (Figure 9.8) for plastic waste management Kolkata. Source segregated waste is collected from households which is transported to the material recycling facility. Since, we are concerned about the dry waste only, we consider the dry waste stream going to the MRF. At the MRF, plastic waste is separated from other dry waste. Now, plastic waste is divided into two sections – a) recyclable and b) non-recyclable. The recyclable part is sent to either respective third-party recyclers or to the upcycling vendors who are usually start-ups working on upcycling. The recyclers convert plastic waste into pellets for reuse or develop new products. On the other hand, the upcycling start-ups can upcycle the plastics into something useful. New products reach the consumers for their dedicated usage, hence reaching the households. This is how the loop is closed in the supply chain adhering to the circular economy principles. The circularity part is shown using dotted arrows. The non-recyclable fractions are usually discarded or landfilled. Here we suggest pyrolysis, an eco-friendly alternative for re-source recovery from non-recyclable plastics. Pyrolysis of plastics produces pyro-char, pyro-oil,

MRF:

{'index': 3, 'previous_hash': '2f57393265866026ebc483af342dfea80cdcb449bebdba2e25ccc0e71f072070', 'current_transaction': {'transactions': [{'timestamp': 1650212592.668153, 'Name': 'Plastic', 'cost per Kg(INR)': 8, 'Characteristics': 'Dry', 'Recyclability': 0.95, 'from': 'MRF central facility', 'to': 'third party recycler ', 'Digital signature': 'approved', 'msg': 'This waste is in good order', 'FLAG': 'N'}, {'timestamp': 1650212592.668153, 'Name': 'Glass', 'cost per Kg(INR)': 10, 'Characteristics': 'Dry', 'Recyclability': 1.0, 'from': 'MRF central facility', 'to': 'glass factory', 'Digital signature': 'not approoved', 'msg': 'This waste is in good order', 'FLAG': 'N'}, {'timestamp': 1650212592.668153, 'Name': 'Organic', 'cost per Kg(INR)': 3, 'Characteristics': 'wet', 'Recyclability': 0.9, 'from': 'MRF central facility', 'to': 'Organic', 'Digital signature': 'approved', 'msg': 'This waste is in good order', 'FLAG': 'N'}, {'timestamp': 1650212592.668153, 'Name': 'E-waste', 'cost per Kg(INR)': 20, 'Characteristics': 'Dry', 'Recyclability': 0.9, 'from': 'MRF central facility', 'to': 'E-waste', 'Digital signature': 'approved', 'msg': 'This waste is in good order', 'FLAG': 'N'}, {'timestamp': 1650212592.668153, 'Name': 'Metal', 'cost per Kg(INR)': 12, 'Characteristics': 'Dry', 'Recyclability': 1.0, 'from': 'MRF central facility', 'to': 'Metal factory', 'Digital signature': 'approved', 'msg': 'This waste is in good order', 'FLAG': 'N'}]}, 'timestamp': 'datetime.now().timestamp()', 'nonce': 37, 'hash': '2d96ed9bca57b6b2d753aaa4bf48322c15d7a4bf1a303644ec3fa3b540fc8780'}

3rd_Party_Recycler:

{'index': 4, 'previous_hash': '2d96ed9bca57b6b2d753aaa4bf48322c15d7a4bf1a303644ec3fa3b540fc8780', 'current_transaction': {'transactions': [{'timestamp': 1650212592.668153, 'Name': 'PET', 'cost per Kg(INR)': 3, 'Characteristics': 'Dry', 'Q.I.': 9, 'from': '3rd party recycler', 'to': 'manufacturing unit', 'Digital signature': 'not approoved', 'msg': 'This waste is not in good order', 'FLAG': 'N'}, {'timestamp': 1650212592.668153, 'Name': 'HDPE', 'cost per Kg(INR)': 4, 'Characteristics': 'Dry', 'Q.I.': 6, 'from': '3rd party recycler', 'to': 'manufacturing unit', 'Digital signature': 'not approoved', 'msg': 'This waste is not in good order', 'FLAG': 'N'}, {'timestamp': 1650212592.668153, 'Name': 'PVC', 'cost per Kg(INR)': 7, 'Characteristics': 'wet', 'Q.I.': 5, 'from': '3rd party recycler', 'to': 'manufacturing unit', 'Digital signature': 'not approoved', 'msg': 'This waste is not in good order', 'FLAG': 'N'}, {'timestamp': 1650212592.668153, 'Name': 'LDPE', 'cost per Kg(INR)': 2, 'Characteristics': 'Dry', 'Q.I.': 8, 'from': '3rd party recycler', 'to': 'manufacturing unit', 'Digital signature': 'not approoved', 'msg': 'This waste is not in good order', 'FLAG': 'N'}, {'timestamp': 1650212592.668153, 'Name': 'PP', 'cost per Kg(INR)': 3, 'Characteristics': 'Dry', 'Q.I.': 5, 'from': '3rd party recycler', 'to': 'manufacturing unit', 'Digital signature': 'not approoved', 'msg': 'This waste is not in good order', 'FLAG': 'N'}, {'timestamp': 1650212592.668153, 'Name': 'PS', 'cost per Kg(INR)': 2, 'Characteristics': 'Dry', 'Q.I.': 9, 'from': '3rd party recycler', 'to': 'manufacturing unit', 'Digital signature': 'not approoved', 'msg': 'This waste is not in good order', 'FLAG': 'N'}]}, 'timestamp': 'datetime.now().timestamp()', 'nonce': 40, 'hash': '7244b64b937324e968bfc3f7de0e81e85ff2ed1bc644439255b0bc728a16ccc0'}

FIGURE 9.8 Outputs of Case 3 Showing Issues Due to Uncertainty.

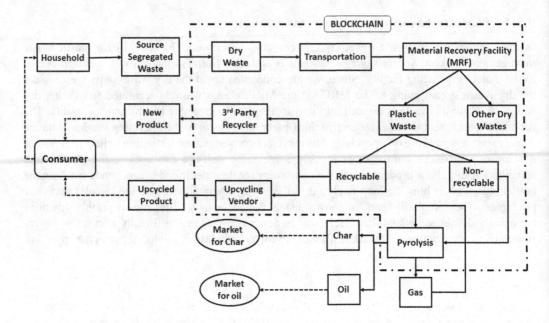

FIGURE 9.9 Proposed Framework for Achieving Circular Economy.

and pyro-gas (Kremer et al. 2021). The gas can be utilized in the process itself to heat up the reactor. The gas has a condensable part that is trapped as pyro-oil. This oil has market value and hence can be sold for industrial usage. Similarly, char derived from this process also has a market, which can be utilized for waste water treatment. In this way, circular economy can also be achieved. Most importantly, the supply chain with major processing of plastic waste, i.e., from collection -transportation to third-party recycler/pyrolysis unit (shown using dashed dot line), is under the supervision of blockchain. Not only does it allows traceability of the supply chain, but also it allows to inspect any uncertainties arising along the supply chain (Figure 9.9).

9.5.5 SUSTAINABILITY ANALYSIS

The recent advancements in the ICT and associated disruptive technologies such as blockchain, the area of supply chain management is being revolutionised towards digitalizing them. However, the question of sustainability is still there. Hence, it is imperative to delve into the sustainability aspects of the proposed system. According to Becker (2011), *"Sustainability is an important concept imperative at this moment and needs to be realized for fruition."* In this study, we are presenting a qualitative sustainability analysis of the proposed system as found in literature (Baidya et al. 2020; Debnath et al. 2022). We have divided our analysis in three sections based on three pillars of sustainability, i.e., environmental, economic and social.

9.5.5.1 Environmental Sustainability

Environmental sustainability can be assessed using the life cycle assessment (LCA) tool (Visentin et al. 2020; Debnath et al. 2021). But it is beyond the scope of this chapter and we present a qualitative analysis from the LCA perspective.

    i. The proposed system focuses on circular economy which will have positive impact on the environment. Closing loops by recycling plastic waste will reduce the carbon footprint (Hahladakis & Iacovidou 2019). The possible affected categories are global warming potential, eutrophication, and water depletion.

ii. Upcycling of plastic waste is not only extending the life cycle of the product but also it is reducing the landfill load as well as stopping them from polluting the marine life. Hence, the overall environment footprint will reduce (Zhao et al. 2022). The expected affected categories are global warming potential, eutrophication, human toxicity, marine eco-toxicity, freshwater ecotoxicity, and natural resource depletion.

iii. A part of the supply chain is under monitoring using the blockchain technology. Using blockchain increases efficiency and reduces the load of physical monitoring, which saves time and money. But this will generate e-waste in the future, which is an environmental issue (De Vries & Stoll 2021). This has to be compensated in future studies.

9.5.5.2   Economic Sustainability

Economic sustainability is the most important aspect of any feasible system. It can be complex but its outcome is important to understand the financial viability of the proposed system (Debnath et al. 2021). The outcomes are summarized below:

i. As mentioned before, implementation of blockchain reduces the load of physical monitoring, which saves money which is a good. On the other hand, the cost of blockchain infrastructure also needs to be considered.

ii. Recycling and upcycling of plastic waste will create more opportunities for cash flow. With the growing awareness of waste management, upcycling business could be the next big thing which is another area to be tapped. Policy recommendations in this regard will be helpful.

iii. Recovery of resources via pyrolysis will recover resources with market value. Some idea can be availed from literature (Fivga & Dimitriou 2018). However, detailed economic analysis on this front is required.

9.5.5.3   Social Sustainability

Social sustainability is often ignored, but it is important to delve into this area as well. Social LCA can be conducted to understand social sustainability. However, we present a generalized qualitative version below:

i. The primary benefit that the society will get is due to creation of new job opportunities in the plastic waste management supply chain network. These could be IT analyst, sustainability manager, to skilled upcycling laborers.

ii. Secondly, with growing jobs in waste management sector, the awareness will grow quite a few fold that one can achieve with typical awareness programs.

iii. Overall, the society will march towards a sustainable and circular ecosystem.

9.6   CONCLUSION AND FUTURE SCOPE

In this study, blockchain has been used to simulate a plastic waste supply chain network of Kolkata city. One base case scenario and two uncertainty scenarios have been simulated. The results show that our simulation identifies the disruptions along the supply chain. One unique point of our study is that the work considers both recyclable and non-recyclable plastics. In addition to that, a novel supply chain framework has been proposed which utilises the blockchain program developed in this work. The framework also proposes closing the loop by utilizing third-party recyclers, up-cycling start-ups and pyrolysis units which confirms to the circular economy principles. It is very important to induce the circular economy principles in plastic recycling value chain, which is achieved by the proposed framework. This work can be treated as prima facie to future body of work in this line.

In the scope of supply chain management industries, this proposed framework can be used as secure tracible chain. The designed framework could be integrated with cloud hosts to make the whole secure network robust and remote. The discussed model can be elevated towards smart network with the help of IoT sensor-based network. IoT helps the plastic waste management supply chain in form of route optimization to efficient garbage collection. It can be useful in financial terms by reducing the costs of fuel, garbage collection, maintenance, etc. The methodical use of pyrolysis can lead to convert waste into green energy, which in turn could help in achieving sustainability. More work is underway that utilizes blockchain simulators and further extension of the chain under consideration.

REFERENCES

Andreoni, A., Barnes, J., Black, A., & Sturgeon, T. (2021). Digitalization, industrialization, and skills development: Opportunities and challenges for middle-income countries. In Andreoni et al. (Eds.). *Structural Transformation in South Africa*. (pp. 261–285), UK: Oxford University Press.

Baidya, R., Debnath, B., De, D., & Ghosh, S. K. (2016). Sustainability of modern scientific waste compacting stations in the city of Kolkata. *Procedia Environmental Sciences*, 31, 520–529.

Baidya, R., Debnath, B., Ghosh, S. K., & Rhee, S. W. (2020). Supply chain analysis of e-waste processing plants in developing countries. *Waste Management & Research*, 38(2), 173–183.

Becker, C. (2011). *Sustainability ethics and sustainability research*. Springer Science & Business Media.

Bekrar, A., Ait El Cadi, A., Todosijevic, R., & Sarkis, J. (2021). Digitalizing the closing-of-the-loop for supply chains: A transportation and blockchain perspective. *Sustainability*, 13(5), 2895.

Bhattacharya, R. R. N., Chandrasekhar, K., Roy, P., & Khan, A. (2018). Challenges and opportunities: Plastic waste management in India. TERI, Available: http://hdl.handle.net/2451/42242

Bhattacharya, S., & Kesar, S. (2018). Possibilities of transformation: The informal sector in India. *Review of Radical Political Economics*, 50(4), 727–735.

Blossey, G., Eisenhardt, J., & Hahn, G. (2019). Blockchain technology in supply chain management: An application perspective. In *Proceedings of the 52nd Hawaii International Conference on System Science* (pp. 6885– 6893).

Brock, J. (2020). Special Report: Plastic pandemic - COVID-19 trashed the recycling dream. Retrieved 5 April 2022, from https://www.reuters.com/article/health-coronavirus-plastic-recycling-spe-idUSKBN 26Q1LO Accessed 15 April, 2022.

Browning, S., Beymer-Farris, B., & Seay, J. R. (2021). Addressing the challenges associated with plastic waste disposal and management in developing countries. *Current Opinion in Chemical Engineering*, 32, 100682.

Dangis, A. (2018). The usage of recycled plastics materials (rPM) by plastics converters in Europe. *Reinforced Plastics*, 62(3), 159–161.

Debnath, B., Das, A., Das, A., Chowdhury, R. R., Gharami, S., & Das, A. (2021). Edge Computing-based smart healthcare system for home monitoring of quarantine patients: Security threat and sustainability aspects. In *Intelligent Modeling, Prediction, and Diagnosis from Epidemiological Data* (pp. 189–210). Chapman and Hall/CRC.

Debnath, B., Ghosh, S., & Dutta, N. (2022). Resource resurgence from COVID-19 waste via pyrolysis: A circular economy approach. *Circular Economy and Sustainability*, 2(1), 211–220.

De Vries, A., & Stoll, C. (2021). Bitcoin's growing e-waste problem. *Resources, Conservation and Recycling*, 175, 105901.

Earth5r.org. (2021). Addressing the Plastic Waste Crisis at Kolkata, India. Retrieved 6 June 2022, from https://earth5r.org/addressing-the-plastic-waste-crisis-at-kolkata-india/

Ferronato, N., & Torretta, V. (2019). Waste mismanagement in developing countries: A review of global issues. *International Journal of Environmental Research and Public Health*, 16(6), 1060.

Fivga, A., & Dimitriou, I. (2018). Pyrolysis of plastic waste for production of heavy fuel substitute: A techno-economic assessment. *Energy*, 149, 865–874.

Hahladakis, J. N., & Iacovidou, E. (2019). An overview of the challenges and trade-offs in closing the loop of post-consumer plastic waste (PCPW): Focus on recycling. *Journal of Hazardous Materials*, 380, 120887.

Hossain, R., Islam, M. T., Shanker, R., Khan, D., Locock, K. E. S., Ghose, A., … & Sahajwalla, V. (2022). Plastic waste management in India: Challenges, opportunities, and roadmap for circular economy. *Sustainability*, 14(8), 4425.

Kataki, S., Nityanand, K., Chatterjee, S., Dwivedi, S. K., & Kamboj, D. V. (2022). Plastic waste management practices pertaining to India with particular focus on emerging technologies. *Environmental Science and Pollution Research*, 29, 24478–24503.

Kremer, I., Tomić, T., Katančić, Z., Erceg, M., Papuga, S., Vuković, J. P., & Schneider, D. R. (2021). Catalytic pyrolysis of mechanically non-recyclable waste plastics mixture: Kinetics and pyrolysis in laboratory-scale reactor. *Journal of Environmental Management*, 296, 113145.

Li, A. Q., & Found, P. (2017). Towards sustainability: PSS, digital technology and value co-creation. *Procedia Cirp*, 64, 79–84.

Malav, L. C., Yadav, K. K., Gupta, N., Kumar, S., Sharma, G. K., Krishnan, S., … & Bach, Q. V. (2020). A review on municipal solid waste as a renewable source for waste-to-energy project in India: Current practices, challenges, and future opportunities. *Journal of Cleaner Production*, 277, 123227.

OECD (2022). *Global plastics outlook: Economic drivers, environmental impacts and policy options*. Paris: OECD Publishing, 10.1787/de747aef-en

Payne, J., McKeown, P., & Jones, M. D. (2019). A circular economy approach to plastic waste. *Polymer Degradation and Stability*, 165, 170–181.

Rafey, A., & Siddiqui, F. Z. (2021). A review of plastic waste management in India–challenges and opportunities. *International Journal of Environmental Analytical Chemistry*, 1–17. 10.1080/03067319.2021.1917560

Rejeb, A., Rejeb, K., Simske, S., & Treiblmaier, H. (2021). Blockchain technologies in logistics and supply chain management: A bibliometric review. *Logistics*, 5(4), 72.

Rosário, A. T., & Dias, J. C. (2022). Sustainability and the digital transition: A literature review. *Sustainability*, 14(7), 4072.

Saberi, S., Kouhizadeh, M., Sarkis, J., & Shen, L. (2019). Blockchain technology and its relationships to sustainable supply chain management. *International Journal of Production Research*, 57(7), 2117–2135.

Shetty, R., Sharma, N., & Bhosale, V. A. (2022). Reverse supply chain network for plastic waste management. In *Emerging Research in Computing, Information, Communication and Applications* (pp. 1009–1025). Singapore: Springer.

Shin, S. K., Um, N., Kim, Y. J., Cho, N. H., & Jeon, T. W. (2020). New policy framework with plastic waste control plan for effective plastic waste management. *Sustainability*, 12(15), 6049.

Silva, A. L. P. (2021). Future-proofing plastic waste management for a circular bioeconomy. *Current Opinion in Environmental Science & Health*, 22, 100263.

The Economic Times. (2021). Over 34 lakh tons of plastic waste generated in FY 2019-20: Govt. Retrieved 6 June 2022, from https://economictimes.indiatimes.com/industry/indl-goods/svs/paper-/-wood-/-glass/-plastic-/-marbles/over-34-lakh-tons-of-plastic-waste-generated-in-fy-2019-20-govt/articleshow/84551047.cms

Vanapalli, K. R., Sharma, H. B., Ranjan, V. P., Samal, B., Bhattacharya, J., Dubey, B. K., & Goel, S. (2021). Challenges and strategies for effective plastic waste management during and post COVID-19 pandemic. *Science of the Total Environment*, 750, 141514.

Visentin, C., da Silva Trentin, A. W., Braun, A. B., & Thomé, A. (2020). Life cycle sustainability assessment: A systematic literature review through the application perspective, indicators, and methodologies. *Journal of Cleaner Production*, 270, 122509.

WHO. (2022). tons of COVID-19 health care waste expose urgent need to improve waste management systems. Retrieved 6 June 2022, from https://www.who.int/news/item/01–02-2022-tons-of-covid-19-health-care-waste-expose-urgent-need-to-improve-waste-management-systems, Accessed 5 June, 2022.

Zhao, X., Korey, M., Li, K., Copenhaver, K., Tekinalp, H., Celik, S., & Ozcan, S. (2022). Plastic waste upcycling toward a circular economy. *Chemical Engineering Journal*, 428, 131928.

Zheng, J., & Suh, S. (2019). Strategies to reduce the global carbon footprint of plastics. *Nature Climate Change*, 9(5), 374–378.

10 Supply Chain Management–Based Transportation System Using IoT and Blockchain Technology

Ahmed Mateen Buttar, Mahnoor Bano, and Amna Khalid
Department of Computer Science, University of Agriculture Faisalabad, Faisalabad, Pakistan

CONTENTS

DOI: 10.1201/9781003264521-10

10.1   INTRODUCTION

In today's world, the market has become more active and demanding, which has intensified the competition. Teamwork, integration, adaptability, and stakeholder trust have become more crucial for supply chains to be able to adapt to the environment's ongoing change. New technical applications have become more significant as a result. Utilizing the most recent specialized software is required to improve supply chain flow management. Blockchain technology is now firmly established as a crucial element of the current competitive scene. Investing in blockchain technology might help businesses quickly respond to shifting market conditions and client demands in today's hectic and competitive business environment. It has been noted that one benefit of using blockchain technology is that it can save costs by streamlining information flows since it makes it simpler to trace items throughout the supply chain. Blockchain technology, which enhances the level of information flow and the security of the information that is communicated, makes these advantages feasible.

Organization of the Chapter: This chapter is organized as: Section 10.2 discusses blockchain and supply chain trends, IoT role in e-commerce, blockchain infrastructure, and importance. Section 10.3 outlines the literature review. Section 10.4 provides discussion and analysis. Section 10.5 highlights a case study regarding how to control expenses using SCM. Section 10.6 concludes the chapter with a future scope.

Supply Chain Management: There are several procedures involved in turning raw materials into completed items, all of which go under the "supply chain management" term supply-side optimization is a strategy that aims to improve customer value and obtain a competitive edge in the market by optimizing the processes involved on the supply side. When a product or service is manufactured, supply chain management is responsible for overseeing the whole production process, beginning with the sourcing of raw materials and continuing through delivery to the end user. In order to get the goods to customers, a firm sets up a network of suppliers. Every product on the market today is the result of the efforts of several firms that form a supply chain. Supply chains have been around for a long time, but organizations have only just begun to see the added value they can provide by using them effectively [1].

In today's competitive corporate climate, supply chain management is generally considered as an indispensable instrument for cost management and economic performance enhancement. However, in light of new issues like the increasing complexity, transparency, and flexibility required by supply chains, companies and industrial sectors must reshape task-based challenges and supply management practices in order to remain competitive [2].

It is challenging to coordinate the supply and demand plans of the businesses that make up the supply chain because of the variances in demand and product portfolios. This is made even more challenging by the fact that these companies are situated in such a broad range of geographical locations [3].

It is important to take measures to protect the product from being harmed in any way throughout the process of transferring sensitive goods from one firm to their final consumers. This includes preventing any damage that may occur to the packaging of the product. Problems like as tampering with items while they are being transported, delays and falsified information, a failure to verify a person's identification, a lack of data management, and inaccuracies in the data itself are all typical in conventional methods of supply chain management [4].

There are a lot of huge modern businesses that have built their own identification systems in order to safeguard and keep the worldwide supremacy of their operations, as well as to hold on to the ability to instruct its suppliers. If not, they are required to depend on big regulatory companies or intermediaries and conform to their rules. Since many years ago, businesses have been able to improve their supply chain operations by employing information technology to their advantage. Originally, the purpose of electronic data interchange (EDI) was to enhance the effectiveness of firms' existing communication infrastructure. Integrating EDI allowed businesses to realize a number of potential advantages, including a quicker and more effective flow of data, flow [5], a reduction in the amount of time needed to finish an order, more distribution flexibility, and enhanced levels of customer service [6].

Over the course of time, technologies like RFID (radio frequency identification) and requirements for use of Internet of Things (IoT) as an example, improved responsiveness and visibility in supply chains. Even though many different modern technology tools are utilized to make supply chain operations easier to use in today's world, problems relating to trust, transparency, and adaptability have not yet been entirely resolved. A lack of clarity in supply chains is root cause of many challenges and complications in the areas of security, traceability, identity identification, and verification. According to the research that has been done on the topic, there is still a substantial amount of work to be done to create an atmosphere in supply chains that is founded on trust, despite the notion that trust is critical to the functioning of supply networks [7].

In addition, when life cycles of a product get shorter and manufacturing periods of time increase longer, it is harder to accurately estimate future demand. At this scenario, the poor demand realization that results from a lack of product availability puts supply networks in danger of having excess capacity. As a consequence of this, the development of flexible supply chains that is very important that businesses can quickly adapt to changes in the market. Because of its robust properties, including decentralization, transparency, and trust, the solution of the problems is potentially provided by the blockchain that have been outlined above. Blockchain has the potential to boost the openness and auditability of material movement, which will lead to a greater use of the technology by supply chains. Blockchain is also an accurate supply chain traceability solution [8]. Utilizing blockchain technology can assist in increasing supply chain efficiency, as well as security and traceability [4].

In addition, the technology behind blockchain may facilitate more cooperation between members of the network, which may result in a reduction of expenses and an improvement in operational effectiveness [7]. The blockchain has the potential to increase consumer trust in businesses because it allows for full product traceability across the entire supply chain. Additionally, blockchain technology offers help for the avoidance of fraud, which has a favourable impact on both cost reduction and efficiency [9].

The use of blockchain technology is a feasible alternative for tracing assets while maintaining data authenticity and security. The information sharing security, real time data collection of product, integrity and transparency in the supply network, and quality management all along the entirety of a product's life cycle are some of the benefits that come with using tracing technology that is based on blockchain technology [10]. Numerous research papers have been written that investigate the possible uses of the technology in positions that involve a lot of physical labor or complex processes. Some examples of these roles include duty delivery, administration of bill, and management of inventory [11].

The use of blockchain technology comes with a number of advantages that has been pointed out is that it can save money by optimizing information flows because it makes it easier to track things along the supply chain. These benefits are made possible by blockchain technology, which improves both the degree of information exchange and the security of the information that is shared. Additionally, blockchain technology may use strengthen security of supply chains by compensating for shortcomings of other components of Industry 4.0 [12]. It is unavoidable for all businesses, for the reasons that were outlined above, to keep a close eye on technological developments and make certain that these advancements are included into their operational procedures. To create a sustained competitive advantage and react to changing market circumstances, in light of the very unpredictable nature of the market, there is a newly introduced standard that has to be satisfied, and that is to take into account the use of reducing technology in supply chain.

10.2 BLOCKCHAIN AND SUPPLY CHAIN MANAGEMENT TRENDS IN A BUSINESS LIFE CYCLE

10.2.1 THE MANAGEMENT OF THE SUPPLY CHAIN PROCESS

SCM consist of five elements, as shown [7] in Figure 10.1.

10.2.1.1 Planning

Develop a strategy and oversee the management of the company's resources in order to meet the demand for the company's goods or services. Once the supply chain has been set up, you will need to devise methods to determine whether or not it is successful, efficient, offers customers value, and contributes to the achievement of corporate objectives.

10.2.1.2 Sourcing

The sourcing process involves selecting suppliers who can provide the necessary goods and services for the production of the product. After that, devise methods for overseeing and managing your connections with your suppliers. Ordering, receiving, keeping inventory up to date, and allowing payments to be made to suppliers are all essential operations.

FIGURE 10.1 Supply Chain Management Process.

10.2.1.3 Manufacturing

Manufacturing is the process of coordinating the processes required to start with the raw ingredients, then you manufacture the product, you test it to ensure its quality, you package it up so it can be sent, and you deliver it on time. The act of delivering a product in accordance with its scheduled release date is also considered part of the manufacturing process.

10.2.1.4 Delivery and Logistics

Orders from customers need to be coordinated, a delivery schedule needs to be created, customer invoices need to be issued, and payments need to be collected.

10.2.1.5 Returning

Build a system or technique for the return of goods that have been damaged, purchased in excess, or are no longer wanted.

10.2.2 BLOCKCHAIN TECHNOLOGY

Transactions are validated, recorded, and distributed on immutable and encrypted ledgers using blockchain technology. This ability is the primary reason for blockchain's growing popularity. Transactions using bitcoin, a decentralized digital currency that operates without the need for a central bank, are made easier thanks to the system. In its most basic form, blockchain technology generates and disseminates the ledger, every bitcoin transaction is sent to tens of thousands, or even millions, of computers connected to networks in different parts of the world. This is called the blockchain, as explained [13] in Figure 10.2.

Since both transactions and ledgers are encrypted, a higher level of security is offered by blockchain technology than the traditional banking model. Additionally, the instantaneous transmission of ledgers and transactions over the Internet excludes the two- to three-day clearance procedure that banks require, as well as the associated fees for transferring money from one account to another. The term "blockchain" refers to a chain that is formed by "blocks" of transactions that have been validated and cannot be changed. These "blocks" are connected together in chronological order to build the chain.

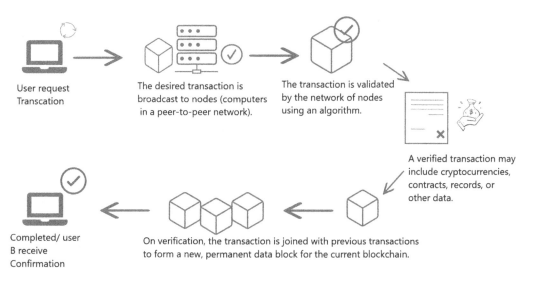

FIGURE 10.2 Blockchain Explained.

10.2.2.1 Background of SCM and Blockchain

One way to think about blockchain is as a distributed blockchain, which is a kind of blockchain that stores digital records of transactions on several computers at the same time. Following the addition of a block of records to the record, the data contained inside the block may be shown to have a mathematical connection to the blocks that follow it. As a direct consequence of this, an unbreakable chain of records will be produced [14].

Because of this mathematical connection, in order to update the information contained in a block, it is necessary to edit all of the blocks. Any change would result in a discrepancy that others would notice if it was implemented [15]. There are both permissioned and permissionless options. Access to permissioned blockchains, such as those used by SC partners, can be restricted to only allow authorized parties. The operation of a permission less blockchain is comparable to that of a shared database. Everyone has access to the entirety of the information. On the other hand, the user has no influence over who is allowed to write. One of the most game-changing uses of the implementation of smart contracts is the focus of blockchain technology [16]. A party's assurance that the counterparty will fulfil their commitments may be provided through a smart contract. The applications and uses of blockchain technology in the medical industry are extremely diverse, as shown [17] in Figure 10.3. The ledger technology makes it possible to transfer patient medical records in a safe manner, regulates the chain of distribution for medicines, and assists researchers in the healthcare industry in decoding genetic information.

Increased supply chain transparency, along with decreased costs and risk, are all potential benefits that may be realized via the use of blockchain technology. To be more specific, advancements in the blockchain supply chain may bring about the following significant benefits:

Primary potential benefits:
- Enhance the material supply chain's capacity to be tracked in order to guarantee compliance with company requirements.
- Decreased losses incurred as a result of gray market and counterfeit goods trade.
- Increasing awareness and ensuring compliance with regard to outsourced contract manufacturing.
- Decrease the amount of paperwork as well as the expenses of administration.

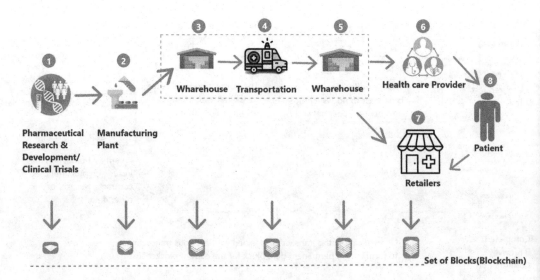

FIGURE 10.3 Supply Chain in Medical Field.

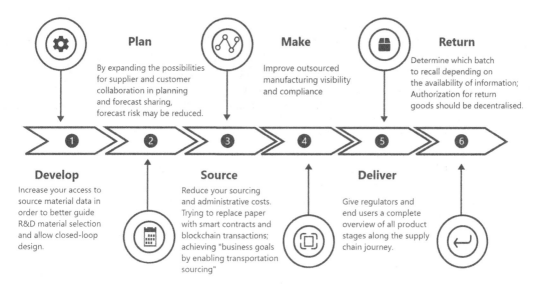

FIGURE 10.4 Driving Value in the Supply Chain.

Secondary potential benefits:
- Improve the company's image by making information about the materials used in goods more transparent.
- Increasing the public's faith in the data presented and its credibility.
- Reduce the danger of supply chain fraud harming your public relations.
- Engage stakeholder.

The efficiency of the supply chain can be affected by its performance in the following six areas: developing, planning, sourcing, making, delivering, and returning products, as shown [18] in Figure 10.4. Companies have the ability to establish and manage these drivers in order to place an emphasis on the appropriate balance between responsiveness and efficiency, taking into account the specific needs of their businesses and their finances.

10.2.2.2 Blockchain Properties
Blockchain has three fundamental characteristics: decentralization, immutability, and cryptography-based authentication [19].

10.2.2.2.1 Decentralization
One may argue that decentralization is at the core of the value proposition that blockchain offers. By enabling decentralized techniques, blockchain has the potential to boost trustworthiness and transparency in activities connected to environmental sustainability. Blockchain technology makes it possible for transactions to take place more rapidly and at a cheaper cost. This is made feasible by the fact that there is no longer a need for the participation of a reliable third party in the transaction of assets. Even among those who are skeptical about blockchain's potential in a wide variety of different markets and applications, there is reason to have optimism over its capacity to foster trust [20].

10.2.2.2.2 Immutability
The word "immutable" originates from object-oriented programming, a kind of computer programming in which the data structure, as well as any operations or functions that may be performed, are specified by the programmer. When something is "immutable," it means that it cannot

be altered once it has been created and saved in the form of computer code. As a direct consequence of this, recorded of transactions on a blockchain can't be deleted or altered in any way. The immutability aspect of blockchain makes it possible to audit transactions, which might contribute to increased transparency. Controlled access to essential data might be offered to a party. For instance, the notion of distributed ledgers that blockchain employs would make it possible for authorities and regulators to have access to vital data and information about sustainability [21].

10.2.2.2.3 Authentication Based on Cryptography

Blockchain systems are required to utilize based on cryptography's modern signatures to authenticate the identities of participants. This is done in order to restrict access to the information to only those individuals who have been granted permission to do so. This keeps the information secure by limiting who may see it. Transactions are signed by users using a private key that is produced for them when they register an account. Typically, a private key takes the form of a lengthy and unpredictable alphanumeric code. Complex algorithms are used in the generation of public keys from private keys by blockchain systems. Utilizing public keys allows for the sharing of information. Because of this feature, it is possible to measure and keep tabs on the outcomes of activities connected to sustainability. If a coffee retailer, for instance, asserts that they pay living wages to coffee producers, the correctness and authenticity of such statements may be established by inspecting payments made to digital wallets that have been given to the farmers.

10.2.3 BLOCKCHAIN-BASED SUPPLY CHAIN

Blockchains have the potential to be a game-changing technology for the supply chain sector in terms of its design, structure, operations, and general administration. Blockchains are decentralized public ledgers that are used to record and verify transactions. The ability of blockchain to ensure the dependability, traceability, and authenticity of information, combined with the potential for smart contractual agreements to create trustless surroundings, all foreshadow a fundamental reevaluation of supply chain logistics and the management of supply networks. Blockchain has the capability of ensuring the dependability, integrity, and accessibility of information. Smart contractual agreements have the potential to create a trustless environment. In this section, we will go further into proposition values of blockchain technology and its applicability to supply chains for commodities and manufacturing, in addition to discussing its structure and perhaps new elements for the management of supply chains. Particularly, we shall concentrate on its applications:

There is still room for interpretation and evolution with regard to how blockchain technology operates within the framework of supply chains. Supply chain networks that are based on blockchain technology could need a situation that is kept secret or hidden, permissioned blockchain with a large number of limited participants. However, there is a possibility that a more open-book approach to relationships is still possible. Choosing the appropriate amount of privacy is one of the first considerations to be made. In supply networks that are based on blockchain technology, there are four primary organizations that perform functions that are not present in conventional supply chains. Registrars, who are responsible for providing actors in the network with distinct identities. Standards organizations are responsible for defining various standards schemes, for example fair trade for more environmentally friendly technical requirements and regulations associated with blockchain technology. Certifiers are individuals or organizations that provide actors the necessary qualifications to become a member of the supplier chain. In order to maintain confidence in the system, some actors, such as manufacturers, retailers, and customers, are needed to get certification from an auditor or certifier that is widely recognized [22].

There are other influences on the flows of products and materials across the supply chain. There is the possibility that every product will have a presence on the blockchain, which would allow all necessary actors to have immediate access to product profiles. Access to a product may be

Blockchain based Supply Chain

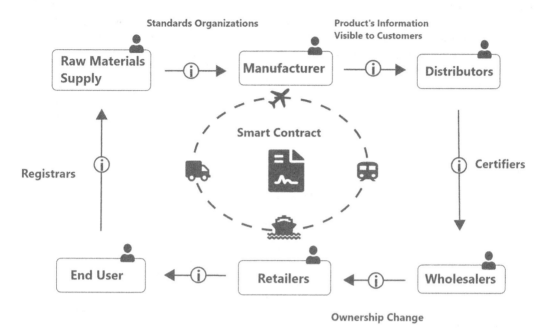

FIGURE 10.5 Supply Chain Transformation.

restricted using security measures, such that only those individuals or organizations who had the appropriate digital keys could use it. [12].

An information tag that is physically affixed to a product can function as an identifier, creating a link between the physical item in the real world and its digital equivalent in the blockchain. The manner in which a certain player "owns" a product or transfers it to another is an interesting facet of structure and flow management that should not be overlooked. It's likely that one of the most important rules will require actors to obtain permission before adding new information to a product profile or beginning a transaction with another party. Obtaining permission might involve entering into agreements via smart contracts and gaining consensus from all of the parties involved. To ensure the legitimacy of the transaction, an electronic contract or the fulfilment of a requirement stipulated by a smart contract might need to be signed by both parties before a good can be handed over (or sold) to a third party. The transaction data are updated on the blockchain ledger as soon as all parties have fulfilled their contractual duties and procedures [17].

At least five essential aspects of a product may be highlighted and specified with the use of blockchain technology. These aspects include the product's nature, its quality, its quantity, its location, and its ownership. Customers can see the whole chain of evidence and transactions, from raw materials to final sale, thanks to blockchain technology, which removes the need for a trusted central body to oversee and carry out this system, as shown [23] in Figure 10.5. As transactions take place across these several blockchain information dimensions, this data is updated and stored in ledgers, where it can be checked for accuracy.

10.2.3.1 The Current Supply Chain Benefits from Using Blockchain Technology

Without using blockchain technology, the overwhelming majority of supply chains are now able to run at scale. Despite this, the parties responsible for supply chain management and information technology have shown an interest in the technology. In addition to this, it has resulted in the publishing of a significant number of publications and motivated major IT players as well as start-ups to undertake potential pilot projects. Additionally, it has resulted in the publication of a huge number of articles. As an example of this, Walmart is now developing an application that authenticates transactions, record keeping that is more accurate and efficient, as well as tracking pork produced in China and produce produced in the United States that is done in order to follow produce and pork in both countries. Blockchain technology is being used in a collaborative effort between Maersk and IBM to provide improvements in the speed and accuracy of international transactions involving many parties. These transactions will take place across borders. DIIP is transitioning away from using spreadsheets in favor of a blockchain-based system to manage samples received from a wide array of suppliers, both internally and outside.

10.2.4 IoT Overview

The term "Internet of Things" refers to the recent increment in the number of devices that can connect to the Internet that also have the capacity for embedded computing. This term refers to a wide variety of technological advancements, includes security cameras and other surveillance systems that can connect to the Internet, as well as industrial equipment and sensors that are networked, as well as commonplace items such as refrigerators and automobiles. The Internet of Things is a technology that is advancing at a lightning-fast rate that has surfaced in the present age of technology. Some applications of Internet of Things such as smart cities, smart homes, and smart transportation, connected devices have made our lives easier to manage. Because the Internet of Things is such a term, there is no one agreed-upon definition for it. Several different organizations and groups of researchers came together to define the IoT. The number of IoT devices has surpassed the total human population around the world. Every day, new software and web-based services can be developed thanks to the numerous benefits provided by the IoT system [24].

10.2.4.1 Characteristics of IoT System

The IoT system is the example of a trying to cut technology that has the potential to bring about significant shifts in our lives, businesses, and economies. The Internet of Things gives rise to an infinite number of digitized services and applications, each of which offers a number of advantages in comparison to previously available options [25]. These applications and services have some characteristics in common, including the following:

1. **Sensing capabilities:** The wireless sensor network is the primary technology that encourages developments in a variety of Internet of Things domains (WSN). WSNs typically take the form of networks of sensors that collect data about their environments and transmit that data to a central location via some form of communication in order to be processed further. The Internet of Things is built on sensors, which are the components that enable the collection of all in genuine and environmentally essential information. This provides decision makers with the ability to make decisions that are accurate and precise in real time.
2. **Connectivity**: As a key component of the Internet of Things, this allows for remote access to billions of devices and things. In addition to this, it makes it possible for different things in our surrounding environment to be linked to and interact with one another through the Internet, which in turn makes it possible to develop new software programs and services.
3. **Intelligence capabilities**: IoT devices are able to make intelligent judgments in a variety of scenarios and intelligently collaborate with other cooperating items because they are equipped with sophisticated hardware, software, and sensor capabilities that allow the collection of a massive amount of contextual data.

4. **Unique identity:** The Internet of Things technology allows a variety of things to connect to one another over the Internet. Only if each piece of hardware had a distinctive identification or identifier, such as an IP address, would it be possible to ensure that it is capable of establishing an Internet connection. For the sake of the Internet system, manufacturers provide a one-of-a-kind identification to each device. This enables the system to upgrade devices to the relevant platforms, which is particularly important in the event of a breach in security.

5. **Big data:** The Internet of Things is comprised of billions of devices, each of which generates its own unique set of data. This massive amount of data is incompatible with the conventional data analysis approaches. The phrase "huge data" is what this is referring to. As a direct outcome of this, the IoT is rapidly becoming one of the most important sources of big data, it generates an enormous amount of data that necessitates the development of novel approaches to data analysis in order to reap the full advantages of IoT data [24].

10.2.5 E-COMMERCE

Electronic commerce is referred to as "e-commerce." Business transactions including the exchange of products and services are carried out through electronic media and the Internet in this context. E-commerce is the act of doing business through the Internet or other electronic means, such as the Electronic Data Interchange protocol (EDI). E-commerce is the activity of doing business over the Internet using a vendor-owned website that enables customers to buy products or services directly from the vendor's website. EFT (electronic fund transfer), credit cards, and debit cards are all accepted forms of payment on the website's digital shopping cart or digital shopping basket. For the goal of creating value between or among businesses, e-commerce involves the use of digital information processing and electronic communications technologies in commercial transactions [26].

10.2.5.1 Best E-Commerce Platforms

10.2.5.1.1 Amazon

One of the most well-known and widely used Internet marketplaces today is Amazon.com. It is best suited for medium-to-large vendors that want exposure to millions of consumers and who are ready to manage the flood of traffic when it arrives. Amazon.com is an online retail company, along with making electronic book readers and offering web hosting services, it has established itself as a leading example of electronic commerce. Customers may shop for anything from books and music to household goods and technology gadgets at Amazon that is large online marketplace. Amazon.com serves as a middleman between the millions of customers of other businesses and the millions of consumers of Amazon.com. "Cloud computing" refers to the practice of renting out data storage and processing power through the Internet. The Kindle e-book reader is a product of this company, which is presently the most popular on the market. Amazon has emerged as a prominent participant in the book publishing market as a result of its promotion of these devices, which has led to a fast growth in the creation of electronic books.

10.2.5.1.2 Walmart

When Walmart first entered the world of e-commerce in the year 2000 with the launch of Walmart.com, the company was already the largest retailer in the entire world. E-commerce sales increased by a staggering 43% in 2018, and there are no indications that this trend will slow down in the near future. The marketplace offered by Walmart is ideal for medium to large sellers who want to increase their exposure and is comparable to that offered by Amazon. To become a seller on the marketplace, you will need to go through an application process, but once your application is accepted, you will have access to over 110 million customers who shop online every single month. Using Walmart's marketplace grants you access to their customer support and safety features, and it also grants you access to a marketplace that integrates with a variety of service providers, allowing you to automate as much of the selling experience as you like.

10.2.5.1.3 Alibaba

One of the most well-known and successful retail and e-commerce companies in the world is Alibaba. As an alternative to charging for membership, Alibaba levies fees for its marketing and technical support services. This helps contribute to a significant and strong market share that is comprised of devoted clients. The majority of Alibaba's revenue comes from advertising and keyword bidding, which accounts for 57% of the company's overall revenues. By scanning a two-dimensional code, users of Alibaba are given the option to either purchase a goods or get customized marketing. Even many firms in the United States have been unable to successfully implement this cutting-edge approach. Additionally, it has more than 2.8 million online stores for different suppliers and more than 5900 different product categories.

10.2.5.1.4 eBay

Since its inception in 1995, eBay has developed from a website that primarily dealt in the sale of previously owned items into one of the most successful and widely utilized forms of online retailing today. This market is ideal for vendors that deal in both new and old goods and are willing to auction off their wares in the hopes of fetching a greater price for their wares. You don't need to fill out any paperwork in order to utilize eBay. Simply make an account by signing up for one, and then post your first listing. Because there are presently over 180 million active users on the platform on a monthly basis throughout the globe, there are a great deal of chances. Even if you are just starting out and do not yet have a website, the eBay Seller Hub is intended to assist you in operating your company in the most efficient manner possible. This is one of the many amazing things about eBay. You are able to create your own mailing labels and even conduct your own campaigns on eBay, for instance.

10.2.5.2 Characteristics of Online Business

The following benefits are provided by electronic commerce:

Payment Methods Other Than Cash: E-commerce enables the use of non-cash payment methods such as credit cards, debit cards, smart cards, electronic money transfers through a bank's website, and other types of electronic payment in addition to the traditional cash payment system.

Service Availability 24x7: The business processes of organizations automated by e-commerce, making it possible for such enterprises to make the services they give to clients accessible around the clock, twenty-four seven. The term "24x7" alludes to the fact that there are 24 hours on each day of the week.

Advertising and Marketing: The reach of advertisements for products and services provided by businesses is greatly increased thanks to e-commerce. It contributes to the improved administration of the marketing of goods and services.

Improved Sales: For allows for the generation of orders for items at any time, from any location, and without the participation of any person. This results in increased sales. In this manner, dependencies to purchase a product are significantly reduced, resulting in a rise in sales.

Support: E-commerce enables businesses to provide a wide variety of pre- and post-sale support options, which helps businesses deliver improved services to their clients.

Inventory Management: Through the use of electronic commerce, the process of managing a company's inventory of items may be automated. Whenever reports are needed, they are immediately created. The management of product inventories is made much more effective and simpler to continue.

Communication Improvement: E-commerce paves the door for improved communication by providing channels that are quicker, more efficient, and more dependable with consumers and partners [27].

10.2.6   IoT's Role in E-Commerce

10.2.6.1   In E-Commerce, the Function of IoT

As is well known, Internet of Things devices connect with one another through the Internet, which enables many retail and e-commerce enterprises to more effectively carry out a variety of operational functions. E-commerce businesses have a significant challenge when it comes to the management of their inventories. The Internet of Things makes it far simpler to keep tabs on a whole inventory than it was before. These Internet of Things sensors, such as RFID tags, make it possible for inventory management to control the flow of goods efficiently. Having smart shelves will lessen the number of unhappy customers you have. It automatically determined the total number of things that were sold and then placed orders for those items to prevent any hassle. Through the use of fully linked appliances, IoT devices help to foster a positive connection between manufacturers and customers. E-commerce stands to gain tremendously from the Internet of Things' capability to gather and transmit pertinent information about a product directly to ERP systems. The degree of interest in IoT among merchants is increasing as a result of increased digital connection. Without logistics channel, the area of e-commerce cannot be considered fully developed. A relatively harmless error in this route may result in significant harm. In these situations, the Internet of Things is essential for ensuring the seamless movement of products from one location to another. E-commerce companies now have the ability to trace the whereabouts of things at any given time thanks to GPS and RFID technology. The Internet of Things makes it easier to learn what customers want and helps to improve their experience.

10.2.6.2   E-Commerce with the Internet of Things for Disabled People

E-retailers and their online consumers and clients rely heavily on their websites as a primary means of communication. E-commerce websites serve as an important communication hub. The goal of businesses that operate e-commerce websites is to provide their customers with a pleasant experience while making online purchases via such websites. According to research by the Globe Health Organization, there are billions of handicapped individuals in the world, which accounts for an estimated 15% of the world's population and prevents them from participating fully in their everyday lives. A person who has a physical disability could find that going shopping makes them feel uncomfortable. Shopping online from the comfort of one's own home is made considerably more accessible to people with disabilities thanks to e-commerce websites. These customers may suffer from a variety of limitations, such as those affecting their senses, their physical skills, or their cognitive ability. The Internet makes it easier for customers with disabilities to purchase or sell things online. The World Material Accessibility Guidelines (WCAG) make it simpler for people with disabilities to access the content that is available on the Internet. WCAG is a set of guidelines that helps websites enhance their accessibility for persons with disabilities such as color blindness, hearing impairments, and vision issues. Insufficient online technology infrastructure is the primary barrier to the accessibility of the web for individuals with disabilities. People who are color blind or blind have a more difficult time exploring the web, owing to accessibility issues on the web. Web accessibility for impaired customers is restricted when dealing with business-to-consumer transactions. The new and burgeoning technology known as the IoT will have a favorable impact on society, particularly on elderly and disabled. IoT connects things that are real with Internet-connected smart gadgets. These intelligent gadgets are able to recognize sense and will reply in the appropriate manner. Opportunities for business, threats to national security, and references architecture for online commercial transactions [28].

10.2.6.3   Internet of Things (IoT): Security Challenges, Business Opportunities for E-Commerce

The Internet has emerged as the predominant medium for communication not just between people but also between people and physical items and between physical objects and other physical objects. According to a forecast by IDC, there will be 30 billion Internet-connected sensor-equipped devices in use by the year 2020. According to an analysis conducted by IBM, 90% of the data produced by smart devices is never used, while 60% of the data loses its value in a matter of seconds after being created. By the year 2020, revenue from products and services related to the Internet of Things will have surpassed $300 billion. The term "Internet of Things" refers to networked, linked smart gadgets that can interact with one another without the intervention of a person. The Internet of Things ecosystem is dependent on many smart items, gadgets, and tablets. Devices are connected to one another via the use of radio frequency identification, quick-response codes, sensors, and other forms of intelligent technology. The Internet of Things makes our everyday lives much more secure, efficient, and convenient. The quantity of devices of Internet of Things and quality of their connection go hand in hand. The Internet of Things comes up against a number of obstacles, the most significant of which is undoubtedly related to data protection and privacy concerns. Data of any kind, including personal, industrial, and organizational, may be saved on apps for the IoT. For instance, Internet of Things apps could save information on a person's health, purchases, habits, location, finances, and company orders, among other things. Therefore, it is essential to protect the sensitive data of customers from things like theft, intruder attacks, alteration, and so on. Due to the fact that these devices create a significant volume of data, technical problems are another obstacle that IoT devices must overcome. Keeping large amounts of data in storage, managing them, and analyzing them may be challenging. According to a poll conducted by Vision Mobile on the mobile developer economy in 2015, more than half (53%) of mobile developers are now engaged in Internet of Things initiatives.

10.3   LITERATURE REVIEW

Niranjanamurthy and Chahar [29] described a different strategy for implementation and the design of security systems within the scope of the Internet of Things scenario. It has been argued that the traditional method of addressing security concerns, which is characteristic of more traditional computer systems and other types of networks, does not take into account all of the facets associated with the newly proposed paradigm of communication, sharing, and actuation. In fact, the IoT paradigm contains new characteristics, mechanisms, and risks that cannot be taken into thorough account by the traditional formulation of security concerns. These new features, mechanisms, and dangers are as follows: The Internet of Things necessitates the adoption of a novel approach to cybersecurity, one that will examine the nature of the security threat from a holistic viewpoint and take into account the new players as well as the dynamics between them. In this work, we investigate the function of each actor and the interactions it has with the other primary players of the proposed scheme, as well as present a systematic method to ensuring the safety of Internet of Things devices.

Talavera et al. [30] discussed applications in the agricultural, industrial, and environmental sectors that make use of the Internet of Things. Discover application in a kind of diverse domains, trends, architectural directions, and obstacles in these two sectors is the impetus behind this project. This research was produced via the process of completing a systematic literature review using academic materials written in English and published in venues that were peer-reviewed between the years 2006 and 2016. The time period covered by this review was between 2006 and 2016. Certain references were collected and organized into four application areas relating to their monitoring, control, logistics, and prediction capabilities. In the process of implementation, the particulars from each of the references that were chosen were pieced together to produce the use and distributions of edge computing modules, communication technologies, storage solutions, and visualization strategies, distributions of sensors, actuators, power sources, and power sources. In the end, the results of the study were compiled into an architecture for the Internet of Things, which displays a wide variety of current solutions in the fields of agriculture, industry, and the environment.

Shafique et al. [31] described Internet of Things (IoT) ideas such as high-resolution video streaming, self-driven automobiles, smart environments, and electronic healthcare demand faster data rates, bigger bandwidths, improved capacities, low latency, and high throughput. In light of these developing ideas, the Internet of Things has completely changed the world by enabling seamless communication across different kinds of networks (HetNets). The ultimate goal of the Internet of Things is to implement plug-and-play technology, which will provide end users simplicity of operation, remote access control, and the capacity to configure their devices. The Technology of Internet of Things is discussed in this article from a high-level perspective, including its statistical and architectural trends, use cases, problems, and future possibilities. In this article, a comprehensive and in-depth assessment of the developing 5G-IoT situation is provided. Cellular networks of the fifth generation, or 5G, provide essential enabling technologies that are necessary for the widespread implementation of the Internet of things. A thorough overview of these vitally crucial supporting technologies, as well as a discussion of the new use cases for 5G-IoT that are evolving as a consequence of advancements in intelligent machines, computer vision, and machine learning.

Abbas et al. [32] presented a number of critical concerns, one of the most significant being privacy and security. The Internet of Things is experiencing a number of issues, some of which include improper device upgrades, a lack of efficient and comprehensive security mechanisms, user unawareness, and the well-known active device monitoring. In this work, we are investigating the history of Internet of Things systems and security measures, as well as identifying various security and privacy issues, methods used to secure the components of IoT-based environments and systems, existing security solutions, and the most effective privacy models that are required and that are suitable for different layers of Internet of Things applications. A novel Internet of Things layered model that is general and flexible with the privacy and security components as well as layer identification was addressed in this paper [32]. The Internet of Things system that is backed by cloud and edge computing has been developed. The bottom layer is symbolized by the Internet of Things nodes that are created by Amazon Web Service (AWS) in the form of Virtual Machines. For the user's information to be protected, security protocols and management sessions were created between each tier of the proposed cloud/edge enabled IoT paradigm. Security certificates were also implemented so that data could be exchanged between levels.

Riahi et al. [33] stated that e-commerce was a topic that was debated security of computer, data security, and other security of information frameworks are all examples of components that might have an effect on e-commerce. Security is a component of the Information Security framework and is applied to these components in order to protect them. E-commerce security has its own special intricacies and is one of the most apparent security components that impact the end user through their regular financial transactions with companies. This is because e-commerce security is closely related to the end user's capacity to make transactions online. The protection of e-commerce assets against unauthorized access, use, alteration, or destruction is what we mean when we speak about e-commerce security. Integrity, non-repudiation, authenticity, secrecy, privacy, and availability are the components that make up the dimensions of e-commerce security. E-commerce has a great potential for the banking industry but also adds a number of new hazards, such as those linked to security. For this reason, information security is a vital managerial and technical necessity for carrying out payment transaction activities through the Internet in an efficient and effective way. Because of the rapid pace at which both technology and business are advancing, it is necessary to match up appropriate algorithmic and technical solutions. In this research work, we covered topics such as an overview of E-commerce security, an understanding of the steps to place an order when shopping online, what the purpose of security is in e-commerce, what the different security concerns are in e-commerce, and how to shop online safely.

Gupta and Dubey [34] addressed the fact that factors such as consumer settling, the information quality of the website, trust, privacy concerns, reputation, security concerns, and the reputation of the firm all have a significant role in Internet customers' confidence in the website. Privacy and safety concerns are major issues that arise for both customers and online retailers while doing business online. Contrary to security, privacy refers to an individual's control over their own personal data, whereas security refers to the prevention of unauthorized users from accessing data. Therefore, the protection of information is a fundamental managerial and technological necessity for any payment transaction activities that take place via the Internet in order to ensure that they are efficient and effective. Integrity, privacy, non-repudiation, authenticity, confidentiality, and availability are the elements of e-commerce security that need to be investigated. E-commerce security is the protection of e-commerce assets against unauthorized access, destruction, modification, or use. In this article, we explore the privacy and security concerns surrounding electronic commerce, as well as its purpose, various security difficulties, and the ways in which it influences customer trust and buying behavior.

Sohaib et al. [28] described the explosion that has taken place in the e-commerce industry as a result of the rapid development and advancement of information technology and the facilities that are related to it in recent years. These developments have been met with unsettling security issues, which have had an impact on both the industry as a whole and all activities that take place online. Since conducting business online and any other activities that can be done online have become essential to our day-to-day lives, it is imperative that research be conducted to determine the dangers, the extent of the damage, and the financial burden that cybercrime places on the individuals and organizations that are victimized. In addition to this, it provides a comprehensive explanation of a variety of illegal online actions that constitute a significant risk to the safety of online business. The popular e-commerce websites Amazon and eBay were utilized as a case study in regard to respondents who are customers of both of these websites for a variety of different types of transactions.

Sarkis [23] examined the concept of blockchain has been gaining traction as an exciting new breakthrough during the 2010s. Many experts believe that this technology has the power to totally change the activities that take place throughout supply chains. Blockchain has been proposed as a way to increase the traceability and transparency of global supply networks from the point of view of SCM. Because of the rapid speed of globalization, supply chain management has been forced to accommodate both complicated disruptions and novel issues. The fact that the information that is collected is frequently kept within a single organization or intermediary is one of the challenges that are currently being faced. This asymmetry of information is related to actors in SCM that do not have sufficient insight into the operations that are being carried out by their partner.

Ndubisi et al. [35] defined that during the last couple of decades, the area of supply chain management (SCM) has developed into a well-established profession. When it was still in its infancy, the SCM idea was organized around optimizing the flow of money, information, and products or services in a way that was both effective and efficient. The external push from customers and other stakeholders for supply chains to become more environmentally friendly further complicates the imbalance of information that already exists.

Gardner et al. [36] discussed the notion of sustainability is constructed using these three pillars as its base: economic development, ecological growth, and social progress. To achieve sustainable development, these building elements are intertwined with one another and reliant on one another. Sustainability in a supply chain is not just the individual responsibility of a company, but also a communal obligation, in which all of the participants in the chain are required to adopt an attitude that is collaborative.

10.4   ANALYSIS AND DISCUSSION

10.4.1   Blockchain Technology in Supply Chain

The use of blockchain technology may make it possible to carry out the specialized services associated with supply chain management. The provision of operating services, reporting services, data administration, and analysis in data services are some examples of these. Furthermore, the technology behind blockchain has the ability to deliver specific business content, such as services for procurement and financial transactions, risk management services for supply chain management, and other services of a similar kind. Businesses need to build a platform for supply chain management that is based on blockchain technology if they want to efficiently satisfy the demand of the supply chain. This platform should be built as soon as possible. Through the use of source tracking, certificate storage, mutual trust, and information transmission, this will make it possible for enterprises to collect huge amounts of data and record information pertaining to the circulation of commodities [37].

In addition, the technology behind blockchain may link businesses that are part of a supply chain, encourage the integration of commodities flow, cash flow, and information flow; cut down on operational expenses; and boost overall quality of operations.

10.4.2   Research Methods and Operational Efficiency

The vast majority of the already available studies on the use of blockchain technology in supply chains tend to focus on qualitative research methods such as theoretical analysis, case studies, and architectural design, as shown [38] in Table 10.1.

TABLE 10.1
Theoretical Applications of Blockchain Technology in Supply Chains

Literature	Research Theory	Research Content and Conclusion
[23]	Supply Chain Objective Theory Innovation Diffusion Theory	Based on the idea of innovation diffusion, this article examines the impact of blockchain on supply chain management goals and concludes that blockchain has a significant advantage over the financial sector in supply chain activities.
[39]	TCA Theory	Transaction costs are reduced and governance expenses and structure are affected by blockchain.
[40]	TAM Theory and SNT Theory	Leaders and managers must provide the appropriate infrastructure and allow the system to optimize the productivity of its users.
[41]	BPR Theory	Process reengineering advocates the use of smart contracts in blockchain technology to automate and disintermediate business processes and foster a multilateral supply chain cooperation network.
[20]	Meaning Construction Theory	New blockchain technology is discussed in this paper using the theory of meaning construction.
[1]	MV Theory	Analyzing global supply chain operational risk in aviation logistics using the MV technique in the age of blockchain.
[8]	PAT Theory TCA Theory RBV Theory NT Theory	A theoretical framework for logistics and supply chain management scholars is presented in this study, which examines the major aspects of blockchain from four perspectives.

10.4.3 Trust and Transparency

When there is a lack of confidence in the traditional supply chain, parties are often unwilling to share information with one another because they believe the other partners to be their competition. Because of this, information asymmetry is ubiquitous throughout the supply chain, which will further add to the bullwhip effect and reduce the efficiency of the supply chain. The phrase "bullwhip effect" refers to the distortion of demand information that is sent as a consequence of poor information exchange between the upstream and downstream of the supply chain. This lack of information exchange will lead to a loss in the supply chain's overall operating efficiency [42]. In addition, as a consequence of the globalization of business and the distribution of production, the issues of information asymmetry and information isolation in the supply chain are becoming an increasingly critical concern. In addition to this, the supply chain that spans international borders has to pay more attention to the requirements imposed by regulatory bodies like customs and quarantine, as well as other departments. Because it is difficult for import regulators to verify the identity of the certificate of origin issuers in cross-border commerce, there is a problem of trust in the supply chain, which is caused by asymmetric information. This information makes it difficult for import regulators [43].

The use of blockchain technology promotes information exchange not only inside an organization but also between the preceding and subsequent phases of a supply chain. This not only helps to effectively establish the trust that already exists within an organization, but it also improves the transparency of the supply chain, fosters greater cooperation between members, and plays an instrumental role in constructively enhancing the capability to carry out operations. An increase in transparency not only ensures that the data are legitimate, reduces the rate of cybercrime, and protects the data of stakeholders, but it also lowers the risk in the operation of food supply chains, which is more favorable to decreasing the uncertainty of supply and demand. These benefits can be attributed to the fact that the increase in transparency also protects the data of stakeholders [8]. To be more specific, improving a company's internal transparency can improve the company's internal monitoring and control ability, eliminate the impact of internal uncertainty, help make better decisions, and earn consumers' trust; improving a company's supply chain transparency can enhance information sharing among business partners, promote the more effective flow of goods and services, trade, and information, and help producers better understand their products; and improving a company's supply chain transparency can help producers better understand their products [41]. Blockchain technology may also be used to verify and audit transactions and product information in real time and at a minimal cost, including the qualifications and reputations of all parties involved and the qualities of the items or services in question [36].

10.4.4 Information Exchange Efficiency

The most expensive part of the supply chain is not the cost of transportation or management; rather, it is the cost of knowledge. To begin, there is a correlation between the length of the supply chain, the complexity of that network, and the amount of trade that takes place. This correlation leads to a rise in the cost of information. Second, since the costs of transportation and monitoring are going down, the fraction of the expenditures that are attributable to information is going increasing [44]. The cost of information encompasses a variety of expenses, including as the cost of signing contracts, the cost of locating trade partners, and the cost of products information incurred as a result of movement of commodities through the supply chain. According to the International Shipping Association (2017), the marine business, which is still considered to be a traditional industry, is responsible for the handling of more than 90% of the products involved in international commerce each year. Along with a significant quantity of papers, a substantial number of shipping containers are moved from port to port all over the globe. It is possible that the expense of processing commercial paperwork will account for as much as one-fifth of the total cost of moving products. In addition, at the moment, the vast majority of businesses solely utilize their systems for the administration of the supply chain.

Because of their isolation, these systems find it difficult to cooperate with one another, which reduces the effectiveness of information transfer [45].

A completely connected ecosystem that is both efficient and transparent may be created via the digitization of the supply chain, which can also eliminate these obstacles. The foundation of supply chain digitalization may emerge from the intersection of blockchain technology with that of artificial intelligence, workflow automation, and the Internet of things. As a result, the blockchain has the ability to advance the trade facilitation agenda, with a particular emphasis on paperless trading across international borders [40]. Blockchain technology and methodologies that are based on smart contracts have the potential to drastically minimize the need for human intervention and automate business activities of supply chain management that now entail manual and paper-based transactions. Because of this, there is a substantial decrease in the amount of business friction, as well as an increase in the effectiveness of the supply chain management service. In addition, given that it is a state machine, the smart contract has the capacity to rapidly monitor the changes in the process state from the point of view of the many stakeholders, including the consumers, distributors, consumers, manufacturers, and logistics service providers. This serves to contribute to an increase in the effective transmission of information [46].

10.4.5 TRACING AND TRACKING

Traceability and the management of data are two of the most significant difficulties associated with the supply chain. However, the vast majority of the present traceability systems lack transparency, the data is mostly held inside the company, and the cost of tampering with the data is quite minimal. The previous centralized management technique makes it easy for supply chain nodes to evade responsibility when product safety or quality concerns develop, and it is difficult to identify the root cause of the problem. When disputes emerge in a company's supply chain, it may be expensive and time-consuming to find the main cause of the problem by conducting a retrospective audit. In addition, the digital commercial world, especially e-commerce, is rife with fake goods. Products that are not authentic make about 2.5% of each year's total worldwide commerce volume, which reached $461 billion in 2018. Particularly damaging to the well-being of customers and the image of brands is the sale of counterfeit goods in grocery and pharmacy shops, as well as issues with the quality of food and the potential for environmental contamination. Traceability solutions are thus urgently required as the primary quality control tool to ensure that items transferred throughout supply chains are safe [47].

The traceability function of the supply chain may be realized with the use of blockchain technology, which can also monitor items or supplies in real time [45]. Decentralized ledgers assist to link suppliers, manufacturers, purchasers, and regulators that are geographically separated from one another owing to the varied methods, regulations, or apps that they use. This, in turn, improves the traceability of the supply chain. The supply chain for food is an ideal use for blockchain technology because it ensures food safety and the ability to track its origins, two factors that may help enhance customer confidence [48]. To be more specific, blockchain technology may be used to monitor the safety of food, aid in the reduction of corruption and waste, and ultimately lower operational expenses. These data are more trustworthy and less susceptible to being tampered with as a direct result of the properties of blockchain. It is possible to efficiently link the data acquired from these nodes in order to guarantee the quality of the food. After an epidemic of an animal or plant illness has occurred, it is possible to track the affected items more rapidly. This helps to improve supply chain risk management. For the supply chain's traceability function, it is also necessary to manage and monitor the items, especially those that are in short supply. Traditional transaction data as well as more complicated data kinds like as temperature, humidity, and dietary requirements may all be tagged on items utilizing blockchain with relative ease [49]; second of all, it is possible to calculate the shelf life and degradation rate of food products in a traceable shelf-life management system, which helps in making decisions about food quality. In addition, the supply chain management system's pricing monitoring component is using blockchain and smart contracts. It is being done to ensure that the product distribution mechanism is completely transparent and to prevent the corporation from chasing excessive profits [16]. Customers have access to accurate and trustworthy pricing information thanks to the system that tracks prices. Both consumers and businesses are able to have a better understanding of one another's preferences because to the availability of pricing information.

10.4.6 OTHER FACTORS

Conducting research on the business model for online commerce based on blockchain and the Internet of Things technologies. Provethings demonstrate both the method of integrating smart contracts based on blockchain technology with traditional e-commerce as well as the protocol's viability from the perspectives of cost analysis and idea [50]. Conversation regarding environmentally friendly supply chain management and blockchain technology, with an emphasis on the idea of leveraging blockchain to create environmentally friendly supply chain management. A study of the effect blockchain technology has on the several risk factors required for the proper functioning of global supply networks [39].

10.5 CASE STUDY

Intel is one of the most well-known companies in the world since it is the leading maker of computer chips. However, after introducing its low-cost "Atom" chip to the market, the business needed to dramatically lower the amount of money it spent on its supply chain. When chips sold for $100 each, supply chain expenses of around $5.50 per chip were manageable, as shown [51] in Figure 10.6. However, the price of the new chip was a fraction of that amount, coming in at approximately $20.

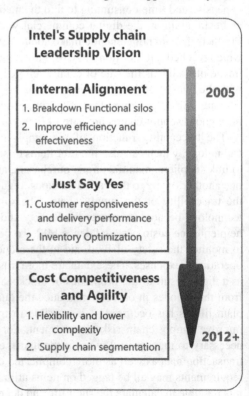

FIGURE 10.6   Intel's Strategy for the Supply Chain.

10.5.1 THE OBSTACLE OF COST REDUCTION ALONG THE SUPPLY CHAIN

Intel needed to find a way to lower the costs associated with the supply chain for the Atom processor, but they only had one area of leverage to work with: their inventory.

Because the processor had to function, Intel was unable to compromise on the quality of the services it provided. Due to the fact that every Atom product is comprised of a single component, there was also no method to lower the amount of duty paid. Because of the high value-to-weight ratio of the chips, Intel was unable to reduce the distribution costs any more despite having previously reduced the amount of packaging to its smallest possible size. The only choice left was to make an effort to cut down on the levels of inventory that had been maintained very high up to that time in order to sustain a nine-week order cycle. The only method Intel could discover to drive down the costs associated with the supply chain was to cut down on the cycle time, and therefore, the amount of inventory they had on hand.

10.5.2 THE WAY TO MAKE COSTS LESS EXPENSIVE

Made to order was somewhat of an outlier in terms of supply chain strategies for the semiconductor industry, but Intel decided to give it a go anyhow. The firm launched its operations with a test run, employing a manufacturer located in Malaysia. They found and removed inefficiencies in the supply chain over the course of many iterations, which allowed for an incremental reduction in the amount of time required to process orders. Additional endeavors for improvement included the following:

- Reducing the test window for the chip assembly from a schedule of five days to one that occurs every two weeks and lasts for two days.
- Implementation of a standardized S&OP planning procedure.
- Switching to a form of inventory that is maintained by the seller wherever and whenever it was feasible to do so.

10.5.3 MANAGEMENT OF COSTS IN THE SUPPLY CHAIN RESULTS

Intel finally brought the order cycle time for the Atom chip down from nine weeks to only two via the use of an iterative approach to cycle time optimization. The business was able to achieve a supply chain cost reduction of more than $4 per unit for the $20 Atom chip as a direct consequence of this, which is a rate that is far more agreeable than the initial number of $5.50, as explained [51] in Table 10.2.

TABLE 10.2
Supply Chain Transformation Enabled by IT

Area	Result
Business Velocity	
• 50% faster ramp up	• Manufacturing process improved 50% more-time, 25% supplier build time and time to place order reduce 95%.
• 65% shorter load times	• 65% reduction in order fulfilment times.
Business Attentiveness	
• Delivery of order 50% faster	• By using VMI hub deliver time 50% increased
• 300% reaction time to customer	• Provide a 3x quicker response to client orders and modification requests.
Business Efficiency	
• 32% inventory reduction	• Inventory reduced by 32% using automation
• 16 to 21 %productivity increase	• 21% increased CPU units and capital dollar invested increased by 16%.

10.6   CONCLUSION AND FUTURE SCOPE

The results of this study will open up a lot of doors for new lines of inquiry in the future, including avenues such as theory construction and idea verification. The discoveries in the fields of e-commerce, IoT, and blockchain will, in fact, need more study in order to be refined and elaborated upon. The findings of the study are eye-opening for the scholars who have an interest in electronic commerce. In this new age of blockchain technology, the key to the successful growth of businesses is to have a firm grasp of the fundamental technological resources of blockchain. In addition, the majority of the pertinent study focuses on the viability of the technology, whereas the investigation of the real-world measurement of the technology's commercial value and the performance of its supply chain is relatively scarce. This may present a challenge to the actualization of the technology. In coming days, it will be necessary to further identify the potential challenges that blockchain may face from both a theoretical and practical perspective. These challenges may include technical complexity, system applicability, cultural impact, issues with collaboration, cost, and security, as well as other aspects, such as the study of the technical limitations of blockchain delay, throughput, size, and bandwidth, version control, hard branch, and multiple branches. Blockchain-based technology paired with employee wellbeing. Second, more research has to be done on the border conditions for sharing quality assurance data among supply chain partners, and a clearer understanding of the appropriate balance between shared data and privacy protection is needed. Additionally, the programmability of blockchain technology makes it possible to conduct automated quality control and performance assessment utilising machine learning-based tools. We might look into the possibility of combining blockchain technology with intelligent technology.

The research area will provide a deeper grasp of the important factors of the most current IoT and e-commerce platform, which will revolutionize their company. The incorporation of a smart device into the Internet of Things necessitates the production of brand-new apps in addition to the expansion of those already in existence. The Internet of Things makes it possible for e-commerce companies to provide more effective and tailored services to their clients in the future. It is helpful for e-commerce enterprises to analyses purchasing patterns in search trends and online browsing in order to offer clients items that are more specifically tailored to their needs. In addition, we may make a connection between some of the traditional application areas and the expansion of our systemic approach in order to bring attention to the limitations and prerequisites that are imposed by the security. In addition, some encryption, protection, verification, and authentication mechanisms will be made available in order to influence consumers' perceptions of security. In conclusion, for the disruptive potential of blockchain to be fully realized, solutions that have the backing of both businesses. At that time, the private sector is in driver's seat when it comes to blockchain initiatives, but government organizations are falling behind. There have been very few attempts that have concentrated on integrating the requirements and demands of government entities. Finding a remedy for the uneven distribution of blockchain resources is of critical importance should be a collaborative effort between the sector and the government. This is done for the simple reason that a marketplace can only be reliable if customers are comfortable conducting business in that setting.

REFERENCES

[1] Kshetri, N. (2018). 1 Blockchain's roles in meeting key supply chain management objectives. *International Journal of Information Management*, 39, 80–89.
[2] Kamilaris, A., Fonts, A., & Prenafeta-Boldú, F. X. (2019). The rise of blockchain technology in agriculture and food supply chains. *Trends in Food Science & Technology*, 91, 640–652.
[3] Kamble, S. S., Gunasekaran, A., Kumar, V., Belhadi, A., & Foropon, C. (2021). A machine learning based approach for predicting blockchain adoption in supply Chain. *Technological Forecasting and Social Change*, 163, 120465.

[4] Dwivedi, S. K., Amin, R., & Vollala, S. (2020). Blockchain based secured information sharing protocol in supply chain management system with key distribution mechanism. *Journal of Information Security and Applications*, 54, 102554.

[5] Hofmann, E. (2016). S. Chopra and P. Meindl: Supply chain management: strategy, planning and operation. *Journal of Purchasing and Supply Management*, 19(3), 212–213.

[6] Ellis, S., Morris, H. D., & Santagate, J. (2015). IoT-enabled analytic applications revolutionize supply chain planning and execution. *International Data Corporation (IDC) White Paper*, 13.

[7] Song, J. M., Sung, J., & Park, T. (2019). Applications of blockchain to improve supply chain traceability. *Procedia Computer Science*, 162, 119–122.

[8] Helo, P., & Hao, Y. (2019). Blockchains in operations and supply chains: A model and reference implementation. *Computers & Industrial Engineering*, 136, 242–251.

[9] Di Vaio, A., & Varriale, L. (2020). Blockchain technology in supply chain management for sustainable performance: Evidence from the airport industry. *International Journal of Information Management*, 52, 102014.

[10] Agrawal, T. K., Kumar, V., Pal, R., Wang, L., & Chen, Y. (2021). Blockchain-based framework for supply chain traceability: A case example of textile and clothing industry. *Computers & Industrial Engineering*, 154, 107130.

[11] Muessigmann, B., von der Gracht, H., & Hartmann, E. (2020). Blockchain technology in logistics and supply chain management—A bibliometric literature review from 2016 to January 2020. *IEEE Transactions on Engineering Management*, 67(4), 988–1007.

[12] Queiroz, M. M., Telles, R., & Bonilla, S. H. (2019). Blockchain and supply chain management Integration: A systematic review of the literature. *Supply Chain Management: An International Journal*, 25(2), 241–254.

[13] Singhal, I., Bisht, H. S., & Sharma, Y. Anti-Counterfeit product system using blockchain technology.*International Journal for Research in Applied Science & Engineering Technology*, 9(12), 291–295.

[14] Yaga, D., Mell, P., Roby, N., & Scarfone, K. (2019). Blockchain technology overview. arXiv preprint https://arxiv.org/abs/1906.11078. https://nvlpubs.nist.gov/nistpubs/ir/2018/NIST.

[15] Kshetri, N., & Voas, J. (2018). Blockchain in developing countries. *IT Professional*, 20(2), 11–14.

[16] Khatoon, A. (2020). A blockchain-based smart contract system for healthcare management. *Electronics*, 9(1), 94.

[17] Kouhizadeh, M., Sarkis, J., & Zhu, Q. (2019). At the nexus of blockchain technology, the circular economy, and product deletion. *Applied Sciences*, 9(8), 1712.

[18] Ali, O., Jaradat, A., Kulakli, A., & Abuhalimeh, A. (2021). A comparative study: Blockchain technology utilization benefits, challenges and functionalities. *IEEE Access*, 9, 12730–12749.

[19] Kshetri, N. (2020). Blockchain could be the answer to cybersecurity. Maybe.*Wall Street Journal*, https://www.wsj.com/articles/blockchain-could-be-the-answer-to-cybersecurity-maybe-1527645960

[20] Dogru, T., Mody, M., & Leonardi, C. (2018). Blockchain technology & its implications for the hospitality industry. *Boston University*.

[21] Vervoort, D., Guetter, C. R., & Peters, A. W. (2021). Blockchain, health disparities and global health. *BMJ Innovations*, 7(2).

[22] Steiner, J., Baker, J., Wood, G., & Meiklejohn, S. (2015). Blockchain: The solution for transparency in product supply chains. *Available at: provenance. org/whitepaper*.

[23] Sarkis, J. (2020). Supply chain sustainability: Learning from the COVID-19 pandemic. *International Journal of Operations & Production Management*. 41(1), 63–73.

[24] Atlam, H. F., Azad, M. A., Alzahrani, A. G., & Wills, G. (2020). A review of blockchain in internet of things and AI. *Big Data and Cognitive Computing*, 4(4), 28.

[25] Lam, K. Y., Mitra, S., Gondesen, F., & Yi, X. (2021). ANT-centric IoT security reference architecture—Security-by-design for satellite-enabled smart cities. *IEEE Internet of Things Journal*, 9(8), 5895–5908.

[26] Vadwala, A. Y., & Vadwala, M. M. S. (2017). E-Commerce: Merits and demerits a review paper. *International Journal of Trend in Scientific Research and Development*, 1(4), 117–120.

[27] Bhat, S. A., Kansana, K., & Khan, J. M. (2016). A review paper on e-commerce. *Asian Journal of Technology & Management Research [ISSN: 2249–0892]*, 6(1).

[28] Sohaib, O., Lu, H., & Hussain, W. (2017, June). Internet of Things (IoT) in E-commerce: For people with disabilities. In *2017 12th IEEE Conference on Industrial Electronics and Applications (ICIEA)* (pp. 419–423). IEEE.

[29] Niranjanamurthy, M., & Chahar, D. (2013). The study of e-commerce security issues and solutions. *International Journal of Advanced Research in Computer and Communication Engineering*, 2(7), 2885–2895.

[30] Talavera, J. M., Tobón, L. E., Gómez, J. A., Culman, M. A., Aranda, J. M., Parra, D. T., & Garreta, L. E. (2017). Review of IoT applications in agro-industrial and environmental fields. *Computers and Electronics in Agriculture*, 142, 283–297.

[31] Shafique, K., Khawaja, B. A., Sabir, F., Qazi, S., & Mustaqim, M. (2020). Internet of things (IoT) for next-generation smart systems: A review of current challenges, future trends and prospects for emerging 5G-IoT scenarios. *IEEE Access*, 8, 23022–23040.

[32] Abbas, K., Tawalbeh, L. A. A., Rafiq, A., Muthanna, A., Elgendy, I. A., El-Latif, A., & Ahmed, A. (2021). Convergence of blockchain and IoT for secure transportation systems in smart cities. *Security and Communication Networks*, 2021, 5597679. 10.1155/2021/5597679.

[33] Riahi, A., Challal, Y., Natalizio, E., Chtourou, Z., & Bouabdallah, A. (2013, May). A systemic approach for IoT security. In*2013 IEEE international conference on distributed computing in sensor systems* (pp. 351–355). IEEE.

[34] Gupta, M. P., & Dubey, A. (2016). E-commerce-study of privacy, trust and security from consumer's perspective. *Transactions*, 37, 38.

[35] Ndubisi, N. O., Zhai, X. A., & Lai, K. H. (2021). Small and medium manufacturing enterprises and Asia's sustainable economic development. *International Journal of Production Economics*, 233, 107971.

[36] Gardner, T. A., Benzie, M., Börner, J., Dawkins, E., Fick, S., Garrett, R., ... & Wolvekamp, P. (2019). Transparency and sustainability in global commodity supply chains. *World Development*, 121, 163–177.

[37] Pan, X., Pan, X., Song, M., Ai, B., & Ming, Y. (2020). Blockchain technology and enterprise operational capabilities: An empirical test. *International Journal of Information Management*, 52, 101946.

[38] Zhao, M. (2020). Blockchain Technology Improves Supply Chain: A Literature Review. *Academic Journal of Computing & Information Science*, 3(4), 73–85.

[39] Cole, R., Stevenson, M., & Aitken, J. (2019). Blockchain technology: implications for operations and supply chain management. *Supply Chain Management: An International Journal*, 24(4), 469–483.

[40] Astill, J., Dara, R. A., Campbell, M., Farber, J. M., Fraser, E. D., Sharif, S., & Yada, R. Y. (2019). Transparency in food supply chains: A review of enabling technology solutions. *Trends in Food Science & Technology*, 91, 240–247.

[41] Kouhizadeh, M., Zhu, Q., & Sarkis, J. (2020). Blockchain and the circular economy: Potential tensions and critical reflections from practice. *Production Planning & Control*, 31(11-12), 950–966.

[42] Fu, Y., & Zhu, J. (2019). Big production enterprise supply chain endogenous risk management based on blockchain. *IEEE Access*, 7, 15310–15319.

[43] Bray, R. L., & Mendelson, H. (2015). Production smoothing and the bullwhip effect. *Manufacturing & Service Operations Management*, 17(2), 208–220.

[44] Allen, D. W., Berg, C., Davidson, S., Novak, M., & Potts, J. (2019). International policy coordination for blockchain supply chains. *Asia & the Pacific Policy Studies*, 6(3), 367–380.

[45] Gao, Z., Xu, L., Chen, L., Zhao, X., Lu, Y., & Shi, W. (2018). CoC: A unified distributed ledger-based supply chain management system. *Journal of Computer Science and Technology*, 33(2), 237–248.

[46] Schmidt, C. G., & Wagner, S. M. (2019). Blockchain and supply chain relations: A transaction cost theory perspective. *Journal of Purchasing and Supply Management*, 25(4), 100552.

[47] Salah, K., Nizamuddin, N., Jayaraman, R., & Omar, M. (2019). Blockchain-based soybean traceability in agricultural supply chain. *IEEE Access*, 7, 73295–73305.

[48] Wang, Q., Zhu, X., Ni, Y., Gu, L., & Zhu, H. (2020). Blockchain for the IoT and industrial IoT: A review. *Internet of Things*, 10, 100081.

[49] George, R. V., Harsh, H. O., Ray, P., & Babu, A. K. (2019). Food quality traceability prototype for restaurants using blockchain and food quality data index. *Journal of Cleaner Production*, 240, 118021.

[50] Russ, H., Muthusam, K., Quah, K. C., Jian, W., & Mark, Z. (2012). Transforming Intel's supply chain to meet market challenges. *IT@Intel White Paper*, 1–8.

[51] Ferrer-Gomila, J. L., Hinarejos, M. F., & Isern-Deya, A. P. (2019). A fair contract signing protocol with blockchain support. *Electronic Commerce Research and Applications*, 36, 100869.

11 Perspective Analysis of Three Types of Services on a Queueing-Inventory System with a Sharing Buffer for Two Classes of Customers

K. Jeganathan
Ramanujan Institute for Advanced Study in Mathematics, University of Madras, Chennai, India

K. Prasanna Lakshmi
Department of Mathematics with Computer Applications, Ethiraj College for Women, Chennai, India

S. Selvakumar
Ramanujan Institute for Advanced Study in Mathematics, University of Madras, Chennai, India

T. Harikrishnan
Department of Mathematics, Guru Nanak College, Chennai, India

D. Nagarajan
Department of Mathematics, Rajalakshmi Institute of Technology, Chennai, India

CONTENTS

DOI: 10.1201/9781003264521-11

11.1 INTRODUCTION

The ultimate aim of any organization is to give service to their satisfaction of the customer ensuring they will always return to the organization for service by delivering the needs of its customers on time and by providing the desired services to the customers satisfaction. The organization distinguishes their customers priorities based on their needs and designation for the profit of the organization by serving them effectively based on their priorities. For example, many call centers give major importance to its customers who are more profitable and at the time organization do not want to lose the customers who are less profitable. Many industrial/social organizations like IT industry, military and Air force, etc., provide service to all its employees but incentives/privileges are given only to its employees in higher cadres. The customers who come to electronic showrooms, automobile showrooms, jewelry showrooms, and mobile showrooms are not only provided with basic services but are also given supplementary or optional services. The customers are prioritized and served based on the above factors.

In the decade, many investigators have actively engaged in stochastic modeling of inventory queuing systems with variety types of multi-priority customers service policy. Sivakumar et al. [1] examined two sorts of customers: type 1 customers may wait during stock-out periods, whereas type 2 customers do not need to wait during stock-out periods. Anbazhagan et al. [2] analyzed the mixed priority service policy of two classes of customer on a queueing-inventory system (QIS) where the HNC will get the basic service along with some optional service, but the LNC demand only with basic service.

Chang et al. [3] investigated a QIS with MAP arrival. For both customer classes, the queue and orbit have the finite capacity. In many real-life situations, dissatisfied customers want servicing on their goods at an indeterminate period. These clients are known as feedback customers in queueing systems. Sivakumar et al. [4] studied the significance of fulfilling the needs of feedback customers combining with a single server QIS. The arrival process of a primary customer follows the Poisson process. In this extension, Amirthakodi et al. [5] studied the same with MAP arrival process.

Anbazhagan et al. [6] presented a QIS with a service facility that includes a limited queue size for priority customers. Jeganathan et al. [7] scrutinized a QIS with non-preemptive priority service. HNC and LNC arrive according to Poisson processes. Jeganathan [8] looked at a perishable inventory model with Erlang-K service and finite source clients. The HNC requests a unit item and asks for service before accepting it, whereas the LNC requests a unit item but does not ask for service. Ning and Lian [9] and Sivakumar and Arivarigan [10] discussed a QIS with two kinds of customers. Regarding service facility, the reader can refer to [11–20].

All these articles have considered that the separate queues are used for separate services. Especially the waiting hall is used for one type of customers service and/or the retrial orbit used for another type of customers' service. To date, all articles have discussed that different types of customers are given separate queues. But this chapter discusses all types of customers will be sharing the same queue. In order to avoid this, all types of customers are made to share a single queue. The customers can be categorized based on their priority/importance, according to which the service policy/pattern of each type of customers can be determined. This will help reduce the waiting time for all customers and will reduce the additional expenses incurred.

Comparing the separate queues with single queue based on service pattern of each type of customers, it is easy to quantify the different types of customers. So every queue is dependent of itself and not depend on the other queues. But in order to differentiate the customers in sharing type is very difficult. Because sharing buffer type will make the queue dependent on each and every other type of customers. It is also difficult to calculate the different measures of system performance based on this. More questions about the inventory system and the topic of perishable commodities may be found by referring here [15, 21–27]. More ordering policies in the queueing inventory system literature are available [12, 28–39]. Several queueing inventory systems with service facilities are documented in the literature [40–44].

Anilkumar and Jose [45] discussed a self-induced interruption in QIS. When a consumer comes, he or she is immediately placed in a high-volume, infinite-capacity queue. The service may be interrupted for personal reasons during his or her service. The interrupted consumer is routed to a lower-priority queue with unlimited capacity. M/M/1 queueing systems with inventory were investigated by Schwarz et al. [46]. Shajin and Krishnamoorthy [47] investigated the advanced reservation, cancellation, and overbooking with impatient customers. More multiple inventory system literature are available [48–50].

Many authors have analyzed the idea of multi priority customers service policy (non-preemptive priorities/preemptive priorities/mixed priority service) with sharing a buffer-type queue in a queueing model in the past [51, 52]. In inventory-queuing literature, there is no article that has been discussed on multi-priority customers with a single sharing buffer-type queue. This is the novelty of this chapter. In the future plan, we have to investigate the sharing buffer inventory queuing system with multi-sever facilities.

Objective of This Chapter

1. This study aims to connect preemptive, non-preemptive, and mixed priorities service polices for a HNC and LNC in a queueing-inventory system using the Markov chain. According to the state of the Markov chain, it computes the infinitesimal generator matrix.
2. Due to the structure of an infinitesimal generator matrix, this study seeks a stationary probability vector to the generator matrix using the matrix analytic method.
3. This study explores the cost analysis of the proposed queueing-inventory setup and provides a numerical illustrations to distinguish the best service polices.
4. The study of priorities between HNC and LNC emphasizes that the mixed priority will give an efficient solution.
5. The policy of adopting the mixed priority will give a balanced service between HNC and LNC. It also helps to reduce LNC lost due to the preemption of HNC.

Organization of This Chapter: The rest of the chapter is organized as: The mathematical model is discussed in Section 11.2. In Section 11.3, the model's steady state analysis is presented. System performance indicators are extracted in Section 11.4. The estimation of total expected cost (TEC) is highlighted in Section 11.5. Section 11.6 presents the analysis of the convexity of the TEC and the implications of priorities. Section 11.7 concludes the chapter with future scope.

11.2 DESCRIPTION OF THE MATHEMATICAL MODEL

We are working on sharing finite queue with a single server and two types of customers: high and low need customers (HLNC). The server serves them with a mixed priority-based service scheme. The capacity of the finite buffer is assumed to be N with the exception of a customer at the service point. let N_1 and N_2 are number of HLNC in the buffer whose sum is $N_1 + N_2 \leq N$. The arrival processes of HLNC are assumed to be independent Poisson processes with rates λ_1 and λ_2, respectively. The servicing time of any class of customer follows an exponential distribution. The commencement of the service of HNC is after selecting the purchased item from the inventory and the LNC is only arrived for the repaired work. Let μ_1 and μ_2 be the rates of service of HNC and LNC, respectively. It is assumed that an item is perishable with exponential rate γ but the item in the service is not perishable so that the items are perishable with the rate $j\gamma$ where j denotes the number items in the stock except the servicing item. The inventory's ordering policy is (s, Q) and $Q > s$, where Q represents the ordering quantity, s represents the reorder point, and $s + Q = S$ represents the maximum stock level. The order's lead time is considered to be exponential with a rate of β.

The server is generally idle in the case of no LNC in the buffer and either no inventory in the system or no HNC in the buffer. At an arrival instant of a HNC, if the server is busy with HNC then the arriving HNC joins the queue behind all the HNC and in front of all the LNC. After finishing the service, a HNC can receive the service by FCFS basis when at least one inventory is available for him. At an arrival instant of a HNC, if the server is busy with LNC and the inventory is available for the arriving customer then he can demand the service with the probability p and the rate of demand is $p\lambda_1$. This is known as a preemptive priority service scheme. Otherwise, he joins the queue behind all the HNCs and in front of all the LNCs in the buffer with the rate $q\lambda_1$, $(p + q = 1)$. This is known as a non-preemptive priority service scheme.

At an arrival instant of a LNC, if the server is busy (high need/LNC), then the arriving customer joins the queue behind all the customers who already occupied. After finishing the service (high need/LNC), a LNC can receive the service immediately by FCFS when either there is no high priority customers in the buffer or no inventory in the stock. Both high and low priority customers are lost when they find the buffer occupied.

11.3   SOLUTION OF THE MATHEMATICAL MODEL

Let $\varpi_1(t) \in \{0, 1, 2, ...,S\}$ be the inventory level. There are three feasible states of the server as follows:

$$\varpi_2(t) = \begin{cases} 0, & \text{server is inactive,} \\ 1, & \text{server is active with HNC,} \\ 2, & \text{server is active with LNC,} \end{cases}$$

Let $\varpi_3(t)$, $\varpi_4(t) \in \{0, 1, 2, ..., N\}$ respectively, indicate the number of HLNC in the buffer with $\varpi_3(t) + \varpi_4(t) \le N$.

Then the 4-dimensional random process $\{h(t) = (\varpi_1(t), \varpi_2(t), \varpi_3(t), \varpi_4(t)), t \ge 0\}$ with state space $F = F_1 \cup F_2 \cup F_3 \cup F_4 \cup F_5$ is given by

$$\begin{aligned} F = {} & \{(\varpi_1, \varpi_2, \varpi_3, \varpi_4) | \varpi_1 = 0, \varpi_2 = 0, \varpi_3 = 0, \varpi_4 = 0, 1,..., N\} \cup \\ & \{(\varpi_1, \varpi_2, \varpi_3, \varpi_4) | \varpi_1 = 0, \varpi_2 = 2, \varpi_3 = 0, 1,...,N, \\ & \varpi_4 = 0, 1,...,N - \varpi_3\} \cup \{(\varpi_1, \varpi_2, \varpi_3, \varpi_4) | \varpi_1 = 1, 2,...,S, \varpi_2 = 0, \\ & \varpi_3 = 0, \varpi_4 = 0\} \cup \{(\varpi_1, \varpi_2, \varpi_3, \varpi_4) | \varpi_1 = 1, 2,...,S, \varpi_2 = 1, \\ & \varpi_3 = 0, 1, 2,..., N, \varpi_4 = 0, 1, 2,..., N - \varpi_3\} \cup \{(\varpi_1, \varpi_2, \varpi_3, \varpi_4) | \\ & \varpi_1 = 1, 2,...,S, \varpi_2 = 2, \varpi_3 = 0, 1, 2,...,N, \varpi_4 = 0, 1, 2,...,N - \varpi_3\}. \end{aligned}$$

More explicitly, the states and their explanation are given below:

The state $(0, 0, 0, \varpi_4) \in F$, $0 \le \varpi_4 \le N$, indicates no item in the inventory, server is inactive, no LNC in the buffer, and ϖ_4 HNC are in the buffer. The state $(\varpi_1, 0, 0, 0) \in F$, $1 \le \varpi_1 \le S$, ϖ_1 items in the inventory, server is inactive, buffer is empty. The state $(\varpi_1, 1, \varpi_3, \varpi_4) \in F$, $1 \le \varpi_1 \le S$, $0 \le \varpi_3 \le N$, $0 \le \varpi_4 \le N$, ϖ_1 items in the inventory, server is active with a HNC, ϖ_3 LNC and ϖ_4 HNC are in the buffer with the condition $\varpi_3(t) + \varpi_4(t) \le N$. The state $(\varpi_1, 2, \varpi_3, \varpi_4) \in F$, $0 \le \varpi_1 \le S$, $0 \le \varpi_3 \le N$, $0 \le \varpi_4 \le N$, ϖ_1 items in the inventory, server is active with a LNC, ϖ_3 LNC, and ϖ_4 HNC are in the buffer with the condition $\varpi_3(t) + \varpi_4(t) \le N$.

Now we denote the levels $<0>$ and $<\varpi_1>$, $1 \le \varpi_1 \le S$ by $<0> = F_1 \cup F_2$ and $<\varpi_1> = \{(\varpi_1, \varpi_2, \varpi_3, \varpi_4) | \varpi_2 = 0, \varpi_3 = 0, \varpi_4 = 0\} \cup \{(\varpi_1, \varpi_2, \varpi_3, \varpi_4) | \varpi_2 = 1, 2, \varpi_3 = 0, 1, 2, ..., N, \varpi_4 = 0, 1, 2, ..., N - \varpi_3\}$. The infinitesimal generator of the Markov chain $\{h(t), t \ge 0\}$ is

$$\Psi = ((h((\varpi_1, \varpi_2, \varpi_3, \varpi_4); (\varpi'_1, \varpi'_2, \varpi'_3, \varpi'_4)))),$$

for all $(\varpi_1, \varpi_2, \varpi_3, \varpi_4)$, $(\varpi'_1, \varpi'_2, \varpi'_3, \varpi'_4) \in F$.

The generator matrix Ψ of the time-homogeneous 4-dimensional Markov chain $\{h(t), t \geq 0\}$ is given by:

$$
\Psi = \begin{array}{c}
<S> \\
<S-1> \\
\vdots \\
<s+1> \\
<s> \\
<s-1> \\
\vdots \\
<1> \\
<0>
\end{array}
\left(
\begin{array}{ccccccccc}
\Phi_S & \Pi_S & & & & & & & \\
& \Phi_{S-1} & \Pi_{S-1} & & & & & & \\
& & \cdots & & & & & & \\
& & & \cdots & \Phi_{s+1} & \Pi_{s+1} & & & \\
\Delta & & & & & \Phi_s & \Pi_s & & \\
& \Delta & & & & & \Phi_{s-1} & & \\
& & \cdots & & & & & \cdots & \\
& & & \Delta & & & & \cdots & \Phi_1 & \Pi_1 \\
& & & & \Delta_1 & & & & \Phi_0
\end{array}
\right)
$$

where $\Delta_1 = [\delta_{1((\varpi_1, \varpi_2, \varpi_3, \varpi_4),(\varpi'_1, \varpi'_2, \varpi'_3, \varpi'_4))}]$, for all $(\varpi_1, \varpi_2, \varpi_3, \varpi_4) \in <0>$ and $(\varpi'_1, \varpi'_2, \varpi'_3, \varpi'_4) \in <Q>$.

$$
\delta_{1((\varpi_1, \varpi_2, \varpi_3, \varpi_4),(\varpi'_1, \varpi'_2, \varpi'_3, \varpi'_4))} = \begin{cases}
\beta, & \varpi'_2 = \varpi_2, \quad \varpi'_3 = \varpi_3, \quad \varpi'_4 = \varpi_4 \\
& \varpi_2 = 0, \quad \varpi_3 = 0, \quad \varpi_4 = 0, \\
& \varpi'_2 = 1, \quad \varpi'_3 = \varpi_3, \quad \varpi'_4 = \varpi_4 - 1, \\
& \varpi_2 = 0, \quad \varpi_3 = 0, \quad \varpi_4 \in V_1^N, \\
& \varpi'_2 = \varpi_2, \quad \varpi'_3 = \varpi_3, \quad \varpi'_4 = \varpi_4, \\
& \varpi_2 = 2, \quad \varpi_3 \in V_0^N, \quad \varpi_4 \in V_0^{N-\varpi_3}, \\
0, & \text{otherwise.}
\end{cases}
$$

$\Delta = [\delta_{((\varpi_1, \varpi_2, \varpi_3, \varpi_4),(\varpi'_1 = \varpi_1 + Q, \varpi'_2, \varpi'_3, \varpi'_4))}]$, $1 \leq \varpi_1 \leq s$, for all $(\varpi_1, \varpi_2, \varpi_3, \varpi_4) \in <\varpi_1>$, $(\varpi'_1 = \varpi_1 + Q, \varpi'_2, \varpi'_3, \varpi'_4) \in <\varpi'_1 = \varpi_1 + Q>$; $\delta_{((\varpi_1, \varpi_2, \varpi_3, \varpi_4),(\varpi'_1 = Q+s, \varpi'_2 = \varpi_2, \varpi'_3 = \varpi_3, \varpi'_4 = \varpi_4))} = \beta$ and otherwise it is zero.

$\Pi_1 = [\pi_{1((\varpi_1, \varpi_2, \varpi_3, \varpi_4),(\varpi'_1, \varpi'_2, \varpi'_3, \varpi'_4))}]$, for all $(\varpi_1, \varpi_2, \varpi_3, \varpi_4) \in <1>$, $(\varpi'_1, \varpi'_2, \varpi'_3, \varpi'_4) \in <0>$.

$$
\pi_{1((\varpi_1, \varpi_2, \varpi_3, \varpi_4),(\varpi'_1, \varpi'_2, \varpi'_3, \varpi'_4))}
= \begin{cases}
\gamma, & \varpi'_2 = \varpi_2, \quad \varpi'_3 = \varpi_3, \quad \varpi'_4 = \varpi_4 \\
& \varpi_2 = 0, \quad \varpi_3 = 0, \quad \varpi_4 = 0, \\
& \varpi'_2 = \varpi_2, \quad \varpi'_3 = \varpi_3, \quad \varpi'_4 = \varpi_4, \\
& \varpi_2 = 2, \quad \varpi_3 \in V_0^N, \quad \varpi_4 \in V_0^{N-\varpi_3}, \\
\mu_1 & \varpi'_2 = \varpi_2 - 1, \quad \varpi'_3 = \varpi_3, \quad \varpi'_4 = \varpi_4, \\
& \varpi_2 = 1, \quad \varpi_3 = 0, \quad \varpi_4 \in V_0^{N-\varpi_3}, \\
& \varpi'_2 = \varpi_2 + 1, \quad \varpi'_3 = \varpi_3 - 1, \quad \varpi'_4 = \varpi_4, \\
& \varpi_2 = 1, \quad \varpi_3 \in V_1^N, \quad \varpi_4 \in V_0^{N-\varpi_3}, \\
0, & \text{otherwise.}
\end{cases}
$$

For $2 \leq \varpi_1 \leq S$, $\Pi_{\varpi_1} = [\pi_{\varpi_1((\varpi_1, \varpi_2, \varpi_3, \varpi_4),(\varpi'_1 = \varpi_1 - 1, \varpi'_2, \varpi'_3, \varpi'_4))}]$, for all $(\varpi_1, \varpi_2, \varpi_3, \varpi_4) \in <\varpi_1>$, $(\varpi'_1, \varpi'_2, \varpi'_3, \varpi'_4) \in <\varpi'_1 = \varpi_1 - 1>$.

$$\pi_{\varpi_1((\varpi_1,\varpi_2,\varpi_3,\varpi_4),(\varpi_1-1,\varpi'_2,\varpi'_3,\varpi'_4))}$$

$$= \begin{cases} \varpi_1\gamma, & \varpi'_2 = \varpi_2, & \varpi'_3 = \varpi_3, & \varpi'_4 = \varpi_4 \\ & \varpi_2 = 0, & \varpi_3 = 0, & \varpi_4 = 0, \\ & \varpi'_2 = \varpi_2, & \varpi'_3 = \varpi_3, & \varpi'_4 = \varpi_4, \\ & \varpi_2 = 2, & \varpi_3 \in V_0^N, & \varpi_4 \in V_0^{N-\varpi_3}, \\ (\varpi_1-1)\gamma, & \varpi'_2 = \varpi_2, & \varpi'_3 = \varpi_3, & \varpi'_4 = \varpi_4 \\ & \varpi_2 = 1, & \varpi_3 \in V_0^N, & \varpi_4 \in V_0^{N-\varpi_3}, \\ \mu_1 & \varpi'_2 = \varpi_2 - 1, & \varpi'_3 = \varpi_3, & \varpi'_4 = \varpi_4, \\ & \varpi_2 = 1, & \varpi_3 = 0, & \varpi_4 = 0, \\ & \varpi'_2 = \varpi_2, & \varpi'_3 = \varpi_3, & \varpi'_4 = \varpi_4 - 1, \\ & \varpi_2 = 1, & \varpi_3 \in V_0^{N-1}, & \varpi_4 \in V_0^{N-\varpi_3}, \\ & \varpi'_2 = \varpi_2 + 1, & \varpi'_3 = \varpi_3 - 1, & \varpi'_4 = \varpi_4, \\ & \varpi_2 = 1, & \varpi_3 \in V_1^N, & \varpi_4 = 0, \\ 0, & \text{otherwise.} \end{cases}$$

$\Phi_0 = [\phi_{0((\varpi_1,\varpi_2,\varpi_3,\varpi_4),(\varpi'_1,\varpi'_2,\varpi'_3,\varpi'_4))}]$, for all $(\varpi_1, \varpi_2, \varpi_3, \varpi_4)$, $(\varpi'_1, \varpi'_2, \varpi'_3, \varpi'_4) \in \langle 0 \rangle$.

$$\phi_{0((\varpi_1,\varpi_2,\varpi_3,\varpi_4),(\varpi'_1,\varpi'_2,\varpi'_3,\varpi'_4))}$$

$$= \begin{cases} \lambda_1, & \varpi'_2 = \varpi_2, & \varpi'_3 = \varpi_3, & \varpi'_4 = \varpi_4 + 1 \\ & \varpi_2 = 0, & \varpi_3 = 0, & \varpi_4 \in V_0^{N-1}, \\ & \varpi'_2 = \varpi_2, & \varpi'_3 = \varpi_3, & \varpi'_4 = \varpi_4 + 1, \\ & \varpi_2 = 2, & \varpi_3 \in V_0^{N-1}, & \varpi_4 \in V_0^{N-(\varpi_3-1)}, \\ \lambda_2, & \varpi'_2 = 2, & \varpi'_3 = \varpi_3, & \varpi'_4 = \varpi_4 + 1 \\ & \varpi_2 = 0, & \varpi_3 = 0, & \varpi_4 \in V_0^N, \\ & \varpi'_2 = \varpi_2, & \varpi'_3 = \varpi_3 + 1, & \varpi'_4 = \varpi_4, \\ & \varpi_2 = 2, & \varpi_3 \in V_0^{N-1}, & \varpi_4 \in V_0^{N-(\varpi_3-1)}, \\ \mu_2 & \varpi'_2 = 0, & \varpi'_3 = \varpi_3, & \varpi'_4 = \varpi_4, \\ & \varpi_2 = 2, & \varpi_3 = 0, & \varpi_4 \in V_0^N, \\ & \varpi'_2 = \varpi_2, & \varpi'_3 = \varpi_3 - 1, & \varpi'_4 = \varpi_4, \\ & \varpi_2 = 2, & \varpi_3 \in V_1^N, & \varpi_4 \in V_0^{N-\varpi_3}, \\ -(\lambda_2 + \bar{\delta}_{\varpi_4 N}\lambda_1 + \beta) & \varpi'_2 = \varpi_2, & \varpi'_3 = \varpi_3, & \varpi'_4 = \varpi_4, \\ & \varpi_2 = 0, & \varpi_3 = 0, & \varpi_4 \in V_0^N, \\ -(\mu_2 + \bar{\delta}_{\varpi_4(N-\varpi_3)} & \varpi'_2 = \varpi_2, & \varpi'_3 = \varpi_3, & \varpi'_4 = \varpi_4, \\ (\lambda_1 + \lambda_2) + \beta) & \varpi_2 = 2, & \varpi_3 \in V_0^N, & \varpi_4 \in V_0^{N-\varpi_3}, \\ 0, & \text{otherwise.} \end{cases}$$

For $1 \le \varpi_1 \le S$, $\Phi_{\varpi_1} = [\phi_{\varpi_1((\varpi_1,\varpi_2,\varpi_3,\varpi_4),(\varpi'_1,\varpi'_2,\varpi'_3,\varpi'_4))}]$, for all $(\varpi_1, \varpi_2, \varpi_3, \varpi_4)$, $(\varpi'_1, \varpi'_2, \varpi'_3, \varpi'_4) \in \langle\varpi_1\rangle$.

$$\phi_{\varpi_1}((\varpi_1, \varpi_2, \varpi_3, \varpi_4),(\varpi'_1, \varpi'_2, \varpi'_3, \varpi'_4))$$

$$= \begin{cases} \lambda_1, & \varpi'_2 = 1, & \varpi'_3 = \varpi_3, & \varpi'_4 = \varpi_4, \\ & \varpi_2 = 0, & \varpi_3 = 0, & \varpi_4 = 0, \\ & \varpi'_2 = \varpi_2, & \varpi'_3 = \varpi_3, & \varpi'_4 = \varpi_4 + 1, \\ & \varpi_2 = 1, & \varpi_3 \in V_0^{N-1}, & \varpi_4 \in V_0^{N-(\varpi_3-1)}, \\ p\lambda_1, & \varpi'_2 = \varpi_2 - 1, & \varpi'_3 = \varpi_3 + 1, & \varpi'_4 = \varpi_4 \\ & \varpi_2 = 2, & \varpi_3 \in V_0^{N-1}, & \varpi_4 \in V_0^{N-(\varpi_3-1)}, \\ (1-p)\lambda_1, & \varpi'_2 = \varpi_2, & \varpi'_3 = \varpi_3, & \varpi'_4 = \varpi_4 + 1, \\ & \varpi_2 = 2, & \varpi_3 \in V_0^{N-1}, & \varpi_4 \in V_0^{N-(\varpi_3-1)}, \\ \lambda_2, & \varpi'_2 = 2, & \varpi'_3 = \varpi_3, & \varpi'_4 = \varpi_4, \\ & \varpi_2 = 0, & \varpi_3 = 0, & \varpi_4 = 0, \\ & \varpi'_2 = \varpi_2, & \varpi'_3 = \varpi_3, & \varpi'_4 = \varpi_4, \\ & \varpi_2 = 1, & \varpi_3 \in V_0^{N-1}, & \varpi_4 \in V_0^{N-(\varpi_3-1)}, \\ & \varpi'_2 = \varpi_2, & \varpi'_3 = \varpi_3 + 1, & \varpi'_4 = \varpi_4, \\ & \varpi_2 = 2, & \varpi_3 \in V_0^{N-1}, & \varpi_4 \in V_0^{N-(\varpi_3-1)}, \\ \mu_2 & \varpi'_2 = 0, & \varpi'_3 = \varpi_3, & \varpi'_4 = \varpi_4, \\ & \varpi_2 = 2, & \varpi_3 = 0, & \varpi_4 = 0, \\ & \varpi'_2 = 1, & \varpi'_3 = \varpi_3, & \varpi'_4 = \varpi_4 - 1, \\ & \varpi_2 = 2, & \varpi_3 \in V_0^{N-1}, & \varpi_4 \in V_1^{N-\varpi_3}, \\ & \varpi'_2 = \varpi_2, & \varpi'_3 = \varpi_3 - 1, & \varpi'_4 = \varpi_4, \\ & \varpi_2 = 2, & \varpi_3 \in V_1^{N}, & \varpi_4 = 0, \\ -(\lambda_2 + \lambda_1 + \varpi_1\gamma + & \varpi'_2 = \varpi_2, & \varpi'_3 = \varpi_3, & \varpi'_4 = \varpi_4, \\ H(s - \varpi_1)\beta) & \varpi_2 = 0, & \varpi_3 = 0, & \varpi_4 = 0, \\ -(\bar{\delta}_{\varpi_4(N-\varpi_3)}(\lambda_1 + \lambda_2)+ & & & \\ \delta_{\varpi_2 1}((\varpi_1 - 1)\gamma + \mu_1)+ & \varpi'_2 = \varpi_2, & \varpi'_3 = \varpi_3, & \varpi'_4 = \varpi_4, \\ \delta_{\varpi_2 2}(\varpi_1\gamma + \mu_2) & & & \\ +H(s - \varpi_1)\beta) & \varpi_2 = 1, 2, & \varpi_3 \in V_0^{N}, & \varpi_4 \in V_0^{N-\varpi_3}, \\ 0, & \text{otherwise.} \end{cases}$$

Matrix	Dimension	Matrix	Dimension
Δ_1	$\left(\frac{N^2 + 5N + 4}{2}\right) \times (N^2 + 3N + 3)$	$\Delta, \Pi_{p_1}, \Phi_{\varpi_1}$, (square matrices) $1 \le \varpi_1 \le S; \ 2 \le p_1 \le S$	$(N^2 + 3N + 3)$
Π_1	$(N^2 + 3N + 3) \times \left(\frac{N^2 + 5N + 4}{2}\right)$	Φ_0, (square matrix)	$\left(\frac{N^2 + 5N + 4}{2}\right)$

11.3.1 STEADY-STATE EQUATIONS

Due to the formation of an infinitesimal matrix "Ψ", the state space of the Markov chain $\{h(t), t \ge 0\}$ is irreducible, aperiodic, and persistent non-null. Let $\nabla = (\nabla^{(0)}, \nabla^{(1)}, ..., \nabla^{(S)})$ be the steady-state probability vector of Ψ of dimension $1 \times (S + 1)$. Then ∇ satisfies

$$\nabla \Psi = 0 \tag{11.1}$$

$$\nabla e = 1, \tag{11.2}$$

where $e = (1, 1, ..., 1, 1)^T$. Note that $\nabla^{(0)}$ and $\nabla^{(\varpi_1)}$, $1 \le \varpi_1 \le S$, are row vectors of dimensions $\left(\frac{N^2 + 5N + 4}{2} \right)$ and $(N^2 + 3N + 3)$, respectively. Each coordinate of the vector $\mathbf{X}$ can be partitioned as follows:

$$\nabla^{(0)} = (\nabla^{(0,0)}, \nabla^{(0,2)}),$$

$$\nabla^{(\varpi_1)} = (\nabla^{(\varpi_1,0)}, \nabla^{(\varpi_1,1)}, \nabla^{(\varpi_1,2)}), \; 1 \le \varpi_1 \le S;$$

Each coordinate of the vectors $\nabla^{(0)}$ and $\nabla^{(\varpi_1)}, 1 \le \varpi_1 \le S$, can be further partitioned into the vectors as follows:

$$\begin{aligned}
\nabla^{(0,0)} &= (\nabla^{(0,0,0)}), \\
\nabla^{(0,2)} &= (\nabla^{(0,2,0)}, \nabla^{(0,2,1)}, ..., \nabla^{(0,2,N)}), \\
\nabla^{(\varpi_1,0)} &= (\nabla^{(\varpi_1,0,0)}), \; 1 \le \varpi_1 \le S; \\
\nabla^{(\varpi_1,\varpi_2)} &= (\nabla^{(\varpi_1,\varpi_2,0)}, \nabla^{(\varpi_1,\varpi_2,1)}, ..., \nabla^{(\varpi_1,\varpi_2,N)}), \; 1 \le \varpi_1 \le S; \varpi_2 = 1, 2;
\end{aligned}$$

Further, each coordinate of the vectors $\nabla^{(0,0)}$, $\nabla^{(0,2)}$, $\nabla^{(\varpi_1,0)}$ and $\nabla^{(\varpi_1,\varpi_2)}, 1 \le \varpi_1 \le S, \varpi_2 = 1, 2$ can be partitioned as follows:

$$\begin{aligned}
\nabla^{(0,0,0)} &= (\nabla^{(0,0,0,0)}, \nabla^{(0,0,0,1)}, ..., \nabla^{(0,0,0,N)}), \\
\nabla^{(0,2,\varpi_3)} &= (\nabla^{(0,2,\varpi_3,0)}, \nabla^{(0,2,\varpi_3,1)}, ..., \nabla^{(0,2,\varpi_3,N-\varpi_3)}), \; 0 \le \varpi_3 \le N; \\
\nabla^{(\varpi_1,0,0)} &= (\nabla^{(\varpi_1,0,0,0)}), \; 1 \le \varpi_1 \le S; \\
\nabla^{(\varpi_1,\varpi_2,\varpi_3)} &= (\nabla^{(\varpi_1,\varpi_2,\varpi_3,0)}, \nabla^{(\varpi_1,\varpi_2,\varpi_3,1)}, ..., \nabla^{(\varpi_1,\varpi_2,\varpi_3,N-\varpi_3)}), \\
& \quad 1 \le \varpi_1 \le S; \varpi_2 = 1, 2; 0 \le \varpi_3 \le N;
\end{aligned}$$

Equation (11.1) gives the following system of linear equations:

$$\nabla^{(\varpi_1)} \Phi_{\varpi_1} + \nabla^{(\varpi_1+1)} \Pi_{\varpi_1+1} = \mathbf{0}, \quad 0 \le \varpi_1 \le Q - 1, \tag{11.3}$$

$$\nabla^{(\varpi_1-Q)} \Delta_1 + \nabla^{(\varpi_1)} \Phi_Q + \nabla^{(\varpi_1+1)} \Pi_{Q+1} = \mathbf{0}, \quad \varpi_1 = Q, \tag{11.4}$$

$$\nabla^{(\varpi_1-Q)} \Delta + \nabla^{(\varpi_1)} \Phi_{i_1} + \nabla^{(\varpi_1+1)} \Pi_{\varpi_1+1} = \mathbf{0}, \quad Q + 1 \le \varpi_1 \le S - 1, \tag{11.5}$$

$$\nabla^{(\varpi_1)} \Phi_{\varpi_1} + \nabla^{(\varpi_1-Q)} \Delta = \mathbf{0}, \quad \varpi_1 = S, \tag{11.6}$$

The key of the Equations (11.3), (11.5), and (11.6) can be expediently described as:

$$\nabla^{(\varpi_1)} = \nabla^{(Q)} \Theta_{\varpi_1}, \quad 0 \le \varpi_1 \le S.$$

For evaluating Θ_{ϖ_1}, we define

$$\prod_{l=k}^{r} c_l = \begin{cases} c_k c_{k-1} \cdots c_r & \text{if } k \geq r \\ 1 & \text{if } k < r \end{cases}$$

$$\Theta_{\varpi_1} = (-1)^{Q-\varpi_1} \nabla^{(Q)} \beta \sum_{r=Q}^{\varpi_1+1} \Pi_r \Phi_{r-1}^{-1}, \quad 0 \leq \varpi_1 \leq Q-1,$$

$$= (-1)^{2Q-\varpi_1+1} \nabla^{(Q)} \sum_{r=0}^{S-\varpi_1} \left[\left(\beta \prod_{v=Q}^{(s+1)-r} \Pi_v \Phi_{v-1}^{-1} \right) \Delta \Phi_{S-r}^{-1} \left(\beta \prod_{u=S-r}^{\varpi_1+1} \Pi_u \Phi_{u-1}^{-1} \right) \right],$$

$$\qquad\qquad\qquad\qquad Q+1 \leq \varpi_1 \leq S,$$

$$= I, \qquad\qquad\qquad \varpi_1 = Q$$

To find the vector $\nabla^{(Q)}$, we can use the following equations:

$$\nabla^{(0)} \Delta_1 + \nabla^{(Q)} \Phi_Q + \nabla^{(Q+1)} \Pi_{Q+1} = 0 \quad \text{and} \quad Xe = 1$$

which gives, respectively:

$$\nabla^{(Q)} \left[\left\{ (-1)^Q \sum_{r=0}^{s-1} \left[\left(\beta \prod_{v=Q}^{(s+1)-r} \Pi_v \Phi_{v-1}^{-1} \right) \Delta \Phi_{S-r}^{-1} \left(\beta \prod_{u=S-r}^{Q+2} \Pi_u \Phi_{u-1}^{-1} \right) \right] \right\} \Pi_{Q+1} \right.$$

$$\left. + \Phi_Q + \left\{ (-1)^Q \beta \prod_{r=Q}^{1} \Pi_r \Phi_{r-1}^{-1} \right\} \Delta \right] = 0,$$

and

$$\nabla^{(Q)} \left[\sum_{\varpi_1=0}^{Q-1} \left((-1)^{Q-\varpi_1} \Omega \prod_{r=Q}^{\varpi_1+1} \Pi_r \Phi_{r-1}^{-1} \right) + I \right.$$

$$\left. + \sum_{\varpi_1=Q+1}^{S} \left((-1)^{2Q-\varpi_1+1} \sum_{r=0}^{S-\varpi_1} \left[\left(\beta \prod_{v=Q}^{(s+1)-r} \Pi_v \Phi_{v-1}^{-1} \right) \Delta \Phi_{S-r}^{-1} \left(\beta \prod_{u=S-r}^{\varpi_1+1} \Pi_u \Phi_{u-1}^{-1} \right) \right] \right) \right] e$$

$$= 1.$$

11.4 SYSTEM CHARACTERISTICS

Here, we explore the necessary systems characteristics measures of the model.

i. Expected inventory level:

$$E[I] = \sum_{\varpi_1=1}^{S} \varpi_1 \nabla^{(\varpi_1,0,0,0)} + \sum_{\varpi_1=1}^{S} \sum_{\varpi_2=1}^{2} \sum_{\varpi_3=0}^{N} \sum_{\varpi_4=0}^{N-\varpi_3} \varpi_1 \nabla^{(\varpi_1,\varpi_2,\varpi_3,\varpi_4)}$$

ii. Expected reorder rate:

$$E[R] = (s+1)\gamma \nabla^{(s+1,0,0,0)} + \sum_{\varpi_3=0}^{N} \sum_{\varpi_4=0}^{N-\varpi_3} (s\gamma + \mu_1) \nabla^{(s+1,1,\varpi_3,\varpi_4)}$$

$$+ \sum_{\varpi_3=0}^{N} \sum_{\varpi_4=0}^{N-\varpi_3} (s+1)\gamma \nabla^{(s+1,2,\varpi_3,\varpi_4)}$$

iii. Expected perishable rate:

$$E[P] = \sum_{\varpi_1=1}^{S} \varpi_1 \gamma \nabla^{(\varpi_1,0,0,0)} + \sum_{\varpi_1=1}^{S} \sum_{\varpi_3=0}^{N} \sum_{\varpi_4=0}^{N-\varpi_3} (\varpi_1-1)\gamma \nabla^{(\varpi_1,1,\varpi_3,\varpi_4)}$$

$$+ \sum_{\varpi_1=1}^{S} \sum_{\varpi_3=0}^{N} \sum_{\varpi_4=0}^{N-\varpi_3} \varpi_1 \gamma \nabla^{(\varpi_1,2,\varpi_3,\varpi_4)}$$

iv. Expected number of HNC in the buffer:

$$E[HNC] = \sum_{\varpi_4=1}^{N} \varpi_4 \nabla^{(0,0,0,\varpi_4)} + \sum_{\varpi_1=1}^{S} \sum_{\varpi_3=0}^{N-1} \sum_{\varpi_4=1}^{N-\varpi_3} \varpi_4 \nabla^{(\varpi_1,1,\varpi_3,\varpi_4)}$$

$$+ \sum_{\varpi_1=0}^{S} \sum_{\varpi_3=0}^{N-1} \sum_{\varpi_4=1}^{N-\varpi_3} \varpi_4 \nabla^{(\varpi_1,2,\varpi_3,\varpi_4)}$$

v. Expected number of LNC in the buffer:

$$E[LNC] = \sum_{\varpi_3=1}^{N} \sum_{\varpi_4=0}^{N-\varpi_3} \varpi_3 \nabla^{(0,2,\varpi_3,\varpi_4)} + \sum_{\varpi_1=1}^{S} \sum_{\varpi_2=1}^{2} \sum_{\varpi_3=1}^{N} \sum_{\varpi_4=0}^{N-\varpi_3} \varpi_3 \nabla^{(\varpi_1,\varpi_2,\varpi_3,\varpi_4)}$$

vi. Expected HNC lost to the system:

$$E[HNCL] = \lambda_1 \nabla^{(0,0,0,N)} + \sum_{\varpi_1=1}^{S} \sum_{\varpi_3=0}^{N} \lambda_1 \nabla^{(\varpi_1,1,\varpi_3,N-\varpi_3)}$$

$$+ \sum_{\varpi_1=0}^{S} \sum_{\varpi_3=0}^{N} \lambda_1 \nabla^{(\varpi_1,2,\varpi_3,N-\varpi_3)}$$

vii. Expected LNC lost to the system:

$$E[LNCL] = \lambda_2 \nabla^{(0,0,0,N)} + \sum_{\varpi_1=1}^{S} \sum_{\varpi_3=0}^{N} \lambda_2 \nabla^{(\varpi_1,1,\varpi_3,N-\varpi_3)}$$

$$+ \sum_{\varpi_1=0}^{S} \sum_{\varpi_3=0}^{N} \lambda_2 \nabla^{(\varpi_1,2,\varpi_3,N-\varpi_3)}$$

11.5   COST ANALYSIS OF THE MODEL

Then the TEC is defined by

$$TC(s, S, L, M) = d_1 E[I] + d_2 E[R] + d_3 E[P] + d_4 E[HNC] + d_5 E[LNC] + d_6 E[HNCL] + d_7 E[LNCL].$$

$$
\begin{aligned}
TC(s, S, L, M) = {} & d_1 \left[\sum_{\varpi_1=1}^{S} \varpi_1 V^{(\varpi_1,0,0,0)} + \sum_{\varpi_1=1}^{S} \sum_{\varpi_2=1}^{2} \sum_{\varpi_3=0}^{N} \sum_{\varpi_4=0}^{N-\varpi_3} \varpi_1 V^{(\varpi_1,\varpi_2,\varpi_3,\varpi_4)} \right] \\
& + d_2 \left[(s+1)\gamma V^{(s+1,0,0,0)} + \sum_{\varpi_3=0}^{N} \sum_{\varpi_4=0}^{N-\varpi_3} (s\gamma + \mu_1) V^{(s+1,1,\varpi_3,\varpi_4)} \right. \\
& \left. \sum_{\varpi_3=0}^{N} \sum_{\varpi_4=0}^{N-\varpi_3} (s+1)\gamma \psi^{(s+1,2,\varpi_3,\varpi_4)} \right] + d_3 \left[\sum_{\varpi_1=1}^{S} \varpi_1 \gamma V^{(\varpi_1,0,0,0)} \right. \\
& + \sum_{\varpi_1=1}^{S} \sum_{\varpi_3=0}^{N} \sum_{\varpi_4=0}^{N-\varpi_3} (\varpi_1 - 1)\gamma V^{(\varpi_1,1,\varpi_3,\varpi_4)} + \sum_{\varpi_1=1}^{S} \sum_{\varpi_3=0}^{N} \sum_{\varpi_4=0}^{N-\varpi_3} \varpi_1 \gamma V^{(\varpi_1,2,\varpi_3,\varpi_4)} \left. \right] \\
& + d_4 \left[\sum_{\varpi_4=1}^{N} \varpi_4 V^{(0,0,0,\varpi_4)} + \sum_{\varpi_1=1}^{S} \sum_{\varpi_3=0}^{N-1} \sum_{\varpi_4=1}^{N-\varpi_3} \varpi_4 V^{(\varpi_1,1,\varpi_3,\varpi_4)} \right. \\
& \left. + \sum_{\varpi_1=0}^{S} \sum_{\varpi_3=0}^{N-1} \sum_{\varpi_4=1}^{N-\varpi_3} \varpi_4 V^{(\varpi_1,2,\varpi_3,\varpi_4)} \right] + d_5 \left[\sum_{\varpi_3=1}^{N} \sum_{\varpi_4=0}^{N-\varpi_3} \varpi_3 V^{(0,2,\varpi_3,\varpi_4)} \right. \\
& \left. + \sum_{\varpi_1=1}^{S} \sum_{\varpi_2=1}^{2} \sum_{\varpi_3=1}^{N} \sum_{\varpi_4=0}^{N-\varpi_3} \varpi_3 V^{(\varpi_1,\varpi_2,\varpi_3,\varpi_4)} \right] + d_6 \left[\lambda_1 V^{(0,0,0,N)} \right. \\
& \left. + \sum_{\varpi_1=1}^{S} \sum_{\varpi_3=0}^{N} \lambda_1 V^{(\varpi_1,1,\varpi_3,N-\varpi_3)} + \sum_{\varpi_1=0}^{S} \sum_{\varpi_3=0}^{N} \lambda_1 V^{(\varpi_1,2,\varpi_3,N-\varpi_3)} \right] \\
& + d_7 \left[\lambda_2 V^{(0,0,0,N)} + \sum_{\varpi_1=1}^{S} \sum_{\varpi_3=0}^{N} \lambda_2 V^{(\varpi_1,1,\varpi_3,N-\varpi_3)} + \sum_{\varpi_1=0}^{S} \sum_{\varpi_3=0}^{N} \lambda_2 V^{(\varpi_1,2,\varpi_3,N-\varpi_3)} \right].
\end{aligned}
$$

where

d_1: holding cost per unit.

d_2: setup cost per order.

d_3: failure cost per unit.

d_4: waiting cost per HNC.

d_5: waiting cost per LNC.

d_6: lost cost per HNC.

d_7: lost cost per LNC.

11.6 NUMERICAL DISCUSSIONS

Generally, a numerical discussion is the most conducive tool to explore the application of the proposed model. In this connection, the parameter analysis is carried out under the assumptions of their values such that $S = 72$, $s = 3$, $M = 10$, $\lambda_1 = 11$, $\lambda_2 = .002$, $\beta = 5$; $\gamma = 4$; $\mu_1 = 12$, $\mu_2 = 3$, $p = 0.5$, $d_1 = 0.01$; $d_2 = 18$, $d_3 = 0.1$, $d_4 = 15$, $d_5 = 2$, $d_6 = 18$, $d_7 = 1$. The main aim of this chapter is priority-based service, so that all the examples are to be concentrated over the priority like the case $p \to 0$, $p \in (0, 1)$ and $p \to 1$ will explain about the non-preemptive, mixed, and preemptive priorities, respectively.

Example 11.1: *This example investigates the total cost with respect to the local convexity and the priority deciding probability p. Indeed, p = 0, p ∈ (0, 1) and p = 1 will explain about the non-preemptive, mixed, and preemptive priorities in Tables 11.1, 11.2, and 11.3, respectively. In all the three cases, there can be seen that a local convexity is obtained and which is indicated as the bold numbers. When comparing these three tables, the optimized minimal total cost for the case p = 1 at $(S^*, s*) = (65, 3)$, the case p = 0.5 at $(S^*, s*) = (70, 4)$ and the case p = 1 at $(S^*, s*) = (80, 5)$ is obtained is 43.67704, 46.3052, and 45.8588, respectively.*

TABLE 11.1
Optimum Values When $p = 0$

S/s	2	3	4	5	6	7	8
65	57.34851	52.93763	51.12191	50.58876	50.77551	51.41108	52.3532
70	56.45734	52.35537	50.67129	50.17538	50.34237	50.9196	51.77435
75	55.80291	51.97299	50.40548	49.94426	50.09709	50.62749	51.41125
80	55.34009	51.75151	50.28785	**49.8588**	50.00142	50.4936	51.21868
85	55.03456	51.66151	50.29093	49.89163	50.02689	50.48741	51.16328
90	54.8598	51.68037	50.39366	50.02189	50.15189	50.58585	51.21993
95	54.79492	51.79041	50.57966	50.23338	50.35966	50.7711	51.36932

TABLE 11.2
Optimum Values When $p = 0.5$

S/s	2	3	4	5	6
55	55 51.244155	48.07111	47.28854	47.61761	48.55241
60	60 50.263931	47.4222	46.73989	47.05342	47.90431
65	65 49.592494	47.02819	46.42986	46.73191	47.51579
70	70 49.158668	46.82965	**46.30252**	46.59577	47.32495
75	75 48.911542	46.78433	46.31822	46.60454	47.28824
80	80 48.813758	46.86139	46.44806	46.72879	47.37409
85	85 48.837294	47.03774	46.67049	46.94663	47.55908

TABLE 11.3
Optimum Values When $p = 1$

S/s	1	2	3	4	5	6	7
50	51.77149	45.75505	44.49171	44.88384	46.02152	47.57555	49.4092
55	50.2858	45.04488	43.9901	44.38121	45.41805	46.81077	48.43814
60	49.248	44.63059	43.73985	44.12815	45.08381	46.34874	47.81434
65	48.54327	44.43606	**43.67704**	44.0615	44.95038	46.1116	47.44709
70	48.09254	44.40923	43.75785	44.13793	44.97087	46.04627	47.27492
75	47.83968	44.51327	43.95122	44.32667	45.11206	46.11529	47.25469
80	47.74384	44.72145	44.2345	44.60527	45.34972	46.29143	47.35517

Example 11.2: *This example discuss the E [HNC], E [HNCL], E [LNC], and E [LNCL] through the graphical representation as shown in* Figures 11.1–11.16. *In such a way that, if the parameter λ_1 increases, both E [HNC], E [HNCL] are increases, whereas E [LNC] decreases and E [LNCL] increases. When the parameter λ_2 increases, E [HNC], E [HNCL] decreases but both E [LNC], E [LNCL] increase. The parameter μ_1 causes an increment in the E [HNC], E [HNCL] and E [LNCL] and decrement in the E [LNC]. Also, the impact of μ_2 has the same relation with E [HNC], E [HNCL], E [LNC], and E [LNCL] as μ_1. When concluding this example, we observe*

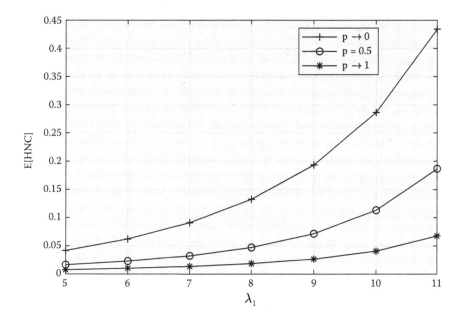

FIGURE 11.1 Expected $E[HNC]$ in the Queue When p vs λ_1.

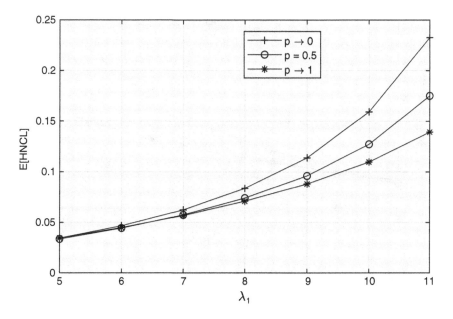

FIGURE 11.2 Expected $E[HNCL]$ in the Queue When p vs λ_1.

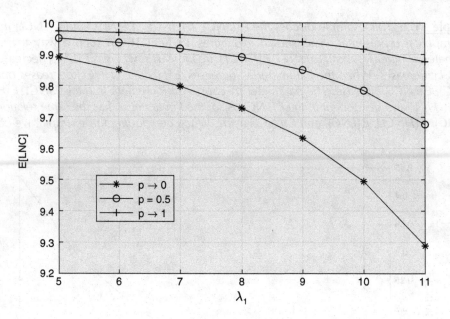

FIGURE 11.3   Expected $E[LNC]$ in the Queue When p vs λ_1.

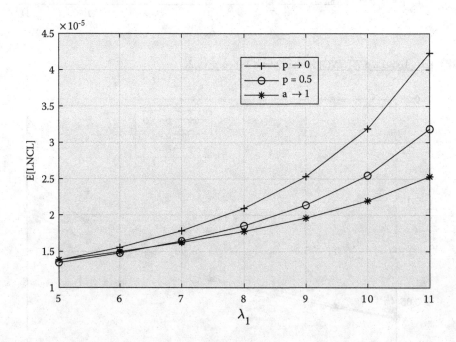

FIGURE 11.4   Expected $E[LNCL]$ in the Queue When p vs λ_1.

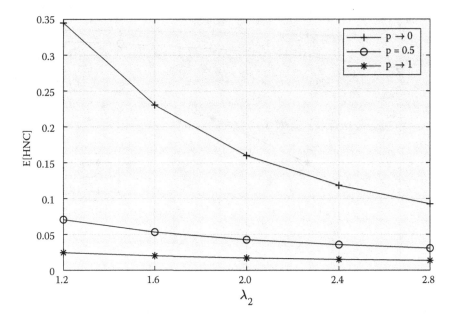

FIGURE 11.5   Expected $E[HNC]$ in the Queue When p vs λ_2.

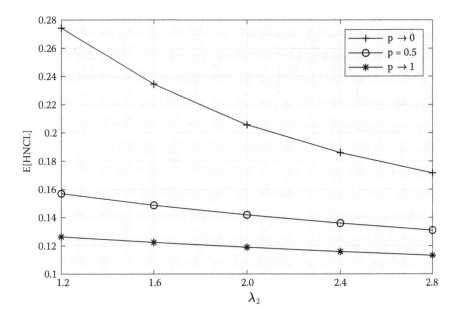

FIGURE 11.6   Expected $E[HNCL]$ in the Queue When p vs λ_2.

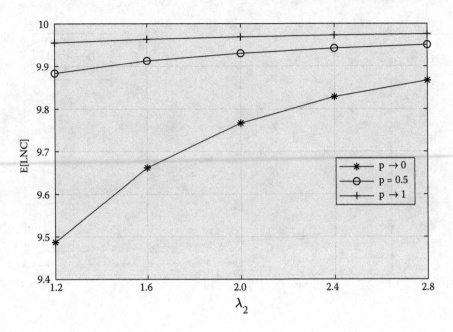

FIGURE 11.7   Expected $E[LNC]$ in the Queue When p vs λ_2.

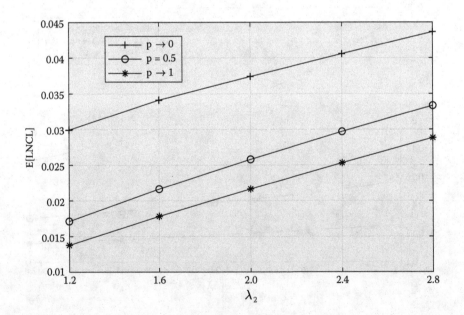

FIGURE 11.8   Expected $E[LNCL]$ in the Queue When p vs λ_2.

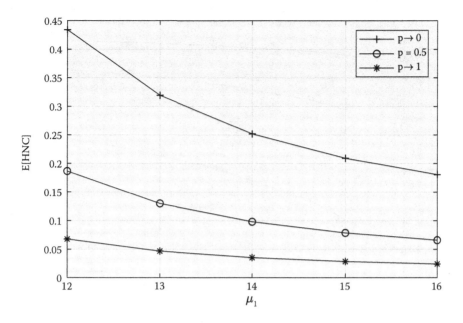

FIGURE 11.9   Expected $E[HNC]$ in the Queue When p vs μ_1.

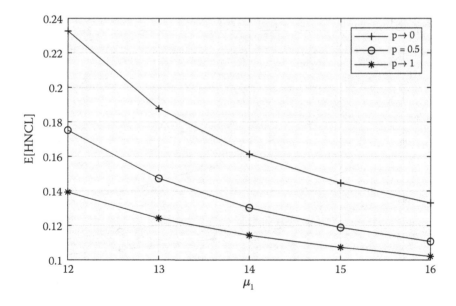

FIGURE 11.10   Expected $E[HNCL]$ in the Queue When p vs μ_1.

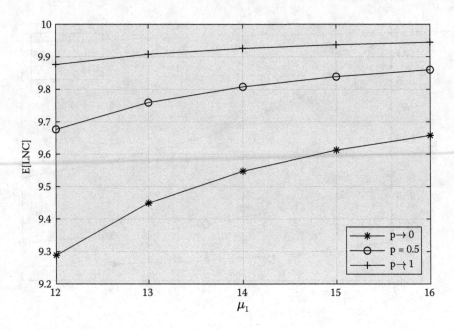

FIGURE 11.11 Expected $E[LNC]$ in the Queue When p vs μ_1.

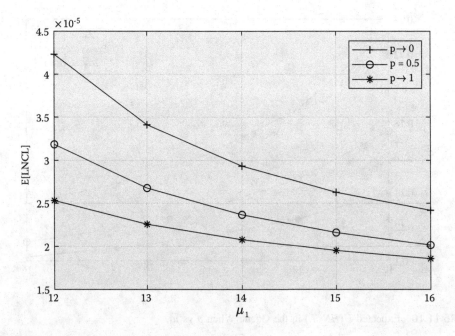

FIGURE 11.12 Expected $E[LNCL]$ in the Queue When p vs μ_1.

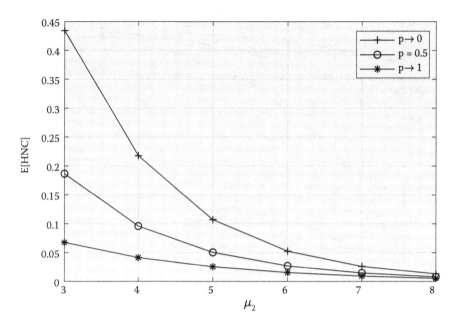

FIGURE 11.13 Expected $E[HNC]$ in the Queue When p vs μ_2.

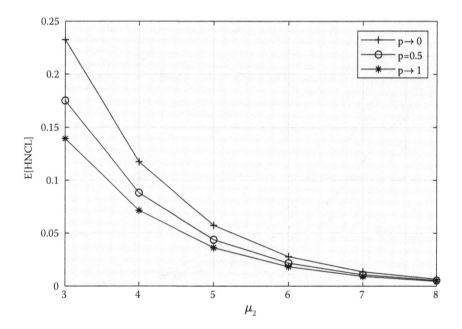

FIGURE 11.14 Expected $E[HNCL]$ in the Queue When p vs μ_2.

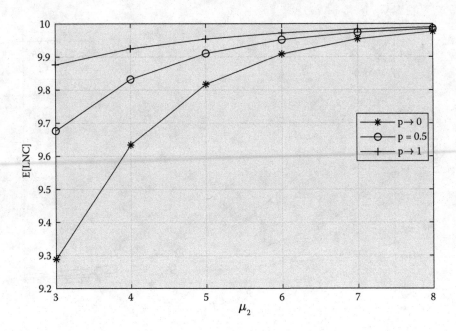

FIGURE 11.15   Expected $E[LNC]$ in the Queue When p vs μ_2.

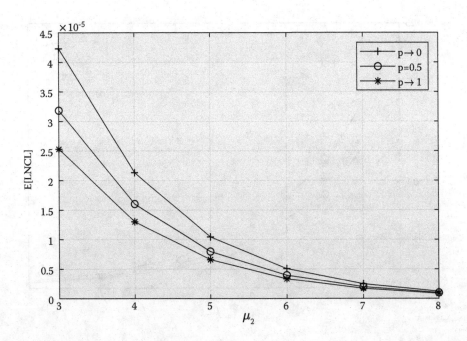

FIGURE 11.16   Expected $E[LNCL]$ in the Queue When p vs μ_2.

that in each graphical representation there is a probability p describes all the preemptive, mixed, and non-preemptive priorities and its impact over λ_1, λ_2, μ_1, and μ_2. From these things, there is a comfort zone for a decision maker to hold a best policy about more profit of the firm when choosing the value of p.

Example 11.3: *In this example, Table 11.4 shows that the impact of p over β, γ with E [HNC], E [LNC], E [HNCL], E [LNCL].*

TABLE 11.4

Impact of p over β, γ with E[HNC], E[LNC], E[HNCL], and E[LNCL]

p	β	γ	E[HNC]	E[LNC]	E[HNCL]	E[LNCL]
p = 0	5	2.5	0.044679	9.924208	0.026834	0.000005
		3	0.117366	9.803318	0.067582	0.000012
		3.5	0.244429	9.595045	0.135526	0.000025
	5.5	2.5	0.034407	9.94126	0.02141	0.000004
		3	0.090616	9.84705	0.054128	0.00001
		3.5	0.189573	9.683427	0.109089	0.00002
	6	2.5	0.027073	9.953531	0.017389	0.000003
		3	0.071482	9.87861	0.044147	0.000008
		3.5	0.150131	9.74761	0.089412	0.000016
	6.5	2.5	0.021683	9.962607	0.014329	0.000003
		3	0.057397	9.902013	0.036535	0.000007
		3.5	0.120977	9.795444	0.07435	0.000014
p = 0.5	5	2.5	0.014759	9.972939	0.01955	0.000004
		3	0.042344	9.923918	0.049465	0.000009
		3.5	0.096303	9.830187	0.100284	0.000018
	5.5	2.5	0.010535	9.980337	0.015648	0.000003
		3	0.030143	9.94482	0.039685	0.000007
		3.5	0.068576	9.876737	0.080661	0.000015
	6	2.5	0.007754	9.985296	0.012742	0.000002
		3	0.022114	9.95883	0.032417	0.000006
		3.5	0.05027	9.908048	0.06609	0.000012
	6.5	2.5	0.005857	9.988735	0.010521	0.000002
		3	0.016648	9.968533	0.026862	0.000005
		3.5	0.037785	9.929777	0.05495	0.00001
p = 1	5	2.5	0.004643	9.990832	0.015948	0.000003
		3	0.013818	9.973477	0.039845	0.000007
		3.5	0.032981	9.93834	0.080056	0.000015
	5.5	2.5	0.003088	9.993684	0.012907	0.000002
		3	0.009072	9.981956	0.032352	0.000006
		3.5	0.021451	9.958453	0.065161	0.000012
	6	2.5	0.002141	9.995473	0.010607	0.000002
		3	0.006212	9.987216	0.026697	0.000005
		3.5	0.014539	9.970857	0.053949	0.00001
	6.5	2.5	0.001538	9.996647	0.008826	0.000002
		3	0.004411	9.990627	0.022315	0.000004
		3.5	0.010217	9.978835	0.045266	0.000008

Case i: *Non-preemptive priority* (p = 0)
When γ increases, E [HNC], E [HNCL], E [LNCL] increase and E [LNC] decreases. The change in β affects E [HNC], E [HNCL], E [LNCL] directly and affects E [LNC] indirectly.

Case ii: *Preemptive priority* (p = 1)
Whenever β increases, E [HNC] and E [HNCL] decrease and E [LNC] increases and E [LNCL] decreases, whereas γ increases the E [HNC], E [LNCL], and E [HNCL] and decreases the E [LNC].

Case iii: *Mixed priority* (p = 0.5)
The parameters β and γ have the same property as in case (i) and case (ii). The important thing from the observation of this discussion provides that mixed priority holds the conducive result.

Example 11.4: *The intent of Table 11.5 is to show the impact of p over β, γ with E [I], E [R], E [P], and TC.*

Case i: *Non-preemptive priority* (p = 0)
Here, γ decreases the inventory level and increase the expected reorder and perishable rates and total cost. The parameter β increases , E [P] and decreases and TC.

Case ii: *Preemptive priority* (p = 1)
At this stage, every change about β and γ holds the same property as said in case (i). Indeed, the result is more significant than case (i).

Case iii: *Mixed priority* (p = 0.5)
Here too, β and γ hold the same property as in case (i). All the time, in practical phenomenon, the flexible service option provide the conducive result because preemptive and non-preemptive service options induce the impatience on the customers' mind. The above argument leads us to the priority will be given to the mixed priority.

11.6.1 MANAGERIAL INSIGHTS

The investigated examples give the result of mixed priorities, which will be more efficient to work with in a single server QIS with priorities. When we provide service to both HLNC by a single server, there will be chance to face the customer lost of any one kind. However, the admission of customer based on the mixed priorities will reduce the LNC lost. Most customer-oriented business services depend the number of customers who come to the system and buys the product from there. Even though a company gives priority to the HNC, the arrival of LNC should be accountable to determine the profit of the business. In such a way, the assumption of mixed priority of this proposed model gave the balanced results between HLNC.

11.7 CONCLUSION AND FUTURE SCOPE

This chapter enhanced the performance of sharing service under the preemptive, non-preemptive, and mixed priorities categories for a finite buffer. The purpose of this investigation will make effective strategies for developing business ideas. Sharing the service of a single server for a preemptive and non-preemptive mode of a customer is discussed in all the examples. Analysis of these priorities explores the advantage of sharing a finite buffer for both low and HNC. If a preemptive priority is preferred, then LNC lost is increased, whereas non-preemptive is preferred, and HNC lost is increased. In the mixed priority case, both high and LNC lost become less. The

TABLE 11.5
Impact of p over β, γ with $E[I]$, $E[R]$, $E[P]$, and TC

p	β	γ	$E[I]$	$E[R]$	$E[P]$	TC
$p = 0$	5	2.5	38.2333	0.116116	95.56422	33.03046
		3	35.32744	0.223056	105.9246	37.54435
		3.5	32.80472	0.375265	114.6813	43.84696
	5.5	2.5	38.30498	0.113293	95.74733	32.78107
		3	35.47108	0.2146	106.3672	36.88187
		3.5	33.04576	0.356033	115.5516	42.46827
	6	2.5	38.36179	0.111179	95.89224	32.60022
		3	35.58329	0.20823	106.7125	36.39933
		3.5	33.23277	0.341413	116.226	41.45699
	6.5	2.5	38.40801	0.109559	96.01001	32.46553
		3	35.67342	0.203327	106.9895	36.03819
		3.5	33.38187	0.330071	116.7631	40.69527
$p = 0.5$	5	2.5	38.32872	0.108402	95.81239	32.43493
		3	35.53488	0.200511	106.5748	35.99539
		3.5	33.16484	0.325261	116.0037	40.69677
	5.5	2.5	38.38149	0.106927	95.94644	32.30351
		3	35.63884	0.195754	106.8935	35.62542
		3.5	33.34056	0.313562	116.6352	39.87508
	6	2.5	38.42405	0.105868	96.05438	32.21157
		3	35.72085	0.192334	107.1443	35.36653
		3.5	33.47691	0.30511	117.1242	39.29896
	6.5	2.5	38.45933	0.105089	96.1437	32.14528
		3	35.78754	0.189811	107.348	35.17958
		3.5	33.58613	0.29885	117.5153	38.88214
$p = 1$	5	2.5	38.37725	0.104461	95.93849	32.1963
		3	35.64877	0.188017	106.9316	35.3054
		3.5	33.37959	0.295032	116.792	39.136
	5.5	2.5	38.41915	0.103789	96.04434	32.12284
		3	35.7274	0.185745	107.171	35.10012
		3.5	33.50841	0.289127	117.2518	38.67612
	6	2.5	38.45386	0.103331	96.13189	32.07167
		3	35.79103	0.184201	107.3644	34.95813
		3.5	33.61039	0.28512	117.6149	38.36063
	6.5	2.5	38.48332	0.103009	96.2061	32.03485
		3	35.84408	0.18312	107.5253	34.85621
		3.5	33.69393	0.282314	117.9117	38.13548

major aim of most inventory businesses is to bring strategic plans to reduce the customers lost. Hence, the proposed model will give such strategies to the upcoming industrialists. This work will be extended in a multi-server QIS setup in the future.

REFERENCES

[1] Sivakumar, B., and Arivarignan, G. A modified lost sales inventory system with two types of customers. *Quality Technology & Quantitative Management*, 5(4):339–349, 2008.

[2] Anbazhagan, N., Jeganathan, K., and Kathiresan, J. A retrial inventory system with priority customers and second optional service. *OPSEARCH*, 53(4):808–834, 2016.

[3] Chang, T.-M., Wang, F.-F., and Bhagat, A. Analysis of priority multi-server retrial queueing inventory systems with MAP arrivals and exponential services. *OPSEARCH*, 54 (4):44–66, 2017.

[4] Sivakumar, B., and Amirthakodi, M. An inventory system with service facility and finite orbit size for feedback customers. *OPSEARCH*, 31(3):9–18, 2002.

[5] Sivakumar, B., Amirthakodi, M., and Radhamani, V. A perishable inventory system with service facility and feedback customers. *Annals of Operations Research*, 31(3):9–18, 2002.

[6] Anbazhagan, N., Yadavalli, V. S. S., and Jeganathan, K. A retrial inventory system with impatient customers. *Applied Mathematics & Information Sciences*, 9(2):637–650, 2015.

[7] Anbazhagan, N., Jeganathan, K., and Kathiresan, J. A retrial inventory system with non-preemptive priority service. *Applied Mathematics & Information Sciences*, 24(2):57–77, 2013.

[8] Jeganathan, K. A perishable inventory model with two types of customers, Erlang-k service, linear repeated attempts and a finite populations. *Mathematical Modelling and Geometry*, 2(3):28–47, 2014.

[9] Ning, Z., and Zhaotong, L. A queueing-inventory system with two classes of customers. *International Journal of Production Economics*, 129:225–231, 2011.

[10] Sivakumar, B., and Arivarignan Karthik, T. An inventory system with two types of customers and retrial demands. *International Journal of Systems Science: Operations & Logistics*, 2(2):90–112, 2015.

[11] Abdul Reiyas, M., Prasanna Lakshmi, K., Jeganathan, K., and Saravanan, S. Two server Markovian inventory systems with server interruptions: Heterogeneous vs. homogeneous servers. *Mathematics and Computers in Simulation*, 155:177–200, 2019.

[12] Abdul Reiyas, M., Selvakumar, S., Jeganathan, K., and Anbazhagan, N . Analysis of retrial queueing-inventory system with stock dependent demand rate: (s, S) Versus (s, Q) Ordering Policies. *International Journal of Applied and Computational Mathematics*, 6(4):1–29, 2020.

[13] Melikov, A. Z., and Molchanov, A. A. Stock optimization in transportation/storage systems. *Cybernetics and Systems Analysis*, 28(3):484–487, 1992.

[14] Saha, S., and Sen, N. An inventory model for deteriorating items with time and price dependent demand and shortages under the effect of inflation. *International Journal of Mathematics in Operational Research*, 14(3):377–388, 2019.

[15] Sangeetha, N., and Sivakumar, B. Optimal service rates of a perishable inventory system with service facility. *International Journal of Mathematics in Operational Research*, 16(4):515–550, 2020.

[16] Sanjai, M., and Periyasamy, S. An inventory models for imperfect production system with rework and shortages. *International Journal of Operational Research*, 34(1):66–84, 2019.

[17] Sigman, K., and Simchi-Levi, D. Light traffic heuristic for anm/g/1 queue with limited inventory. *Annals of Operations Research*, 40(1):371–380, 1992.

[18] Suganya, C., and Sivakumar, B. MAP/PH(1), PH(2)/2 finite retrial inventory system with service facility, multiple vacations for servers. *International Journal of Mathematics in Operational Research*, 15(3):265–295, 2019.

[19] Ushakumari, P. V. A retrial inventory system with an unreliable server. *International Journal of Mathematics in Operational Research*, 10(2):190–210, 2017.

[20] Sivakumar, B., Arivarignan, G., Yadavalli, V. S. S., and Adetunji, O. A finite source multi-server inventory system with service facility. *Computers and Industrial Engineering*, 63:739–753, 2017.

[21] Jose, K. P., and Reshmi, P. S. A production inventory model with deteriorating items and retrial demands. *OPSEARCH*, 14:1–12, 2020.

[22] Manuel, P., Sivakumar, B., and Arivarignan, G. A perishable inventory system with service facilities, MAP arrivals PH-service times. *Journal of Systems Science and Systems Engineering*, 16(1):62–73, 2007.

[23] Manuel, P., Sivakumar, B., and Arivarignan, G. A perishable inventory system with service facilities and retrial customers. *Computers and Industrial Engineering*, 54(0):484–501, 2008.

[24] Reshmi, P. S., and Jose, K. P. A map/ph/1 perishable inventory system with dependent retrial loss. *International Journal of Applied and Computational Mathematics*, 6(6):1–11, 2020.

[25] Sivakumar, B., Anbazhagan, N., and Arivarignan, G. A two-commodity perishable inventory system. *ORiON*, 21(2):157–172, 2005.

[26] Sivakumar, B., Elango, C., and Arivarignan, G. A perishable inventory system with service facilities and batch markovian demands. *International Journal of Pure and Applied Mathematics*, 32(1):33, 2006.

[27] Yadavalli, V. S. S., Adetunji, O., Sivakumar, B., and Arivarignan, G. Two-commodity perishable inventory system with bulk demand for one commodity articles. *South African Journal of Industrial Engineering*, 21(1):137–155, 2010.

[28] Anbazhagan, N., and Arivarignan, G. Two-Commodity continuous review inventory system with co-ordinated reorder policy. *International Journal of Information and Management Sciences*, 11:19–30, 2000.

[29] Anbazhagan, N., and Arivarignan, G. Analysis of two-commodity Markovian inventory system with lead time. *The Korean Journal of Computational and Applied Mathematics*, 8(2):427–438, 2001.

[30] Anbazhagan, N., and Arivarignan, G. Two-commodity inventory system with individual and joint ordering policies. *International Journal of Management and Systems*, 19:129–144, 2003.

[31] Krishnamoorthy, A., Artalejo, J. R., and Lopez-Herrero, M. J. Numerical analysis of (s,S) inventory systems with repeated attempts. *Annals of Operations Research*, 141(1):67–83, 2006.

[32] Kaplan, E. H., Berman, O., and Shimshak, D. G. Deterministic approximations for inventory management at service facilities. *IIE Transactions*, 141(25):98–104, 1993.

[33] Sivakumar, B., Keerthana, M., and Manuel, P. An Inventory system with postponed and renewal demands. *International Journal of Systems Science: Operations and Logistics*, 0(0):1–19, 2018.

[34] Krishnamoorthy, A., and Jose, K. P. Comparison of inventory systems with service, positive lead-time, loss, and retrial of demands. *Journal of Applied Mathematics and Stochastic Analysis*, 2007(Article ID 37848):1–23, 2007.

[35] Sivakumar, B. Two-commodity inventory system with retrial demand. *European Journal of Operational Research*, 187(1):70–83, 2008.

[36] Sivakumar, B. An inventory system with retrial demands and multiple server vacation. *Quality Technology & Quantitative Management*, 8(2):125–146, 2011.

[37] Ushakumari, P. V. On (s, s) inventory system with random lead time and repeated demands. *Journal of Applied Mathematics and Stochastic Analysis*, 2006, 2007.

[38] Veinott, Jr, A. F., and Wagner, H. M. Computing optimal (s, s) inventory policies. *Management Science*, 11(5):525–552, 1965.

[39] Wagner, H. M., O'Hagan, M., and Lundh, B. An empirical study of exactly and approximately optimal inventory policies. *Management Science*, 11(7):690–723, 1965.

[40] Dshalalow, J. H., Abolnikov, L., and Dukhovny, A. On stochastic processes in a multilevel control bulk queueing system. *Stochastic Analysis and Applications*, 10:155–179, 1992.

[41] Knessl, C., Choi, D., II, and Tier, C. A queueing system with queue length dependent service times, with applications to cell discarding in ATM networks. *Journal of Applied Mathematics and Stochastic Analysis*, 12(1):35–62, 1999.

[42] Fakinos. The G/G/1 (LCFS/P) queue with service depending on queue size. *European Journal of Operational Research*, 59:303–307, 1992.

[43] Gebhard, R. F. A queueing process with bilevel hysteretic service-rate control. *Naval Research Logistics*, 14:303–307, 1967.

[44] Harris, C. M. Some results for bulk-arrival queue with state-dependent service times. *Management Science*, 16:313–326, 1970.

[45] Anilkumar, M. P., and Jose, K. P. A Geo/Geo/1 inventory priority queue with self induced inter-ruption. *International Journal of Applied and Computational Mathematics*, 6(4):1–14, 2020.

[46] Schwarz, M., Sauer, C., Daduna, H., Kulik, R., and Szekli, R. M/m/1 queueing systems with inventory. *Queueing Systems*, 54(1):55–78, 2006.

[47] Shajin, D., and Krishnamoorthy, A. On a queueing-inventory system with impatient customers, advanced reservation, cancellation, overbooking and common life time. *Operational Research*, 21(2):1229–1253, 2021.

[48] Alscher, T., and Schneider, H. Resolving a multi-item inventory problem with unknown costs. *Engineering Costs and Production Economics*, 6:9–15, 1982.

[49] Elango, C. N., and Kumaresan, V. Analysis of two-commodity inventory system with compliment for bulk demand. *Mathematics Modelling and Applied Computing*, 2(2):155–168, 2011.

[50] Yadavalli, V. S. S., Arivarignan, G., and Anbazhagan, N. Two commodity coordinated inventory system with markovian demand. *Asia-Pacific Journal of Operational Research*, 23(04):497–508, 2006.

[51] Bondi, A. B. An analysis of finite capacity queues with priority scheduling and common or reserved waiting areas. *Computers & Operations Research*, 16(3):217–233, 1989.

[52] Feng, W., and Masataka. Analysis of a finite buffer model with two severs and two non preemptive priority classes. *European Journal of Operation Research*, 192:151–172, 2009.

12 Ensuring Provenance and Traceability in a Pharmaceutical Supply Chain Using Blockchain and Internet of Things

Anna N. Kurian and P.P. Joby
Computer Science and Engineering, St. Joseph's College of Engineering and Technology, Palai, Kerala, India

Tomina Anoop
M.Sc Scholar, Tesside University, Middlesbrough, United Kingdom

Allen Mathew
Financial Analyst, X L Dynamics, Kochi, Kerala, India

CONTENTS

DOI: 10.1201/9781003264521-12

12.1 INTRODUCTION

Today, in the health-conscious society, everyone depends on drugs that supplement their life. The pharmaceutical drug supply chain, the backbone of the healthcare sector, carts life-saving medications from laboratory to market. Pharmaceutical drug research and development took a long time to complete. It needs to ensure correct proportion of active ingredients of the right quality, approval from regulatory agencies, manufacturing, and quality assurance before distribution in the market. Unfortunately, most pharmaceutical companies own complex supply chains but they are unutilized and inefficient [1]. Low-income nations like India are considered an open route for the supply of counterfeit medications. In 1992, World Health Organisation (WHO) and International Federation of Pharmaceutical Manufacturers and Associations jointly defined counterfeit drugs as:

> A counterfeit medicine is one which is deliberately and fraudulently mislabelled with respect to identity and/or source. Counterfeiting can apply to both branded and generic products and counterfeit products may include products with the correct ingredients or with the wrong ingredients, without active ingredients, with insufficient active ingredient or with fake packaging. [2].

Later in 2017, after complex discussions, WHO member states jointly decided to define counterfeit and substandard drugs into a new category as "substandard/spurious/falsely-labelled/falsified/counterfeit medical products" (SSFFC) [3,4].

Circulation of fake drugs needs to be paused because it directly and indirectly affects the human life. Not only a falsified drug supply is a concern, there are some temperature-sensitive drugs, especially vaccines, that need a controlled environment to prevent deterioration. Due to the lack of accurate temperature monitoring and record keeping across the supply chain, many deteriorated drug cases go unreported. Tamper-proof records resolve the issues of counterfeit drug supply, compliance, and traceability of pharmaceutical supply chain in an effective way.

Counterfeit drugs are a recent issue, but it was first identified as an emerging problem by WHO in 1985. Measuring the scale, the problem has intensified substantially. Counterfeit drugs have invaded the legitimate supply chain, the Avastin – a drug for treatment of cancer, case reported in United States [5]. Also, the Internet pharmacy significantly contributes to the circulation of counterfeit drugs [6]. and the situation became worse because it opens a number of distribution channels. Poor quality drugs or substandard drugs can also come from the legitimate pharmaceutical company due to the negligence in the manufacturing and improper transportation. Even in the COVID-19 pandemic situation, there is a huge circulation of counterfeit hand sanitizers, low quality PPE kits, and other essentials raise the situation to a worse case, the Authentication Service Provider's Association (ASPA) reported.

From the perspective of a pharmaceutical company, counterfeit drug supply results in financial loss and ruins public trust over genuine medicines and the national healthcare system. Speaking from the other perspective of common medication consumers, the counterfeit drugs directly and indirectly affect their health. The risk management of a pharmaceutical drug supply chain has to be extensively studied to combat the circulation of counterfeit and deteriorated drugs across the supply chain.

The pharmaceutical sector needs an immediate solution to pause the circulation of counterfeit and deteriorated drugs in the market, which causes adverse effects on life. We need a practical solution to ensure provenance and traceability in the pharmaceutical drug supply chain in an effective way.

12.1.1 Objectives

Initially, a study on pharmaceutical ecosystem was conducted, which divides the ecosystem into three categories – drug discovery, drug manufacturing, and drug distribution. Concentrating more on risk management in the supply chain, circulation of counterfeit and deteriorated drugs in the market is identified as problems that require an immediate solution. In a developing country like India, it is estimated that 25% of drugs on the market are counterfeit ones. Many cases remain unreported; the deterioration of drugs during its transport from laboratory to market due to temperature variation is a major threat for public health and need immediate concern.

12.1.2 Trust and Transparency

In the supply chain management, from the manufacturer's end and customer's end, both the former who produces the drugs and the latter who consumes the produced drugs should be able to track the path of the drug across the supply chain. Manufacturers should be able to ensure the supplied drugs at any point of time, who are distributors, packagers, and also finally reach the intended customers without any deterioration. Customers should be satisfied that they can authenticate the originality and quality of drug that they intake [7].

12.1.3 Visibility and Data Privacy

Supply chain visibility can be defined as the capacity to track various merchandise goods and/or products in transit, giving a clear view of the inventory and activity [8]. Visibility and data privacy really contradict, ie., if one wins other lose. The data should be kept private and at the same time should be open for public verification. Manufacturer's private data should be kept secret and the unique product code, active ingredients, dosage to be taken, manufacturing date, and date of expiry should be visible for public verification [8].

12.1.4 Secured Tamper-Proof Records

Conventional paper-based drug sheets and e-pedigrees are not tamper proof. Each company maintains their own e-pedigree within their database and, for authentication purposes, each has to contact the previous custodian, which creates additional overhead on the network. Tamper-proof records reveal all the information about the drug and its quality, which would be accessible to the consumers is an effective remedial measure to combat the counterfeit and deteriorated drugs supply.

Organization of the Chapter: The rest of the chapter is organized as: Section 12.2 delivers a systematic review on the Pharmaceutical Ecosystem includes the drug discovery, drug manufacturing, and drug discovery, as well as the risks involved in the pharmaceutical sector. Section 12.3 elaborates the study on existing visibility technologies and temperature sensors used for cold chain monitoring. Section 12.4 discusses case studies on the global threat – counterfeit drug supply, and analysis of its impact on public health, society, and economy. Section 12.5 covers blockchain, a distributed digital ledger technology, data structures, reference model, and basic concepts has raised the potential to provide tamper-proof records of all logistics activities undertaken in the supply chain. Section 12.6 elaborates discussion and open challenges and section 12.7 concludes the chapter with future scope.

12.2 PHARMACEUTICAL ECOSYSTEM

A pharmaceutical ecosystem is a combined effort of drug manufactures, researchers, regulatory authorities, distributors, wholesalers, retailers, pharmacies, physicians, and patients. Everyone has to collaborate and cooperate to keep the ecosystem alive. Starting from the new drug production, dynamic demand of drugs due epidemic outbreak, R&D cost, intellectual property and patent rights, issues of generic as well as the branded drugs, the quality of raw materials used, regulatory acts passed to be followed, production cost involved, and supply of counterfeit and deteriorated drugs in the distribution chain are major concerns reflected in the pharmaceutical sector. To properly comprehend an effective solution for the alarming pharmaceutical sector needs a systematic and thorough study of these issues. The categorization of pharmaceutical ecosystem is shown in Figure 12.1.

12.2.1 Drug Discovery

Pharmaceutical companies target the discovery and development of medications that are approved by regulatory authorities, patients, healthcare professionals, and healthcare service providers [9]. All of the stakeholders need efficacious and safe drugs that improves patient's life expectancy. Although, they also expect new drugs to be more cost efficient (i.e., "health economical"). In addition, pharma companies also need to produce sustainable revenues and return for the investment from shareholders involved. Therefore, speed, quality, and cost have become a mantra for all drug discovery and developmental activities. Focusing in this direction, Elebring et al. [9] in 2012 published a feature addressing a crucial question – "What is the most important approach in current drug discovery: are we doing the right things or doing things right?" Mignani et al. [10] completely agreed with Elebring et al. [9], that the right selection of pharmacologic target and chemical lead ensures an improvement in R&D productivity and implementation of robust pipelines. There occurs a sudden spread of communicable diseases without any alarm and requires research and development of new drugs to cure the epidemic.

The dynamic demand for certain drugs also creates fluctuations in the market. New drug discovery takes several years to complete as it is time consuming – thorough study about disease, research on active ingredients required, approval from the regulatory authorities, and finally the test sample on the combination of several active ingredients to match the hypothesis. A recently

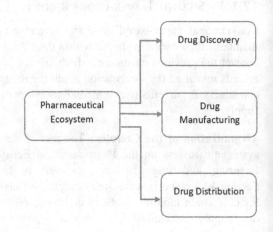

FIGURE 12.1 Categorization of Pharamaceutical Ecosystem.

reported epidemic outbreak was in Kerala – Nipah Virus, a brain-damaging virus. The Kerala State Government bought Ribavirin, an antiviral drug from Malaysia, to tackle down the Nipah menace because the Nipah virus was initially identified in Malaysia [11]. The H1N1 case reports in October 2018, nearly 32 deaths reported in Kozhikode district itself. Oseltamivir, a drug which is sold under the brand name Tamiflu, being used for the treatment and prevention of H1N1 influenza [12].

For the development of new drug and carting it into the market, a company requires nearly $300 million to $1000 million including all the risk in R&D. So, no company is ready to display patent for their product without any revenue returns. Hence, the patent right should be preserved. This will not be a problem for short-duration development drugs. But menace for drugs evolved from biotechnology path especially from gene [13]. Many strategies have been developed by companies to tackle down the cost and achieve trade advantage include outsourcing R&D activities, forming R&D partnerships, and establishing alliance [14].

New technologies alter the drug economy. Biopharmaceuticals took a chair with a goal to produce personalized medications for everyone as their genome is mapped onto a single chip. The main issue of concern is the storage of this personal database [15].

The TRIPS Agreement (Agreement on Trade – Related Aspects of Intellectual Property Rights) of World Trade Organisation (WTO) attempts to have a common internationally accepted rule [16]. But acceptance of TRIPS brought changes to Indian pharmaceutical industry. The Patent Act in 1970, supports exponential growth in generic drug industry in India [17]. Hence drugs are easily accessible with cheaper rate. Local firms in India started manufacturing copies of drugs unit, and send to authorities abroad for patent recognition. Then from 1 January 2005, with the implementation of TRIPS, product patent era came. Indian firms cannot manufacture patented drugs without permission from the patented owner [17]. TRIPS ends up the generic drug production but opened a route for new drug discovery. Many regulatory authorities passed acts to control and coordinate the pharmaceutical sector in a proper and efficient manner.

12.2.2 DRUG MANUFACTURING

Counterfeit drugs can be classified as: (1) Products do not contain specified active ingredients even though mentioned over the packaging, (2) products contain active ingredients that is not mentioned over the packaging, (3) product contain the active ingredients in correct proportion but the source is same as the declared one, (4) products contain active ingredients but not in quantity as specified and also contain impurities or contaminants [18].

Food and Drug Administration (FDA), defined Good Manufacturing Practices (GMP) standards in countries with pharmaceutical manufacturing operations to ensure that the products are produced according to quality standards [19]. It covers includes all aspects from the raw material, environment, machinery and equipment, as well as personal hygiene of staff. The law should always support post-surveillance (Pharmacovigilance). In 2002, WHO defined pharmacovigilance as "the science and activities relating to the detection, assessment, understanding and prevention of adverse effects or any other drug-related problems" [19].

The raw materials used in the manufacturing drugs should be from authenticated source and of good and specified standard. Improper proportion of active ingredients results in falsified drugs. Recently, the UN General Assembly released political declaration and endorsed all member nations to end tuberculosis worldwide by 2035. Tuberculosis is easily diagnosable and curable but required affordable quality diagnosis and drugs [20]. Production cost for each drug is estimated to be $300 million. The cost gets added to the final product. Civica Rx, a new generic drug company in the United States, to help patients and seek medication with affordable price [21]. It does not aim for profit.

Medicines can change their compound structure when exposed to higher or lower temperatures than the specified one. The traceability is an issue in cold supply chain. During the transportation of vaccines, the temperature should be continuously monitored and should be recorded to verify the vaccine is a life-saver or life-destructor [22]. Title II of Drug Quality and Security Act (DQSA) passed in the United States, in 2013, clearly mentions that there should be an inter-operable monitoring system to track the temperature of drugs during its transportation [23].

In some cases, drugs may be mislabelled. An antibiotic Rofact, generic name Rifampicin [24], case was reported in Canada in 2009; a batch of antibiotics were mislabeled. The actual content in the bottles were anti-epileptic clonazepam. Another case in Canada, in 2011, minocycline was mislabeled as amlodipine [25]. In USA, 2011 [23], bottles with finasteride were labeled as cita-lopram. Zopiclone as a substitute for furosemide in France, 2013 [26].

12.2.3 DRUG DISTRIBUTION

Drug distribution is the tail end of the pharmaceutical ecosystem. The pharmaceutical supply chain is vulnerable to the circulation of counterfeit drugs that are extremely dangerous for health [55].

Adverse effects of counterfeit drugs on health can be both directly or indirectly. Indirectly means those drugs does not contain active ingredient in correct proportion to kill the disease-causing agents and on consumption they create drug resistant strains which makes the consumption of genuine drugs useless. In terms of directly, drugs may or may not contain active ingredient, if contained it may not be in correctly specified quantity otherwise, it may also contain impurities that may cause health problems [27]. Counterfeit drugs are not new; earlier in 1985, WHO had been identified as a problem that needs to be rectified. As a remedial measure in 2006, WHO initiated to build the "International Medical Products Anti-Counterfeiting Taskforce" (IMPACT). In 2008, formation of the Central Drugs Standard Control Organisation (CDSCO) is India's first step to tackle the alarming threat.

There are many supporting literature in history that mark counterfeit drugs – a global threat. In Haiti, reports in 1995, 95 people died with the intake of paracetamol cough syrup which was con-taminated with diethylene glycol [18]. In Pakistan, in 2006, 60 adults of two different cities died with consumption of large dosages of cough syrups as a drug addiction that contains levomethorphan – a dangerously powerful drug, but has the same molecular formula as dextromethorphan [4].

Incidents are reported on the distribution of contaminated drugs in the market. In India, in 2009, a paclitaxel formulation – Albupax contains excessive endotoxin level and seized from market [28]. The presence of 2, 4, 6-tribromoanisole found in Tylenol, Motrin, Rolaids, and Benadryl cases were reported in the USA [29]. Generic Clopidogrel, marketed in India, contains methyl chloride, which damages hepatic, renal, and nervous systems [30]. In 2011, Tanzania Foods and Drug Authority banned Methyldopa 250 mg tablet [31] because the identification can be easily detached from the package; "vivid fungal growth" present on the tablets.

Counterfeited drugs enter into the legitimate supply chain – as an example, Avastin [4], a cancer treatment drug. Other sources can be Internet pharmacy-purchasing medicines online that bring health risks. Rogue is an online pharmacy [26] that sells "lifestyle" medicines to "lifesaving" medications. These pharmacies offer fast delivery services, with lower prices than market prices, "without any prescription" [26]. This results in a large inflow of illicit drugs – the case of heroin in the family of opioids drug addicts in Punjab [32] and abuse of oxytoxin [33], harmful to animals and human beings.

A worldwide effort, Pangea by INTERPOL, the main aim is to eliminate illicit pharmaceutical products from circulation and raise awareness of the dangers associated with buying medicines online. With Operation Pangea XIV (conducted in May 2021), fake and unauthorized COVID-19 testing kits and all medical devices seized and a massive number of unregulated websites were shut down [34].

In India, Bhagirath and Chandni Chowk in New Delhi are a main hub for the production of counterfeit and spurious drugs [35]. Northeastern states Bihar, West Bengal, Uttar Pradesh, and Gujarat are local market for spurious drugs. Reselling expired medication – another prompting issue in the distribution side. Reverse logistics is a crucial aspect for pharmaceutical industry. In

1998, the European Working Group on Reverse Logistics (REVLOG) clearly defined reverse logistics: "The process of planning, implementing and controlling the flow of raw materials, in process inventory and finished goods from a manufacturing, distribution or use point to a point of recovery or point of proper disposal" [36].

12.3 EXISTING TECHNOLOGIES

Diversion of legitimate product outside the authorized track of distribution channels is another serious issue. The pharmaceutical industry should enforce appropriate measures to eliminate counterfeiting of drugs because it poses serious health hazards and also decline the pharmaceutical revenue. The industry requires more sophisticated anti-counterfeiting technologies because the counterfeiters became more advanced. The existing technologies, holograms, radio frequency identification (RFID), barcodes, and different temperature monitoring sensors, were studied.

12.3.1 VISIBILITY TECHNOLOGIES

12.3.1.1 Holograms

Holograms are laser created otical image device bearing diffractive elements due to the surface relief. They have variable optical information that exhibits rainbow colors and image content that can be animated or stereoscopic. Holographic technique a proactive option that provide the consumers a belief that the medications they intake are authenticated ones [37]. The only advantage of using a hologram is that it supports item-level verification. In holograms, details of the drugs get stored inside. It can be read using the laser light. It is impossible to imitate the same hologram printing, but there is a deterrent that the traceability of hologram orgination machine.

Hologram provides multi-level security features: overt features, covert features, and forensic features. Figure 12.2 depicts clearly the three levels of security features of holograms.

Level 1 includes the overt features or the visible authentication as it can be verified even by an untrained eye without any instruments, commonly the end consumers, distributors, agents, and customs officials. Level 2 includes the covert features or the hidden information, and these features are not visible to the naked eye, but can be verified with some instruments or a reader. Holograms can be with micro text, nano text, and CLR, with raster images. The hidden informations are normally read by the drug inspectors or the higher officials and not accessible to the end consumer. Level 3 includes the foresensic information that are invisible to the naked eye and it can be read or verified only with the microscope or spectrometers in the laboratory.

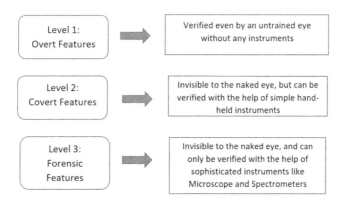

FIGURE 12.2 Three Levels of Security Features of Holograms.

If the hologram gets tampered, the entire data gets lost. The cost or rate changes according to the complexity of the hologram used. Hologram counterfeiting can be done. Several cases are in the article [38]. The problems of using holograms:

- Not an effective solution for long-term use.
- It can be easily counterfeited.
- It does not provide the brand owner the e-pedigree, so tracking of medicines is not possible.

12.3.1.2 Radio Frequency Identification (RFID)

It can be defined as a wireless data collection technology that deploys radio waves to identify the objects [37]. The principle behind the technology is that an electronic chip with transponders emits messages and is read using an RFID reader. The RFID tag can be hidden, which prevents tampering. The advantage is that a reader can be placed in its proximity; no line of sight is required. The RFID tag can be placed in a batch of medicine. Figure 12.3 shows the workings of RFID.

The RFID Reader deploys an antenna part, transcevier part, and a decoder part. It always remains powered on with an external power source. During the ON state, the oscillator inside it generates a signal with a desired frequency; if frequency fades, it is amplified by the amplifier circuit and for a long distance, modulation can be done using a modulator circuit. The signal has to be transmitted with the antenna, which converts the electrical signal into an electromagnetic signal.

When a RFID tag (can be active and passive, for a pharmaceutical sector tag is passive), comes to nearby RFID reader, the tag contains a coil that detects the reader's signal and converts the received RF signal into an electrical signal, which alone is enough to power up the microchip inside the tag. Once the microchip gets powered up, it sends the data (unique ID) stored in it.

In the transceiver part, when the signal comes through, the antenna is fed into the demodulator and to a decoder to obtain original data and then to perform specific task, later fed into a microcontroller or a microprocessor.

Problems of using RFID are:

- RFID tag requires a reader to authenticate the medicine.
- RFID tag cost get added to the medicines, which may not be economically feasible.
- Using RFID tags only requires batch level authentication. Consumers always look for item-level verfication.

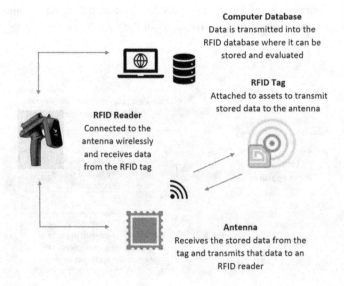

Computer Database
Data is transmitted into the RFID database where it can be stored and evaluated

RFID Tag
Attached to assets to transmit stored data to the antenna

RFID Reader
Connected to the antenna wirelessly and receives data from the RFID tag

Antenna
Receives the stored data from the tag and transmits that data to an RFID reader

FIGURE 12.3 Workings of RFID.

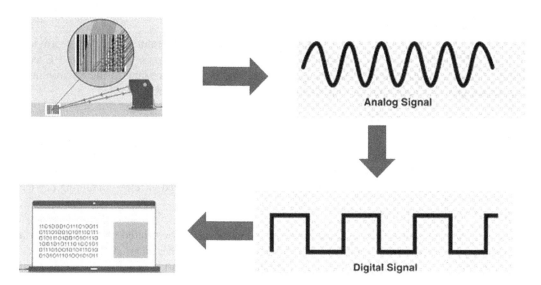

FIGURE 12.4 Working of Barcodes.

12.3.1.3 Barcodes

Barcodes are small images of lines (bars) and separated with spaces. Barcodes typically consist of: (1) quiet zone, (2) start character, (3) data characters, (4) stop characters, (5) another quiet zone. Every product is given a unique digital identity that is generated by a software-based encryption engine and the same algorithm is used for decryption [37]. There are different kinds of barcodes: linear barcodes, 2-D matrix barcodes, 2-D quick response barcodes. The most preferable one is 2-D matrix barcodes that contain a 16-bit alphanumeric code.

The working of barcode technology: When the laser beam incidents on a mirror/prism, it is directed towards left to right of the barcode. The dark portion absorbs the light incident on it, and the white portion reflects the light. The photo diode generates voltage waveform, similiar to the bars and the space pattern of barcode. The photodiode measures the light reflected and gives out analog electrical signal. When the corresponding barcode is read, the signal is then converted into a digital signal. There are different types of barcode scanners: pen scanners, laser scanners, mobile scanners, Charge Coupled Device (CCD). Figure 12.4 shows the working principle of barcodes.

The advantage of using barcodes are:

- It allows each strip of tablets or vial of vaccines verification.
- Science is software based; no specialized reader is required. Barcodes can be scanned using mobile phones.
- Cost effective and secure encryption of data.
- Supports track and trace with e-pedigree.

The problem of using barcodes is:

- If the barcode gets a scratch over it, the entire data is lost.

12.3.2 TEMPERATURE MONITORING SENSORS FOR COLD SUPPLY CHAIN

Vaccines can be termed as biological products that lose their potency when exposed to extreme heat and/or freezing conditions. Monitoring devices can be placed in cold boxes or in vaccine carriers.

12.3.2.1 Electronic Freeze Indicator

The electronic freeze indicator [39] is a single alarm electronic temperature indicator. It provides an irreversible temperature if exposed to temperature relative to alarm setting of $<-0.5°C$ continuous duration of 60 minutes, "X" changes to "X," indicating the change in the environment. The device provides active monitoring and recorded data are irreversible.

12.3.2.2 Cold Chain Monitor

A cold chain monitor is a paper-based temperature monitoring device [40]. It uses time-temperature functions. The indicator strips are attached to the instructions for use in the packet. If exposed to change in temperature, depending on the time, it changes its color irreversibly on a constant rate. The primary purpose is for the shipment of freeze-dried vaccine containing dry ice.

12.3.2.3 Vaccine Vial Monitor

A vaccine vial indicator [40] is a round sensitive indicator. It can be printed on the vaccine vial label or may be fixed on the top of the vial. It contains heat-sensitive material. The inner square portion is initially colored a light shade and it gradually becomes darker in color when exposed to heat or change in temperature. Health workers compare the color differences and determine whether vaccine has lost its potential or not.

12.3.2.4 Electronic Shipping Indicator

Electronic shipping indicators [40] are electronic temperature indicators that have a temperature-sensitive electronic circuit associated with a liquid crystal display (LCD). If the temperature is within a specified range, then "OK" will appear; otherwise, out-of-range means an "ALARM" sign appears.

12.4 COUNTERFEIT DRUG SUPPLY – A GLOBAL THREAT

A case: [4]. in Paraguay on September 2013, 44 children felt breathing difficulties and were admitted to the hospital. Six out of 44 were affected the worse. Physicians were not able to identify the cause of the sudden outbreak. The condition started with a common- cold and the parents treated it with locally available cough medicines. The National Medicine Regulator reported this case to World Health Organization Substandard and Falsified Medical Products Group in Geneva. A similar case was reported earlier; 60 adults in two different cities of Pakistan died with consumption of a large dosage of cough syrup because they were addicted to it. The Pakistan government suddenly suspended the production of the cough syrup. On detailed investigation, both manufacturers had changed their raw material used in the manufacture of cough syrup to a cheaper one. Levomethorphan has the same molecular formula but is extremely more powerful than dextromethorphan [4]. India was the raw material supplier.

In the investigation in Paraguayan progresses, the import records for dextromethorphan had taken, here again the levomethorphan – raw material supplier was the same India supplier who supplied in Pakistan. Two similar cases occurred within the gap of ten months. The case underlines that there is a sudden requirement of real-time reliable mechanism for people, to check the authenticity of medicines before consumption. World Health Organization received a case – WHO's Substandard and Falsified Medical Products Surveillance database from 2013–2017 [39].

There exist many assumptions that high-income countries with strong regulations can prevent the entry of substandard drugs in their market. But the scenario today was entirely different. There is no distinguishing between the high-income or low-income country; every region over the world is in the adverse of counterfeited drugs. The lighter portion in the map does not mean that there is no issue or impact of substandard drugs. Sometimes the case was not reported. Anti-malarial drugs and antibiotics are most commonly counterfeited ones. The aim of counterfeiters is to gain profit and destroy the economic stability.

12.4.1 Impact of Substandard/Falsified Drugs Supply

The circulation of substandard/falsified drugs marks its effect on public health, social, and economic areas [41].

12.4.1.1 Impact on Public Health

Mortality and morbidity: Consumption of falsified drugs that may contain impurities or large doses of active pharmaceutical ingredients than the prescribed standard is hazardous for human life. The fake medication produces drug-resistant strains in our body that may suppress the action of genuine drugs.

Disease prevalence: Fake drugs cannot cure diseases. Rather, they produce drug resistance. Even then consuming genuine medication, it will be useless. With globalization an exponential growth in population and widespread disease-causing microbes. If diseases are not cured, then a sudden epidemic outbreak will extend to non-endemic regions.

Anti-microbial resistance: Anti-microbial drugs kill the pathogens effectively, only with high proportion of active pharmaceutical ingredient [42–44]. Since 2013, the World Health Organization received 1500 cases of counterfeiting drugs. Out of 1500 antimalarial and antibiotic medicines are more counterfeited ones.

Loss of confidence: Supply of fake drugs in the market creates mistrust among the people about the drugs they intake to save their life. Spurious drugs are not life-savers; they are life destructors. Common people lost their trust and confidence in availing medication.

12.4.1.2 Impact on Economy

Increased out-of-pocket spending: From both manufacturers' and patients' perspectives, the counterfeit drug supply is a major threat. Patients lose their money and their diseases are not cured or rather the situation become worse. Manufacturers spent more on resources and revenue for R&D, production of drugs, and purchase of raw materials of good quality. With the counterfeits, all these efforts go out.

Wasted resources: Patented owners for various drugs spent resources for research and development, manufacturing, licensing their distribution, and so on. Counterfeits in the intermediate make a profit. The entire resource spent by genuine owners get wasted.

Economic loss: Counterfeited drugs cause loss of credibility and affect the stability of financial economy.

12.4.1.3 Socioeconomic Impact

Lost productivity: Patented manufacturers are masked by counterfeits. There is a sharp decline in productivity.

Lost income: Counterfeits suppress the income for genuine manufactures. Income is proportional to socioeconomic development.

Increased poverty: People spend their money for medications to save their life. But in the present scenario, the consumption of falsified drugs makes the disease state worse and the have to go for re-treatment. Loss money increases poverty.

Lack of social mobility: Counterfeited drugs are available at a low price. The illicit drugs are supplied at a large amount through online pharmacies without any prescriptions. These activities uproot the socioeconomic stability.

12.5 BLOCKCHAIN – A DISTRIBUTED DIGITAL LEDGER

Sathoshi Nakamoto in 2008 published a white paper on Bitcoin, the digital cryptocurrency [45]. Bitcoin is a decentralized peer-to-peer network of transactions. The Bitcoin network as it is decentralized and a public network faces the problem of double spending. That means the same

Bitcoins are transferred to two different people. For example, suppose A has 50 coins and A decided to transfer the 50 coins to friend B. At the same time, A tries to have another friend C. Here, A double spends the money. To solve this problem, blockchain technology was introduced in a Bitcoin network.

Blockchain cuts the role of intermediators in the network with peer-to-peer trust. The trust is ensured with cryptographic hash functions. The chronological order chain of blocks is distributed to all peers in the network as ledger. All transactions in the Bitcoin are recorded in blocks and with Proof-of-Work (PoW), a mining mechanism blocks are added to the existing network; it can neither be deleted nor modified. The new data can be updated by creating a new block.

Iansiti and Lakhani in 2017 gave a clear definition for a blockchain [45]. "Blockchain is an open distributed ledger that can record transactions between two parties efficiently and in a verifiable and permanent way." The definition contains many keywords: open (everyone in the network can access and verify the information), distributed (every peers in the network have a copy of public ledger), efficient (works a fast and scalable protocol), verifiable (everyone in the network should take the validity of the information), and permanent (information entering in the blockchain, is persistent, or tamper-proof, if you want to update the transaction then make a new transition, and the old transition is invalid). Whatever transaction already committed, a committed transaction cannot be rolled back and someone cannot change that transaction. The fundamental base of blockchain is a cryptographic hash algorithm. Hash algorithms are one-way functions and have a avalanche effect. The cryptographically secured chain of blocks are formed with the addition of hash of a previous block to the next block, again for hashing mechanism.

Blockchain is capable to do much more than the financial transactions. In 1996, Nick Szabo, first coined "smart contract." He claimed that a smart contract can be reliazed with a public ledger. The notion of smart contracts took blockchain to the next pace. A smart contract is an automated computerized protocol over a decentralized platform for digitally verifying or enforcing the negotiation or performance of a legal contract devoid of intermediates and directly validating the contract. It ensures immutability, and executes business processes over the network securely.

12.5.1 Data Structures

In a typical blockchain architecture, every node over the network maintains a local copy of a global data sheet. A block is defined as a container data structure that contains a series of transactions. The ledger consists of two data structures:

1. Blockchain – Each block over the network forms a linked list that describes a set of transitions. And they are immutable.
2. World State – The most recent state of smart contracts/output of transactions are stored in a traditional database (e.g., key-value store). Data elements can been added, modified, and deleted; all recorded as transactions on a blockchain.

12.5.2 Reference Model and Concepts

In [45], proposed a six-layered reference model. The model characterized as well as standardized the architecture and major components of the blockchain platform. A complete blockchain system has been decoupled into six layers [46–54].

1. **Data Layer:** The data get stacked into chained blocks in a network with asymmetrically encrypted, hashed, and time-stamped Merkle tree data structure. A typical Bitcoin block

consists of: (1) header, (2) body part. The header has all the meta-information while the body part shows a Merkle tree of verified and hashed data (e.g., double SHA256 algorithm). The blocks are hash chained one by one, chronologically forming the entire transaction history from the genesis block to the newly generated one.

2. **Network Layer:** As the blockchain system is decentralized, emergent, and with bottom-up control all participants over the network are equally privileged. These nodes listen to the network, verifying the broadcasted data and blocks created according to pre-defined checklists. Invalid blocks will be discarded and the valid will get forwarded towards the neighboring nodes.

3. **Consensus Layer:** The blockchain network deploys a variety of consensus algorithms to guarantee data consistency and the fault-tolerant ability of the shared ledger among distributed nodes. In traditional application, typically in closed ecosystems where the entities trusting each other, algorithms such as PAXOS are sufficient to reach consensus efficiently. But the blockchain models focus on open and dynamic environments where a mass of trustless entities with possible Byzantine failures, so more complex algorithms are needed, such as practical Byzantine fault tolerance for semi-open environments and proof-of-X (POX) type consensus for open environments (e.g., the cryptocurrency markets). Proof-of-Work (PoW) nodes have to repeatedly do a mathematically difficult computation and validate the data. The winning node will be given the chance to append its block over the ledger; while proof-of-stake (PoS) requires the node with the largest amount of predefined stakes (e.g., coins) to create the new block; delegated PoS (DPoS), proof-of-movement, etc. are other algorithms.

4. **Incentive Layer:** The layer incorporates the economic rewards into the blockchain network. The data verification and the process of creating blocks are driven by consensus mechanisms and the participating nodes that contribute their computing power. Bitcoin and ETH are examples of rewards. The mechanism: once a new block is created, a certain amount of cryptocurrencies will be issued as rewards and allocated to the winning node to motivate the entire network, continuing their efforts in data verification and block creation.

5. **Contract Layer:** The contract layer is packed with smart contracts, mechanisms, and algorithms, to serve as the high-level business logics. Once a group of parties consent to a set of predefined terms or rules, they can codify them as a smart contract, cryptographically sign it, and broadcast over the P2P network for verification. Once preconditions are triggered, the stipulations and associated actions will be activated and self-executed without human interventions.

6. **Application Layer:** All possible application scenarios and use cases of blockchain.

12.5.3 Types of Blockchain Implementation

The following are the types of Blockchain:

1. **Permissionless Blockchain:** The blockchain is open and public. Anyone can join the network without permission. There is pseudo-anonymity of all peers in the network. Bitcoin and Etheruem are the examples. Eg: Ethereum

2. **Permissioned Blockchain:** The blockchain is private and closed for an organisation or a group of organisation. Anyone can join the network but needs prior permission. The identity of all peers in the network are known to everyone in the network. Hyperledger was originally an open blockchain project developed by IBM. Table 12.1 tabulates the difference between permissionless and permissioned blockchain.

TABLE 12.1

Comparison of Permissionless and Permissioned Blockchain

Factors	Permissionless Blockchain	Permissioned Blockchain
Access	Open read/write access to database	Permissioned read/write access to database
Scale	Scale to large number of nodes, but not in transaction throughput	Scale in terms of transaction throughput, but not to a large number of nodes
Consensus	Proof of Work/Proof of Stake	Closed membership consensus algorithms
Identity	Anonymous/Pseudonymous	Identities of nodes are known, but transaction identities can be private/ pseudo-anonymous
Assets	Native assets	Any asset/data/state

A smart contract gears up the blockchain technology to next level. With smart contracts, risks of many companies are reduced to a great extent.

12.5.4 Requirements for Enterprise Blockchain Application

Everyone in the network knows about the piece of information, and immediately get access to this information. In other words, every node in a blockchain network stores a replica of all information that is transacted. Consider each exchange of information as a transaction and this transaction gets recorded on the blockchain. Hence, every participant of the blockchain records it on their ledger.

From an enterprise context, all the participants in the network store these transactions on completely immutable append-only log. Participants can only add transactions to the ledger, without tampering the previous transactions. With the notion of consensus mechanism, all transactions get added on to the ledger, every parties agrees upon legitimate transactions and get added to the ledger. The immutable log improves the auditability and ensures provenance. Along with consensus, the security of the system must be soundly pronounced, in the case for an enterprise application. The notion of a smart contract makes the enterprise blockchain different from a Bitcoin blockchain.

In an enterprise application, rather than exchanging information it is important to conduct business processes. Smart contracts execute accordingly with business logic set for them. Comparing Bitcoin blockchain and enterprise blockchain, the former executes a static smart contract while the latter executes pluggable contracts (we are able to write our own business contracts). Smart contracts in Bitcoin determine who the owner is of a particular Bitcoin or who transferred how many Bitcoins between people. Hence, the asset management in transfer can be done. For an enterprise application, we need benefits of distributed application and at the same time with data privacy and better scalability. Bitcoin world is completely censorship resistant; no authority is present there to control illegal transactions as it is a public permissionless blockchain network. Enterprise blockchain focuses on scaling transaction throughput, whereas Bitcoin blockchain focuses on scaling the number of nodes.

12.5.5 Blockchain Strenghthens IoT

Blockchain, a decentralized framework, eliminates the prerequisite for a trusted third party by permitting participants in the network to check correctness of the data and to ensure immutability. IoT can utilize the blockchain to enroll themselves and organize, store, and share streams of information effectively and with reliability. Blockchain is a very flexible innovation, as it creates a digital ledger in which each transaction into the ledger is encrypted, verified, and signed by trusted entities. This creates the primary advantage where each transaction is confirmed and affirmed by the cryptographic keys.

Another advantage, all old transactions recorded can always be traced back; a chain of trusted transactions that cannot be modified without breaking the cryptographic code. The smart contract controls the access to the encrypted data, as every access to the data is logged on and transparent for the user. Blockchain technology can be deployed as the key to solve scalability, privacy, and reliability problems in connection to the IoT.

12.6 OPEN CHALLENGES AND DISCUSSIONS

The notable challenges regarding adopting the blockchain-based provenance and traceability solutions towards the pharmaceutical industry.

1. **Stakeholder reluctant:** All the involved pharmaceutical supply chain stakeholders store their core and sensitive private business data on a distributed ledger, a blockchain network. Many potential stakeholders show reluctance towards participating in the networks where different business competitors exist in the same supply chain, as this would lead to losing their competitive advantage.

2. **Interoperability:** Interoperability can be defined as the adoption of massive business software and platforms across different organizations to provide efficient and effective integration. As there are no standardized solutions, to ensure integration, adaptability, and implementation easier, the existing medication traceability solutions as well as blockchain-based solutions and platforms lack complete interoperability. But, different blockchain platforms are trying to find solutions for the issues, and to provide complete interoperability, ensure maximum scalability, and adaptability for enabling internal and external communication among the business organizations.

3. **Cost of Implementation:** One of the leading challenges faced by the majority of the surrounding enterprises, including the supply chains of pharmaceutical sector, is the implementation and computational energy costs. Considering the example, Hyperledger Fabric can execute more than 3500 transactions per second and its power consumption is significantly reduced when compared to Ethereum, depending on the different consensus protocols used.

4. **Attacks and security vulnerabilities:** Phishing scams, technology vulnerabilities, implementation exploits, and malware, due to lack of standards and procedures, are the serious challenges to be addressed. And recently, many cybersecurity reports highlight several security risks involved, such as bad actors and the middleman attacks, being in the blockchain network, and exposing the networks to vulnerabilities. The definition of the blockchain states its advantage as its resilience against various types of security attacks, including cyberattacks. The current blockchain implementations are left inherent to vulnerabilities and bugs due to the development of immature processes and systems.

5. **Lack of standardized regulations:** The role of drug regulatory authorities includes monitoring the quality, safety, and efficacy and post-market surveillance of the pharmaceutical products becomes more pertinent and complex with the blockchain-based solutions, as it becomes hard and difficult for the agencies to define the legal boundaries and environment for blockchain technology.

12.7 CONCLUSION AND FUTURE SCOPE

A pharmaceutical supply chain is an interdependency between different groups of people includes drug manufacturers, research and development, suppliers, wholesalers, retailers, doctors, regulatory authorities, and patients. Taking the risk management part of the pharmaceutical sector, elimination of counterfeit and fake drugs supply, quality assurance of medications are important challenges. Recently, the World Health Organization (WHO) released that one out of ten drugs produced in India are counterfeits. On the ground of evidence and analysis on the impact of counterfeit drugs in public health, social, and economic areas, alarming global threats need a reliable solution. The integration of two emerging technologies, blockchain and Internet of Things (IoT), in a pharmaceutical supply chain creates a tamper-proof record about each drug, starting from the time of its discovery throughout its flow over the supply chain. Hence, the provenance and traceability of drugs in a pharmaceutical supply chain shall be ensured. Some of the future directions are: The ability of a smart contract to be extended to include financial transactions and the payment services can be enabled. The potential of machine learning algorithms can be exploited, as the data from the IoT module causes a data flood problem. The machine learning algorithms can be used to extract the most valuable data and could be a great direction of research.

REFERENCES

[1] Simon. (2011). Pharma 2020: Supplying the future which path will you take? [Accessed: 5-July-2020]. https://www.pwc.com/gx/en/pharma-life-sciences/pdf/pharma-2020-supplying-the-future.pdf

[2] WHO. (2010). Counterfeit Medical Products. Annual Report [Accessed: 5-July-2020].

[3] WHO. (2010). Assessment of medicines regulatory systems in sub-Saharan African Countries: An overview of findings from 26 assessment reports. *Annual Report* [Accessed: 5-July-2020].

[4] WHO. (2017). Global Surveillance and Monitoring System for Substandard and Falsified Medical Products. Annual Report [Accessed: 5-July-2020].

[5] Weaver, C., & Whaler, J. (2012). [Accessed: 5-July-2020]. How fake cancer drugs entered US. The Wall Street Journal. https://www.wsj.com/articles/SB10001424052702303879604577410430607090226

[6] Liu, R., & Ludin, S. (2016). Falsified medicines: Literature review. *Working Papers in Medical Humanities*, 2(1), 1–25.

[7] The path to supply chain transparency: https://www2.deloitte.com/us/en/insights/topics/operations/supply-chain-transparency.html (Accesse: 1 July 2021).

[8] Cooksey, B. Why supply chain visibility is so important. https://www.chrobinson.com/blog/why-is-supply-chain-visibility-so-important/ (Accessed: 1 July 2021).

[9] Elebring, T., Gill, A., & Plowright, A. T. (2012). What is the most important approach in current drug discovery: Doing the right things or doing things right?. *Drug Discovery Today*, 17, 1161–1169.

[10] Mignani, S., Huber, S., Thomas, H., Rodrigues, J., & Majoral, J. P. (2016). Why and how have drug discovery strategies in pharma changed? What are the new mindsets? *Drug Discovery Today*, 21(2), 239–249.

[11] Kerala brings anti-viral drug from Malaysia after Nipah claims 11 lives. India Today, May 25, 2018. Article [Accessed: 5-July-2020]. https://www.indiatoday.in/india/story/kerala-brings-anti-viral-drug-from-malaysia-after-nipah-claims-11-lives-1241287-2018-05-25

[12] 32 H1N1 cases in Kozhikode. The Hindu, October 02, 2018, Article [Accessed: 5-July-2020]. https://www.thehindu.com/news/cities/kozhikode/32-h1n1-cases-three-deaths-reported-in-sept/article25100564.ece?utm_source=kozhikode&utm_medium=sticky_footer

[13] Saha, C. N., & Bhattacharya, S. (2011). Intellectual Property Rights: An overview and implications in pharmaceutical industry. *Journal of Advanced Pharmaceutical Technology Research*, (2), 88–93.

[14] Mrdula, B. S., & Durgadevi. (2009). Intellectual Property Right pinpoint at IPR spotlights coveted RD. *Drug Invention Today*, 2, 197–201.

[15] Glasgow, L. J. (2001). Stretching the limits of intellectual property rights: Has the pharmaceutical industry gone too far? *IDEA Journal of Law and Technology*, 41, 227–258.

[16] Nair, M. D. (2010). TRIPS, WTO and IPR: Counterfeit drugs. *Journal of Intellectual Property Rights*, 10, 269–280.

[17] Verma, S., Kumar, R., & Philip, P. J. (2014). The business of counterfeit drugs in India: A critical evaluation. *International Journal of Management and International Business Studies*, 4, 141–148.

[18] Sheth, P. D., Siva Prasada Reddy, M. V., Narayana, D. B. A., Regal, B., Kaushal, M., & Sen, K. (2007). Extent of spurious (counterfeit) medicines in India [Accessed: 5-July-2020]. https://www.researchgate.net/publication/268519536_EXTENT_OF_SPURIOUS_COUNTERFEIT_MEDICINES_IN_INDIA

[19] Management Science for Health. (2012). Pharmaceutical Legislation and Regulation [Accessed: 5-July-2020]. https://pdf.usaid.gov/pdf_docs/PA00M3BX.pdf

[20] Ending TB. The Hindu, September 7, 2018. Article [Accessed: 5-July-2020]. https://www.thehindu.com/opinion/op-ed/ending-tb/article24884403.ece

[21] Accessible and Affordable. The Hindu, September 16, 2018, Article [Accessed: 5-July-2020]. https://www.thehindu.com/education/accessible-and-affordable-education-to-all/article33372379.ece

[22] IBM. (2009). Leveraging Track Trace in the Pharmaceutical Industry. *An Industry Whitepaper* [Accessed: 5-July-2020]. http://frequentz.com/wp-content/uploads/2014/10/Leveraging-track-and-trace-pharma.pdf

[23] DQSA Act. (2013) [Accessed: 5-July-2020]. https://www.fda.gov/drugs/drug-supply-chain-integrity/drug-supply-chain-security-act-dscsa

[24] Health Canada. (2009). A small number of bottles of the antibiotic Rofact (rifampin) may contain a different drug [Accessed: 5-July-2020]. http://www.healthycanadians.gc.ca/recall-alert-rappel-avis/hc-sc/2009/13384a-eng.php

[25] Health Canada. (2011). Immediate Recall by Mylan Pharmaceuticals: potential serious risk due to mislabeling of products [Accessed: 5-July-2020]. http://www.healthycanadians.gc.ca/recall-alert-rappel-avis/hc-sc/2011/13584a-eng.php

[26] Johnston, A., & Holt, D. W. (2013). Substandard drugs: A potential crisis for public health. *British Journal of Clinical Pharmacology*, 78(2), 218–243.

[27] Roy, S. F., & Jerreny, M. (2009). African Counterfeit Pharmaceutical Epidemic: The road ahead. *Anti-Counterfeiting and Product Protection* [Accessed: 5-July-2020]. https://a-capp.msu.edu/article/africas-counterfeit-pharmaceutical-epidemic-the-road-ahead/

[28] Deccan Herald. (2009). Govt declares breast cancer drug Albupax sub-standard [Accessed: 5-July-2020]. https://www.deccanherald.com/content/66392/govt-declares-breast-cancer-drug.html

[29] Kavilanz, P. (2010). Tylenol recall: FDA slams company. CNN Money [Accessed: 5-July-2020]. https://money.cnn.com/2010/05/04/news/companies/tylenol_recall_fda_inspection_report/index.htm

[30] Zoler, M. L. (2010). Methyl chloride contamination found in generic clopidogrel [Accessed: 5-July-2020]. https://www.mdedge.com/internalmedicine/article/19222/cardiology/methyl-chloride-contamination-found-generic-clopidogrel

[31] Ernest, S. (2011). Tanzania: TFDA stops sale of nine types of drugs [Accessed: 5-July-2020]. https://www.tmda.go.tz/pages/disposal-of-unfit-medicines

[32] Punjab's New Addicts. The Hindu, September 8, 2018, Article [Accessed: 5-July-2020]. https://www.thehindu.com/news/national/other-states/punjabs-new-addicts/article61535183.ece

[33] Unease at labor rooms over drug curbs. The Hindu, July 1, 2018. Article [Accessed: 5-July-2020]. https://www.thehindu.com/news/national/kerala/unease-at-labour-rooms-over-drug-curbs/article24301510.ece

[34] Pharmaceutical Crime operations: https://www.interpol.int/en/Crimes/Illicit-goods/Pharmaceutical-crime-operations [Accessed: 5-July-2020].

[35] Jagadeesh, K., & Abhijith, L. M. (2018). Spurious drugs: A tragedy in health system?. *RGUHS Journal of Medical Science*, 8, 6–9.

[36] Ali, C., & Abdelsalam, A. (2017). Analyzing pharmaceutical reverse logistics barriers an interpretive structural modelling approach. *The Business and Management Review, International Conference on Business and Entrepreneurship, Supply chain Management and Information Systems*, 8, 88–99.

[37] Hall, C. (2012). Technology for combating counterfeit medicine. *Pathogens and Global Health* [Accessed: 5-July-2020]. https://www.ncbi.nlm.nih.gov/pmc/articles/PMC4001489/

[38] McGrew, S. P. (April 1990). Hologram counterfeiting: Problems and solutions. Proc. SPIE 1210, Optical Security and Anticounterfeiting Systems, [Accessed: 5-July-2020]. http://www.nli-ltd.com/publications/hologram_counterfeiting.php

[39] IMS Health. (2017). IMS health forecasts global drug spending to increase 30 percent by 2020, to $1.4 trillion, as medicine use gap narrows [Accessed: 5-July-2020]. https://www.iqvia.com/-/media/iqvia/pdfs/institute-reports/global-medicines-use-in-2020

[40] UNICEF Supply Division. (2014). Guidelines: Temperature Monitoring Devices. Annual Report [Accessed: 5-July-2020]. https://www.unicef.org/supply/media/4391/file/E006-temperature-monitoring-devices-procurement-guidelines.pdf

[41] WHO. (2017). A study on the public Health and Socio-economic impact of Substandard and falsified medical products. Annual Report [Accessed: 5-July-2020]. https://www.who.int/publications/i/item/9789241513432

[42] WHO. (2010). Assessment of medicines regulatory systems in sub-Saharan African Countries: An overview of findings from 26 assessment reports. Annual Report [Accessed: 5-July-2020]. https://joppp.biomedcentral.com/articles/10.1186/s40545-020-00255-x

[43] Peyraud, N., Rafael, F., Parker, L. A., Quere, M., Alcoba, G., Korff, C., et al. (2017). An epidemic of dystonic reactions in central Africa. Lancet Global Health [Accessed: 5-July-2020]. https://www.thelancet.com/journals/langlo/article/PIIS2214-109X(16)30287-X/fulltext

[44] Pisani E. Antimicrobial resistance: What does medicine quality have to do with it? [Accessed: 5-July-2020]. https://amr-review.org/sites/default/files/ElizabethPisaniMedicinesQualitypaper.pdf

[45] Nakamoto, S. Bitcoin: A peer-to-peer electronic cash system. Available: http://bitcoin.org/bitcoin. pdf [Accessed: 3-June-2020].

[46] Lansiti, M., & Lakhani, K. R. (2017). The truth about blockchain. *Harvard Business Review*, 95(1), 118–127.

[47] Shortage of personal protective equipment endangering health workers worldwide. https://tinyurl. com/v5qauvp [Accessed: 3-June-2020].

[48] Corrado, C., Antonucci, F., Pallottino, F., Jacopo, A., David, S., & Paolo, M. (2013). A review on agri- food supply chain traceability by means of RFID technology. *Food and Bioprocess Technology*, 06(3), 353–366.

[49] Huang, Y., Wu, J., & Long, C. (2018). Drugledger: A practical blockchain system for drug trace-ability and regulation. IEEE Conference on Internet of Things.

[50] Arsene, C. (2019). Hyperledger project explores fighting counterfeit drugs with blockchain. Available online: https://healthcareweekly.com/blockchain-in-healthcare-guide [Accessed: 5-July-2020].

[51] Muniandy, M., & Ong Tze Ern, G. (2019). Implementation of pharmaceutical drug traceability using blockchain technology. *INTI Journal*, 2019: 035. eISSN:2600-7920.

[52] Nizamuddin, N., Salah, K., Azad, M. A., Arshad, J., & Rehman, M. H. (2019). Decentralized document version control using ethereum blockchain and IPFS. *Computers & Electrical Engineering*, 76, 183–197.

[53] Andrychowicz, M., Dziembowski, S., Malinowski, D., & Mazurek, L. (2015). On the malleability of bitcoin transactions in the proceedings of Financial Cryptography and Data Security, Pages 1–18.

[54] U.S. Food and Drug Administration. Drug Supply Chain Security Act. Available: https://fda.gov [Accessed: 3-June-2020].

[55] Uddin, M., Salah, K., Jayaraman, R., Pesic, S., & Ellahham, S. (2021). Blockchain for drug trace-ability: Architectures and open challenges. *Health Informatics Journal* doi: https://doi.org/10. 1177/14604582211011228.

Index

Note: Page numbers followed by *f* or *t* indicate figures or tables

273

Printed in the United States
by Baker & Taylor Publisher Services